The New York City Draft Riots

The New York City Draft Riots

Their Significance for American Society and Politics in the Age of the Civil War

IVER BERNSTEIN

New York Oxford
OXFORD UNIVERSITY PRESS
1990

Oxford University Press
Oxford New York Toronto
Delhi Bombay Calcutta Madras Karachi
Petaling Jaya Singapore Hong Kong Tokyo
Nairobi Dar es Salaam Cape Town
Melbourne Auckland

and associated companies in
Berlin Ibadan

Published by Oxford University Press, Inc.,
200 Madison Avenue, New York, New York 10016

Library of Congress Cataloging-in-Publication Data

Bernstein, Iver.
The New York City draft riots : their significance for American
society and politics in the age of the Civil War / Iver Bernstein.
p. cm.
Bibliography: p.
Includes index.
ISBN 0-19-505006-1
1. Draft Riot, New York, N.Y., 1863. I. Title.
F128.44.B47 1990 89-2858
974.7'103—dc 19

Printed in the United States of America

For Kay

Preface

I began this project in 1980 on a suspicion that New York's ugly riot against the Civil War draft might tell us much about the intricate and often obscure processes that gave rise to modern urban America. This riot was one of those unusual events important in its own right—it mattered in the war and in the life of the city—and important for its illumination, like a flash of lightning, of a darkened historical landscape. Veiled alliances and animosities, rarely articulated definitions of authority and justice and configurations of power were all disclosed. The Civil War context of the draft riots made the social and political tableaux exposed by the event stand out in bold relief. Because of the wartime preoccupation with national loyalty, any form of participation in or association with the violence was tainted with treason. Inaction was suspect: one had to declare against the rioters to demonstrate loyalty to the government and the war. This charged political climate explained the extraordinary self-consciousness of New Yorkers during the violence. All had to clarify their views and proclaim allegiances. Here, it seemed, a contentious world lay open to view.[1]

The long journey from that suspicion in 1980 to this book was made possible by generous funding from the Graduate School of Yale University, the Mrs. Giles M. Whiting Foundation, the Yale Council on West European Studies, the Stephen Charney Vladeck Fellowship at New York University, the Division of the Social Sciences at the University of Chicago, the American Council of Learned Societies, the National Endowment for the Humanities and the Faculty Research Grant Committee of Washington University, St. Louis.

I am also grateful for the patient attentions and creative suggestions of the librarians and staff of Sterling Memorial Library, Yale University; the New York Public Library; the New-York Historical Society; the New York Municipal Archives and Records Center; the Rare Books and Manuscripts Divisions of Butler Library, Columbia University; Houghton Library, Harvard University; the Tamiment Institute at New York University; the New York State Library, Albany; the National Archives; the Manuscripts and Archives Division of the Library of Congress; the Huntington Library, San Marino, California; and the International Institute for Social History, Amsterdam, the Netherlands. Special appreciation is reserved for Idilio Gracia-

Peña and Kenneth Cobb at the New York Municipal Archives and Records Center, Susan Davis of the Manuscripts Division of the New York Public Library, Jane Reed, at the library of the Union League Club of New York, and Thomas Dunnings of the New-York Historical Society. Miriam Frank of the Tamiment Library provided expert assistance with some questions of translation and interpretation in German-language manuscript sources. Without the aid of Christine Smith and her staff in the Interlibrary Loan Office of Olin Library, Washington University, this project surely would never have been completed.

Trudi and Isaac Behar and Joan and Walter Kornbluh made a research trip to Los Angeles far more pleasurable with their special hospitality. Amy Stanley and Craig Becker offered me the comforts of their Washington, D.C. home one hot week several springs ago. My mother-in-law Shirley Sherman's cheerful willingness to serve as a New York host allowed hundreds more hours in city archives during the final stages of the project.

Sheldon Meyer of Oxford University Press saw the possibilities early on and helped to shape a dissertation into a book. Stephanie Sakson-Ford prepared the manuscript with precision and a keen ear for the right word. Scott Lenz maintained uncompromisingly high standards through to the end of the editorial process.

David Montgomery encouraged my inclination to wrestle with the most dangerous implications of a history of the draft riots and to conceive of the study in its broadest dimensions. His commitment to the project and patience with its author have been extraordinary. My debt to him can barely be put into words, nor can it easily be repaid. David Brion Davis helped to nurture an early graduate seminar paper into a dissertation. He inspired me to discover the relations between ideas and material life that form the architecture of this book. Howard R. Lamar has tried to teach me how to think about the past with the example of his own refined historical imagination. His erudition, counsel, and unfailing good humor were instrumental in bringing this book to completion; more than once he provided the key insight that allowed me to move past a conceptual barrier. Emilia Viotti da Costa was of enormous help and gave searching and exhaustive criticisms to drafts of early chapters. John L. Thomas cultivated in me a deep respect for the craft of historical writing. As a scholar and a teacher he has served as a model.

This project would have scarcely been conceivable without the pioneering work of many historians of New York City who shared their ideas and passion for the material. Joshua Brown, Peter Buckley, Graham Hodges, Bruce C. Levine, David Scobey, James P. Shenton, Michael Wallace, and Sean Wilentz have all provided intellectual and moral support. Certain friends have also taken the trouble to read and comment on versions or chapters: I thank Jeanie Attie, Kathleen Neils Conzen, Ileen DeVault, John D. French, Lori Ginzberg, James R. Grossman, Steven Hahn, Jane

Melnick, Grace Palladino, James T. Patterson, Carl E. Prince, Paula Shields, Amy Stanley, and R. Jackson Wilson for their help and encouragement. Participants in the Workshop on Metropolitan Dominance of the Social Science Research Council Committee on New York City, the Social History Seminar of New York University, the Chicago Area Labor History Group at the Newberry Library, and the Urban History Seminar of the Chicago Historical Society examined earlier guises of several chapters to my great advantage. Joseph O. Losos shared with me one of his many expertises, in the history of American business. Extended conversations with Thomas Bender, Elizabeth Blackmar, and Michael E. McGerr helped me to develop and refine many of my ideas; as well, each gave painstaking attention to the manuscript and improved it considerably. James Henretta magnanimously read the penultimate version and had some extremely useful advice. Eric Foner introduced me to sources and gave me the benefit of his superb editorial eye and incomparable knowledge of mid-nineteenth-century America.

My colleagues in the Department of History at Washington University have provided a rare and very precious sort of stimulation and camaraderie. Especially I would like to thank Rowland Berthoff, Richard W. Davis, George C. Hatch, Jr., Derek M. Hirst, Peter Riesenberg, and Marc Saperstein for their readings of an earlier draft of the manuscript, and Henry Berger, Gregory Claeys, David T. Konig, Mark Kornbluh, and Kenneth H. Ludmerer for their generosity of intellect and spirit.

Harriet and Leon Bernstein have been dedicated to this project from the outset. They fostered my interest in American history. Through them and their past, I became obsessed with understanding the aspirations of New Yorkers and the ways the myriad parts of New York City fit together. Their courage, commitment to hard work, and meticulous honesty have, I hope, informed my efforts. The loyal support of Sally Germain has been a deeply valued resource over the years.

Benjamin Sherman Bernstein arrived when this manuscript was already well advanced, but no doubt some of his zeal for life found its way into these pages. Kay Sherman has helped more than anyone else. She has listened, read and reread. She has shared all the burdens of this endeavor with boundless patience and humor. Whatever clarity and vision can be found in that which follows result from her involvement and her example.

St. Louis I. B.
June 1988

Contents

Abbreviations

AICP	New York Association for Improving the Condition of the Poor
AHR	*American Historical Review*
CU	Rare Books and Manuscripts, Butler Library, Columbia University, New York
HL	Henry E. Huntington Library, San Marino, California
HU	Houghton Library, Harvard University
IISH	Marx-Engels Archive, International Institute for Social History (*Internationaal Instituut voor Sociale Geschiedenis*), Amsterdam, the Netherlands
IWA	International Workingmen's Association
JAH	*Journal of American History*
LC	Manuscripts and Archives Division, Library of Congress
LUBA	Laborers' Union Benevolent Association
LUBS	Longshoremen's United Benevolent Society
MARC	New York Municipal Archives and Records Center
MVHR	*Mississippi Valley Historical Review*
NA	National Archives
NYH	*New York History*
NYHS	New-York Historical Society
NYHSQ	*New-York Historical Society Quarterly*
NYPL	Rare Books and Manuscripts Division, New York Public Library
NYSL	Manuscripts and Special Collections, New York State Library, Albany
SDPK	Society for the Diffusion of Political Knowledge
Strong Diary	Allan Nevins and Milton H. Thomas, eds., *The Diary of George Templeton Strong* (New York, 1952), 4 vols.
Tribune	New York *Daily Tribune*
WOR	U.S. War Department, *The War of the Rebellion: A Compilation of the Official Records of the Union and Confederate Armies* (Washington, D.C., 1899)

The New York City Draft Riots

Introduction

The New York City draft riots had deep and troubling significance for mid-nineteenth-century Americans. For five days in July 1863, armed mobs interrupted enforcement of the first federal conscription and struggled with authorities for sway over the nation's manufacturing and commercial capital. What began on the morning of July 13 as a demonstration against the draft soon expanded into a sweeping assault against the local institutions and personnel of President Abraham Lincoln's Republican Party, as well as a grotesque and bloody race riot. The awesome destruction of property and life stunned a generation of urban Americans well familiar with street violence.[1]

In the context of the Civil War, the draft riots gave sudden focus to controversial questions that Northerners intent upon sectional unity would have preferred to ignore. National issues included the fate of federal conscription, the authority of the new Republican government in Washington, and the future of the post-Emancipation Proclamation war effort. At stake in New York were the social and political order of the city and, related to that, the meaning of this complex event for the status and aspirations of the city's diverse social classes. During the riots national and local issues became joined as Americans waited anxiously to see whether, or on what terms and under whose auspices, the powerful metropolis with its Democratic majorities and vast immigrant poor would participate in the war and the Republican project of nation-making. It is fair to speculate, as some Northerners did, that if Confederate General Robert E. Lee had pushed north from Gettysburg at the moment rioting erupted in New York City, European intervention would have stalemated the war. As it happened, Lee escaped south across the Potomac on the night of July 13. On July 16 and 17, five Union Army regiments ordered back from Gettysburg suppressed the riots, kept New York City behind the war effort, and preserved the ascendancy of the Republican Party in national politics.[2]

But it remains unclear whether the Republican Party or any other group emerged wholly victorious from the New York upheaval. The participants, victors, victims, origins, dynamics, and consequences—indeed the full significance—of the draft riots are still obscure. We do have a wealth of secondary literature describing the events of July as well as several accounts by contemporaries. These narratives nearly all regard the revolt as a series of colorful scrimmages and fail to assess its broader political and historical meaning for the contesting groups and the nation. We have yet to understand why the draft riots occurred and what their implications were for the creation of modern urban America.[3]

This poverty of analysis may be explained in part by the refusal of any of the participants to parade the riots as part of their heritage. A nervous gentry sooner pretended the revolt had never occurred, only acknowledging it as if by some painful reflex during subsequent moments of social crisis. The labor movement disowned a grisly race riot with overtones of treason as well as an episode that revealed many wage earners uniting with their employers at the workplace to shoot down fellow workers in the crowd. Trade union leaders did not want their infant organizations linked with violence and destruction of property in the public mind and repudiated the riots. The immediate victims, the black community and the white poor who opposed the draft and fought federal troops, did allow versions of their story to filter into newspaper accounts, court affidavits, coroner's reports, and comptroller's office petitions. But this victims' record was fragmentary, well buried and embarrassing to the many metropolitan wage earners who came to prefer more orderly forms of protest in the decade following the riot. When in late 1863 Northern military victory began to appear imminent, an anachronistic reading back of national unity in a grand cause encouraged many Northerners to repress further the recollection and meaning of draft resistance in New York and around the country. What gradually ensued was an instance not of what historians call collective memory but rather of collective amnesia.[4]

Nor was this an event that future reform movements would readily incorporate into their usable past. As a communal uprising against the power of an expanding and centralizing federal government, the New York upheaval recalled American traditions of popular resistance to tyrannical regimes dating back to the Stamp Act riots and the colonial agitation against British rule. But an episode so tainted with treason and implicitly or overtly linked with sympathy for the Southern slaveowning aristocracy would not attract the attention of subsequent popular movements against excessive or intrusive governmental power. Protestant moral reformers would find the legacy of the riots equally ambiguous. These men and women saw the quashing of the violence as a crucial homefront victory for a righteous Northern cause. But the riots also exposed a cultural chasm between the reformers and their immigrant clientele. The benevolent empire, the riots made clear, might occasionally have to underwrite its program of moral uplift for the poor with coercion and bloodshed. Socialists would also shun or ignore the legacy of July 1863. Karl Marx saluted the Paris Communards of 1871 as Prometheans who dared to storm the gates of heaven to liberate all humanity.

He could not have so praised the late-week draft rioters, with their murderous attacks on black workingmen.

To the extent the riot has been noticed, its meaning has been obscured by a tendency to classify it according to antebellum precedents.[5] Drawing on the work of European scholars George Rudé and Edward P. Thompson, American students of crowd behavior have pointed to a home-grown tradition of republican mob vigilantism that substituted the *posse comitatus* for the state when community interests were threatened.[6] Rioting in this tradition was legitimated in the name of the state and was largely non-violent. But the 1863 insurrection seems almost un-American, reminiscent of European revolutionary violence such as the bloody Parisian June days of 1848. Like the Parisian workers, at least some of the draft rioters directed their violence against the state.[7] The number of casualties and loss of property in 1863 dwarfed the figures for ethnic and anti-abolitionist mobs in the forties and fifties. The most deadly riot in pre-Civil War Philadelphia, the nativist Southwark riot of July 1844, killed twelve, while New York's exceptionally violent Astor Place riot of 1849 cost twenty-two lives. The death toll from the draft riots was at least 105, nearly five times higher.[8] For some draft rioters, the revolt against conscription may have remained within the confines of the disciplined artisan republicanism that infused the New York City strike movements of the 1830s and early 1850s. But for many others, the riots provided an outlet for a violent style of social and political resistance that was, at least in an American context, entirely new.[9]

The draft riots violated other conventions of antebellum crowd behavior. Hostile immigrant mobs had assailed black workers in many Northern cities before the Civil War.[10] In seaport New York, interracial violence dated back to the early eighteenth century. In more recent memory, many Civil War-era New Yorkers could recall the attacks on black homes and churches during the eight-day anti-abolitionist riot of July 1834. Beginning in 1854, organized Irish dockworkers fought to drive black strikebreakers from the waterfront.[11] These antebellum incidents were localized and almost always limited to the pillaging of black property. By comparison, the draft rioters' horrifying slaughter of black men suggested a far more extreme, citywide campaign to erase the post-emancipation presence of the black community.

Urban violence among street gangs and fire companies in the forties and fifties revolved around the neighborhood, as each ethnic group "fought to regulate who lived near them, who socialized at their pubs and taverns, and which companies serviced their people."[12] This ethnic street violence could become political when it involved the defense of immigrant culture and political organizations against the enforcement of middle-class reform laws.[13] By contrast, the draft riots encompassed both neighborhood and workplace, and were in a sense both mob action and strike. The closing of factories, destruction of telegraph lines, railroad and streetcar tracks, and threats to ferries and bridges implied not merely a fight to control ethnic-group territory, but a bid to gain sway over an entire city. No doubt for some participants the draft riots were another contest for political power between Democratic immigrants and Republican reformers. But for many others the riots were charged with the ideological tensions of a civil war and became a full-throated challenge to the rules of a new politics and society emerging during the war years.

Since the New York draft riot represented a turning point in the Civil War and a critical moment in Republican efforts at centralization, and because it involved an entire city and featured wholly new forms of urban social and political conflict, it demands a broad and multilayered explanation. We cannot regard the riots, as many of the narratives do, as a simple one-dimensional episode of Irish Catholic ethnic hatred, white lower-class racism, Confederate or Copperhead sympathy, or resentment of the poor toward the inequities of the Conscription Act—though each of these themes accounts for important parts of the event. The unprecedented scope, audacity, and violence of the insurrection must also be explained. To comprehend all aspects, we must situate the draft riots in the context of an ongoing process of urban change beginning in the early 1850s and concluding in the 1870s, with ramifications at the regional and national levels. A new complex of social, cultural, and political relations made the new antagonisms of July 1863 possible.

The book begins with a narrative of the draft riots, first from the viewpoint of the rioters themselves, then from the perspective of the "better classes" and authorities who sought to restore order to the city. Part One, "Draft Riots and the Social Order," identifies the riot week as a moment of political crisis because two formidable questions had to be settled at the same time: who was to rule the nation and who was to rule New York City? This section then distinguishes an array of metropolitan groups whose actions and debates defined the contours of the crisis. These groups are the dramatis personae of the book. The volatile and politicized relations among these groups become the central fact of the draft riots to be explained.

Part Two, "Origins of the Crisis, 1850s and 1860s," analyzes the changing experiences and outlooks of each of the participant groups identified in the opening section, with particular attention to the decade before the Civil War. Beginning in the fifties, these groups began to debate how they could best impose order on a city and a nation ravaged by competition, abruptly changing markets, and rampant individualism. The tendency of some workers (and other groups) to blur distinctions between "economic" and "political" action in their pre-war efforts to impose a style of consolidation on the city's topsy-turvy economic growth foreshadowed the highly politicized conflict of the 1863 riots. A sharp disagreement among the "better classes" over how best to rule the city also helped to create the possibility for the draft riots and an era of surprising popular political power in New York City.

Republican enforcement of the first federal conscription provoked a riot and a crisis in New York City because it crystallized and focused these politicized urban conflicts.

Part Three, "Resolutions of the Crisis, 1860s and 1870s," begins with an examination of Boss William M. Tweed's Tammany Hall, the most successful attempt to create a political rapprochement among the various groups at odds during the riot week. After the war, many New Yorkers looked to the Tweed Democracy—committed to loyalty to the nation, Americanization of the Irish, debt financing, metropolitan expansion, local self-government, and white supremacist popular rule—as an instrument of coherence in the industrializing city. But some of the intractable conflicts of 1863 persisted through the Tweed years and eventually, in summer 1871, undid Tweed's political settlement. Part Three will show that the political crisis exposed by the draft riots was not fully resolved until the cataclysmic strikes for the eight-hour day in spring 1872, which forced a reunification of middle- and upper-class groups around certain basic issues of social and political rule and a repudiation of popular political influence in the name of "the best men." In this denouement, it can be said, was the draft rioters' final moment of defeat. After 1872, the possibility of volcanic political violence on the scale of July 1863 largely disappeared. The ultimate significance of the draft riots, the book will conclude, was their situation at the center of a contentious era, an era of politicized social conflict. For a time, such conflict allowed popular definitions of justice to attain enormous power and legitimacy in urban life.

The federal Conscription Act was passed by Congress in March 1863 to reinforce the Union Army at a low point in Northern military fortunes. Lincoln and his counselors had much cause for concern that spring. The war had been raging for nearly two years. The army was making uncertain progress on the battlefield and its ranks were thinned from the illegal absence of more than one hundred thousand persons. Before long, the three-year volunteers of 1861 who had fought the war thus far would come to the end of their enlistment terms. Horrifying tales of carnage at the front circulated through the North. Campaigns to raise new volunteers were flagging, and state drafts attempted the previous year in Maryland, Pennsylvania, and Wisconsin had met popular resistance and widespread evasion. The anticipation of such a response had probably helped to spare New York from a state conscription.[14] This grim scenario convinced many War Department officials that only a new and powerful federal draft bureau could accomplish the related tasks of raising an effective army and monitoring loyalty on the homefront. The "Act for Enrolling and Calling Out the National Forces" created a national Provost Marshal Bureau directed by a Provost Marshal General responsible solely for enforcing the draft. The sweeping provisions of Section 5 empowered a federal provost marshal in each congressional district to make summary arrest of draft evaders and resisters, deserters and spies. The law was an attempt to sustain the flow of men into the army and keep them there once enlisted.

The fielding of an effective army was inseparable from a larger issue, the legitimacy of the Republican federal government and its wartime policies. The Republicans' primary task from the beginning of the war was to reunify the nation. As it became clear after a few months of fighting that a simple police action would not subdue the rebellion, the Republican government resorted to a series of extreme measures to enhance and centralize its authority. Two of these, federal taxation and the Legal Tender Act, were needed to finance the unprecedented cost of the Northern war effort. The third, the Emancipation Proclamation, was a drastic attempt to revive Northern military fortunes by interesting four million slaves in the Union cause. The expansion of federal powers entailed in these acts was deeply controversial in the North. Black emancipation provoked an especially heated debate: it lent the war the dignity and urgency of a moral crusade for some Northerners but antagonized many others. In November 1862, the Northern Democratic Party, increasingly critical of Republican conduct of the war, revived at the polls when it captured thirty-five Republican-held congressional seats and a host of statewide offices. In this climate of growing opposition to the Lincoln administration, the Conscription Act entitled the Provost Marshal Bureau to bypass hostile Democratic state and local governments to discipline individuals directly; district provost marshals were largely free to define treason as they pleased. The most drastic example of Republican centralization, federal conscription was a political measure designed to contain opposition to the Washington regime.[15]

The new law made all men between the ages of twenty and thirty-five and all unmarried men between thirty-five and forty-five liable to military duty. Names were procured through a laborious house-to-house enrollment conducted by government agents. Then a lottery in each congressional district determined who would go to war. Drafted men who presented an "acceptable substitute" or paid three hundred dollars were exempted.[16] The burden of the March Act fell most heavily on the poor; among the poor it singled out young men, and women and families economically dependent on the young male conscript. More than any other Republican wartime legislation, the new draft brought the presence of the federal government into the community, into the waged workplace, and into the household—into nearly every corner of working-class life.

The provisions of the law highlighted three explosive issues in mid-century New York City: relations between the wealthy and the poor, between blacks and whites, and between the city and the nation. Even before the war began, the idea that the hardships of an all-out sectional struggle might exacerbate class tensions in Northern cities and drive the poor to desperate acts had crossed the minds of both Southerners and Northerners. In late 1860 the Virginia secessionist Edmund Ruffin envisioned a sectional war in which New York City was ravaged by riots and Washington became the capital of a victorious Southern republic.[17] That same fall the New York Unitarian minister Henry Whitney Bellows confided to his sister fears that secession would trigger an urban revolution, "driving our populace into panic for bread and violence toward capital and order. . . ."[18] In New York the enthusiastic patriotism of the early months of the war only briefly dispelled a growing sense that a treasonous outbreak of violence was possible. In its 1861 report, the city's leading moral reform organization congratulated itself on the loyalty of the immigrant poor to the war effort and the absence of homefront disturbances.[19] Soon wage earners were groaning under the weight of wartime inflation; in spring 1863 the local labor movement revived to protest the political and economic effects (either experienced or anticipated) of the Republican Legal Tender Act, taxation, and black emancipation. Workers who felt that these measures reinforced the advantages of the well-to-do could still justify them as necessary for winning the war. The Conscription Act was less easily tolerated: the baldly inequitable three-hundred-dollar commutation clause seemed a naked exercise of class power by the wealthy. For their part, the Republican framers of the draft law considered the commutation clause a democratic measure that would place a ceiling on the price of a substitute and bring exemption within the reach of poor men (and studies of the operation of the draft in some Northern communities suggest that commutation may have ultimately done just that). But in the spring and early summer of 1863, before the draft was enforced and before draft insurance clubs, factory owners and political machines mobilized to aid the needy and unwilling conscript, the March Act appeared highly mischievous to its intended beneficiaries. Here, it seemed, was yet another Republican scheme to transfer the burden of the war to the poor. Many among the metropolitan poor were now convinced that the wartime government sought only to aggravate the exploitative social relations of the industrializing city. In turn, local leaders braced for the anticipated mob challenge to Republican rule.[20]

There was no explicit mention of race in the Conscription Act, but because the law pertained to "able-bodied male *citizens* of the United States" (my emphasis), only whites were subject to the draft. The relative status of black and white labor was especially controversial in New York City. As a Northern city with longstanding ties to Southern slavery and as the national capital of both the abolitionist and anti-abolitionist press, New York was a place acutely sensitive to racial issues. New York was also a sprawling seaport, host to what was no doubt the nation's most fluid labor market. Here it was easy for any group, no matter how well established, to feel threatened by the daily flood of new arrivals; poor and often unskilled immigrant workers felt especially vulnerable. Many of New York's poor immigrants believed the Democratic orators who in spring and early summer 1863 predicted that emancipation would bring north low-wage black freedmen to compete with white laborers for employment (a prediction never borne out). The Conscription Act, with its three-hundred-dollar commutation clause, seemed to compound the threat posed by black emancipation. Now the Republican government was not only privileging black labor but subjecting poor whites to special risks to life and livelihood. Maria Lydig Daly noted that the white "laboring classes" thought their status had been lowered by the draft: "[They] say that they are sold for $300 whilst they pay $1000 for negroes." Many white New Yorkers felt that the consequence, if not the explicit intent of the draft law, was to exacerbate racial tensions and degrade the status of white labor.[21]

The new law also stirred debate over New York City's relations with the Republican government in Washington. New York's Democratic leaders had enjoyed vast influence in the Democratic White House of the antebellum period, notwithstanding occasional quarrels over patronage.[22] The election of Lincoln in 1860 and the elaboration of a wartime bureaucracy over the next two years gave city Republicans new authority and distanced Democratic leaders from the seat of national power. Many of those Democrats, committed in principle to local self-government, now bristled against Washington's real and imagined interference with New York's prerogatives. In this setting, the Conscription Act's subordination of New York City's and State's Democratic authorities to an external federal draft bureau appeared to be a Republican stratagem to undermine New York City's rights and interests. New York State Governor Horatio Seymour and other Democratic officials were certain that federal authorities set the state's draft quotas disproportionately high, with the brunt falling upon the heavily Democratic metropolis. While there is no evidence that the Lincoln administration actually conspired against the city, Provost Marshal General James Fry acknowledged years later that New York State had not received sufficient credit for its volunteers when its share of the conscription was calculated. Such unfairness—and few Republicans would admit to it in 1863—was irrelevant to supporters of the President. "Affairs had reached such a point in the spring of 1863," wrote Fry, "that . . . there was no neutral ground between supporting the enforcement of the Enrollment Act and the prosecution of the war as part and parcel of the same thing, and opposing them." Republicans who sought above all to strengthen the wartime government saw Seymour's grievances as treason. They believed (quite rightly) that the New York governor was philosophically opposed to federal conscription as unfair interference with state and local rights and would probably balk at enforcing a more equitable law. The anticipated refusal of Seymour to execute the draft was, in the view of Republican financier John Jay, "the last great card . . . of the rebellion."[23]

It is not hard to see, then, how the Conscription Act—biased against the poor, magnifying white racial fears, and involving the federal government as never before in local affairs—galvanized ongoing conflicts in the city. By late spring 1863 the stage for the draft riots was nearly set.

While in many Northern districts resistance to federal conscription involved a protracted series of confrontations between federal officials and local communities, in New York City draft-related conflict was a spontaneous eruption, volcanic and short-lived. Enrollment was completed with relatively few incidents of collective violence and almost no resort to military force. An auctioneer named Thomas Gaffney was arrested by the provost marshal on May 27 for forcibly resisting an enrolling agent, and three days later, nine men were arrested in a Bowery saloon for giving fictitious names. In the latter incident, the twenty-four-man Provost Marshal Guard had come to the saloon to arrest an army deserter. The Guard's attempt to assist a Bureau agent in enrolling hostile bystanders led to the giving of false names and the arrests. The fictitious name cases posed an early problem for the Provost Marshal Bureau in New York when it became clear that the loosely worded Section 25 of the March 3 Act did not include obstructing enrollment among those acts of draft resistance punishable by arrest. But resistance to the June enrollment remained more a legal than a political problem in New York City. By the middle of the month, New York's Acting Assistant Provost Marshal General Robert Nugent wrote his Washington superior Colonel Fry that enrollment in the city was nearing completion without serious incident.[24]

With Democrats controlling the state house and city legislature, many New Yorkers wondered whether a draft would ever come to pass in the city. Governor Seymour promised to satisfy the state's draft quota with volunteers, and, failing that, he expected the courts to declare the federal conscription law unconstitutional. If Seymour could not stave off the draft from Albany, then many city Democrats saw the protection coming from local sources. While pro-war Tammany Democrats in the Common Council might block the more invidious provisions of the draft law, the greater obstacle to enforcement was the thinly veiled threat of violent resistance from the Peace Democrats. The Peace or Copperhead faction of the Democracy believed that the increased power and centralization of the Republican wartime government had brought hard times and political tyranny to the North; poor white men had paid for the war with their blood and toil, while blacks reaped the benefits of misguided Republican philanthropy; normalcy could be restored only by a negotiated settlement between the sections. June 1863 was the high moment of local Peace Democrat influence. Now the Copperhead themes sounded at a vast "Peace Convention" on the 3rd were being echoed at the newly formed artisan trades congress and along the waterfront, among the city's striking longshoremen. Even Democratic editor of the *Herald* James Gordon Bennett, hardly a Peace sympathizer, predicted that Coppery leader Fernando Wood would ride the swelling national peace movement into the

White House in 1864 with an armistice to follow. "Peace," Bennett forecast during the first week of June, "is henceforth the platform of the democracy." The *Herald*'s pronouncements augured a new respectability for Copperhead sentiments among city Democrats. In this political climate, resistance to enrollment might well have seemed premature. On June 16, the day Lee pushed north across the Potomac River, the Peace journal *Daily News* best expressed this wait-and-see attitude: "No further attempts are being made to avoid the enrollment, people generally seeming to think it is quite time enough to offer resistance when the draft is put into operation." If and when mobs took to the streets to protest Republican policies, many New Yorkers thought, Copperhead orators would lead the way.[25]

The Lincoln administration followed political developments in New York City but concluded that the Peace movement there offered no real threat to Republican rule. We know this from an unusual source inside the White House, one T. J. Barnett of Indiana, a minor official in the Department of the Interior and Washington correspondent for one of New York's Democratic weeklies. The Hoosier journalist charmed his way into Lincoln's inner circle with a mixture of brass and country humor; he was also an intimate friend and business associate of Samuel L. M. Barlow, the New York corporate lawyer and backstage Democrat who engineered the presidential candidacy of General George B. McClellan in 1864. Barnett's clandestine role as Barlow's source in the Lincoln White House (and conceivably Lincoln's source in the New York Democracy) said much about the distant and tense relationship between the leaders of the nation and the city. Barnett wrote Barlow on June 10,

> I am convinced now that the President apprehends nothing from the Peace Party. He looks upon it as an amalgam of elements of discontent in New York, and of folks apprehensive of the personal effect of the Conscription act, seized upon to enure the advantage of such men as the Woods, as so much capital in his [*sic*] hands, whereby to demand a price of Tammany Hall—and so he regards the same movements in England, as hobbies. He has no belief that such an issue could be made to prevail even in the City of New York—and he thinks that "it will give the Democrats far more trouble than it will anybody else." He . . . will keep "pegging away at the Rebels" wholly satisfied that New York demonstrations are Pickwickian, and that his head will not be brought to the block. . . .[26]

Not long after Barnett made these observations, thousands of troops were dispatched from New York to stem the Confederate advance across the Mason-Dixon Line. The city was left virtually undefended in late June as it completed enrollment and readied for the draft lottery. Lincoln's faith in the loyalty of New York City may help explain why Republican officials proceeded with the draft under such unpromising conditions.

How to conduct the draft without military supervision or reinforcements?

Provost Marshal General Fry advised New York subordinate Robert Nugent and officers in neighboring cities to proceed with the lottery one congressional district at a time since it was "not probable that sufficient military force can be furnished . . . to suppress successfully at one and the same time a resistance to the draft in your State should such occur simultaneously in any great number of the districts under your charge." Nugent, a New Yorker, a War Democrat, and an Irishman, knew the volatility of the situation. He hoped to orchestrate a quiet selection by careful location and timing of the lotteries around the city. On July 9 he informed Fry that he would begin drawing in the remote areas of Upper Manhattan, Queens, and Suffolk counties so "that in case there should be any trouble or opposition to the Draft it may be confined to special localities and not become general over the city and Long Island." Nugent's strategy did have the stated advantage of beginning the draft in the isolated districts on the city periphery. But in Manhattan, this ploy placed the lotteries in the heart of the uptown tenement and shanty district, among the poor immigrant workers most suspicious of Republican policies.[27]

Not until the weekend of July 11 did New Yorkers fully realize that Democratic officials would fail to shelter them from the draft. Before that date, there is little evidence of the planning of a riot. At an Independence Day rally, Peace Democrats denounced the draft and military arrests as despotic assaults on popular rights and liberties, and Governor Seymour warned that Republican policies might provoke mob violence.[28] But it was one thing to predict an insurrection or even to invite one as a last resort, quite another to plot a riot; there is no evidence that even the most fiery Copperheads contemplated the latter course. Union victories at Gettysburg and Vicksburg on July 3 and 4 relieved the threat to Northern coastal cities, confirmed the Mississippi as a Union river, and, we can imagine, altered the political climate in New York. Lee's uncertain position in Northern territory could have worked to raise expectations on all sides: of decisive Confederate defeat, for Union supporters, or of an eleventh-hour rally, for Peace Democrats and Confederate sympathizers. The dominant feeling that the tide of the war was turning in favor of the North may have eroded Peace support. On Saturday, July 11, the first names were drawn in the uptown Ninth District draft without incident. New Yorkers now saw that Governor Seymour and city Democrats had not been able to prevent or postpone conscription.[29]

The first actual plans for resistance seem to have been formed on Saturday night and Sunday, as working people gathered in saloons, streets, and kitchens to discuss their own remedies for the lotteries to be resumed Monday morning. "Those who heard the scattered groups of laborers and mechanics who congregated in different quarters on Saturday evening, and who canvassed unsparingly the conscription law," wrote the editor Bennett, "might have reasonably augured that a tumult was at hand." That weekend the sense of imminent crisis was palpable—a journalist for the New York

Sun reported on uptown shipyard blacksmiths' attitudes toward the draft as if he knew exactly where Monday's revolt would start. Upper West Side Police Captain George Washington Walling thought "the growing discontent among certain classes" serious enough to merit remaining at the station house Sunday evening. At Monday dawn, an Upper Manhattan "gentleman" was informed by his coachman that railroad lines would be disturbed that day. The riot New Yorkers had so long imagined was beginning.[30]

PART I

Draft Riots and the Social Order

CHAPTER 1

A Multiplicity of Grievances

On Friday, July 17, 1863, the last day of the draft riots, Peace Democrat Congressman and newspaper editor James Brooks published a brief article entitled "The Riot—Its History." By the 17th, New Yorkers had developed their own versions of the riot-week events and interpretations of the rioters' motives.[1] But Brooks, a popular uptown figure intimately familiar with the attitudes of his constituents, was one of the few observers to draw up a calendar of the rioters' activities:

> *Sunday*—A day of leisure, thousands of Workingmen pondering upon the draft of Saturday.
>
> *Monday*—The Conscription Riot, developed in attacks upon the Provost Marshals and their places, etc.
>
> *Tuesday*—The Riot of Thieves, not only from New York—but from Philadelphia, Boston, and all quarters, who rushed here to steal.
>
> *Wednesday*—Not a Conscription Riot nor a Thief Riot—but the consequences of the collisions of the military and the mob.[2]

Brooks's calendar discriminated among the rioters' targets and chronicled the day-to-day progression of the violence. He did not figure the riot's various forms of racial violence into his account and his notion that rioters who looted were marauders from other cities seems farfetched—he doubtless wished to rehabilitate the event as a pure revolt against Republican Party centralization. But his observation that the insurrection went through phases and that each phase reflected the prominence of different rioters and targets is supported by much of the available evidence. The draft riots involved diverse groups of workers and a multiplicity of grievances against Republican rule. Each group had its own understanding of what the strike against conscription meant.

Monday, July 13

The draft riots began Monday morning not at the hour of the draft but at the hour of work. Between six and seven o'clock, four hours before the Ninth District draft selection was scheduled to begin, employees of the city's railroads, machine shops, and shipyards, iron foundry workers, laborers for an uptown street contractor, and "hundreds of others employed in buildings and street improvements" failed to appear at their jobs.[3] By eight o'clock, many of these workers were streaming up Eighth and Ninth avenues, closing shops, factories, and construction sites along the way and urging workmen to join the procession. After a brief meeting at Central Park, the crowd broke into two columns and, with "No Draft" placards held aloft, marched downtown to the Ninth District Provost Marshal's Office, at Third Avenue and Forty-seventh Street, scene of the draft lottery to be held later that morning.[4] On the way, some rioters cut telegraph poles and committed the first acts of theft, breaking into a hardware store to steal broadaxes.[5] On Third Avenue, a crowd had begun to hack down telegraph poles and lines. Not long afterwards, Irish women used crowbars to pull up the tracks of the Fourth Avenue railway, and soon crowds had stopped the Second and Third Avenue cars.[6] Rioters also attacked several police officers. Superintendent of Police John A. Kennedy, out of uniform, was spotted by a crowd, dragged through the mud and beaten on the head until "unrecognizable."[7] Kennedy was later rescued by John Eagan, a Tammany politician in the Nineteenth Ward.[8] During the course of the morning, rioters who attended the Central Park meeting may have encountered others who had assembled further downtown to close Allaire's Works, the Novelty Works, and other factories along the East Side waterfront. The convergence of rioters from different parts of the city was more likely a result of rumor and circumstance than explicit and coordinated design.[9]

At ten-thirty, draft selection began in the Ninth District Office while the now sizable crowd milled outside. After fifty or so names had been drawn, the Black Joke Engine Company Number Thirty-three arrived dressed in full fire company regalia. One of their men had been selected in Saturday's lottery. Many firemen thought their traditional exemption from militia service should extend to the federal draft, and over the weekend the Black Joke Company had resolved to halt Monday's Ninth District proceedings.[10] A pistol shot rang out, and the Black Joke men burst into the office, smashed the selection wheel, and set the building on fire. A deputy provost marshal who tried to persuade the Black Joke men to fight the flames was beaten to the ground.[11] The throng outside listened to a Virginia lawyer and Confederate sympathizer named John U. Andrews deliver an anti-draft speech; some mistook Andrews for the Peace journalist Benjamin Wood.[12] As the blaze spread to tenements next door, Chief Engineer John Decker of the Fire Department urged the crowd to let his men through and spare the belongings of poor workingmen. He was seconded by Black Joke foreman, uptown building contractor, and Democratic alderman Peter Masterson, who approved the destruction of the draft office but hoped the rioters would now let Decker and his company do their job. The crowd applauded Masterson and cleared a path, but a mob returning to the scene after a battle with police drove Decker and his men away. By eleven-thirty that morning, orders were given to suspend the draft and transfer government papers to Governor's Island for safekeeping.[13]

By this time, the rioters had virtually halted the business of the city, particularly in the uptown wards. In one reporter's account, "men left their various pursuits; owners of inconsiderable stores put up their shutters; factories were emptied, conductors or drivers left their cars, employees at railroad depots all added formidable accessions to the depots. . . ." The crowds on the Upper East Side avenues had now swelled into a "concourse of over twelve thousand" that included men, women, and children of every social grade who had put down their work to discuss the Conscription Act or merely to watch the disturbance and ponder what direction it would take. As most of the city turned its attention toward the rioters and their activities, the draft became Monday's "all-absorbing subject."[14]

Rioters now began to gather at sites emblematic of the political choices of the day to voice their approval or objection. One cohort paid a friendly visit to Democratic General George B. McClellan's house on East Thirty-first Street.[15] Shortly after noon, rioters appeared at Printing House Square, the journalistic thoroughfare, where one could always find the makings of a crowd awaiting the latest war news. They huzzahed McClellan and the buildings of the Peace Democrat *Daily News* and *Weekly Caucasian* and groaned at the offices of the Republican *Times* and *Daily Tribune*. To many New Yorkers, the round, bespectacled face, drooping figure, and telltale white overcoat of Horace Greeley, editor of the *Tribune,* embodied the Republican Party and the anti-slavery cause. James H. Whitten, a neighborhood barber known for his extreme Confederate views, challenged Greeley to show himself and threatened to kill the abolitionist and gut the office. The crowd resisted Whitten's urgings and left the building alone when Greeley failed to appear. After chasing a policeman across City Hall Park, the rioters dispersed and Whitten returned to his barber's chair.[16]

During the morning, divisions among the rioters began to emerge. One committee that closed factories on the East Side explicitly limited its aims to halting the draft. James Jackson, owner of an iron works, reported a conversation with a delegation from an assemblage of one hundred men and boys outside his factory between nine and ten o'clock: "The leaders said they wanted the shop to close and [I] asked how long it must close. They said I might go to work the next day. They stated that their only object was to make a big show to resist the draft. They said they had no other motive than to have the men join them to put down the draft."[17] Employee Charles Clinch heard the leader of the deputation, cartman Thomas Fitzsimmons, say "he did not wish to injure Mr. Jackson, but wished the mob to go away."[18] Jackson closed his shop and his property was left unharmed, while the crowd marched off to nearby Franklin's Forge.[19] For Fitzsimmons and his committee, the "strike" against conscription was a one-day affair. They sought to interrupt the draft but wanted no wholesale onslaught against private property. Their notion of a peaceable "demonstration" may have been shared by the rioters who counseled restraint at the Provost Marshal's Office and the *Tribune* building. Fitzsimmons's action was already different in style and scope from that of the rioters who committed the early morning thefts, interfered with transportation and communication lines or encouraged the spread of flames from the Ninth District office to adjacent buildings.

After Monday noon the revolt against conscription began to expand beyond the limited protest of rioters like Fitzsimmons and his associates. By midday enrollment officers around the city had received orders to suspend the draft; soon there was a popular awareness that the draft had been interrupted and "all its machinery [put] speedily beyond reach of any such undertaking as that of the Black Joke."[20] The deserted Eighth District Provost Marshal's Office at Broadway and Twenty-ninth Street was burned at five o'clock by Patrick Merry, an Irish cellar digger who lived in the neighborhood, and two or three hundred men and boys. Before firing the draft office, Merry ordered a gang of marble cutters at a Broadway construction site to stop work and join his "band" and then attempted to close down John W. Onderdonk's hardware shop next door to the draft office.[21] Meeting resistance at Onderdonk's, Merry and his comrades then sacked the Provost Marshal's Office and set it and the hardware shop ablaze. While the fire spread down the block, the crowd went on a looting spree. Gold bracelets and brooches were taken from a jewelry store and valuables grabbed from upstairs apartments. Finally, the crowd attacked a black fruit vendor and stole his produce.[22]

Here was the closing of work sites and shops, the assault on federal property—the same sort of strike against the Republican draft—that Thomas Fitzsimmons and most Monday morning rioters would have found familiar. But the Merry episode had a new twist. The strike was aimed more at purging the neighborhood of the draft apparatus, since the Eighth District lottery had been interrupted hours earlier. Now, too, the rioters' anti-draft action was joined with looting, destruction of property, and an assault on a black man, activities soon to be repudiated by some of the participants in the revolt.

By five o'clock Monday, the list of the rioters' targets had lengthened considerably. Beginning Monday, homes suspected of harboring policemen were burned by the crowd. Policemen caught by the rioters were often stripped of their clothing and literally defaced—beaten on the face and head until unrecognizable.[23] Bitter assaults against the police would continue through the week. By mid-afternoon Monday, hostility toward the police was combined with an animus against well-dressed gentlemen and the houses of wealthy Republicans.[24] Rioters tore through expensive Republican homes on Lexington Avenue and took—or more often, destroyed—"pictures with gilt frames, elegant pier glasses, sofas, chairs, clocks, furniture of every kind, wearing apparel, bed clothes. . . ."[25] At three o'clock, a nine-year-old black boy was attacked by a downtown mob at the corner of Broadway and Chambers Street.[26] Later in the afternoon, a crowd threatened the Eighteenth Ward Police Station, raided an armory in search of weapons, and drove off the police sent to protect government property. When police reinforcements arrived, the crowd torched the building.[27] Another crowd left the charred site of the Ninth District draft building to burn Allerton's Bull's Head Hotel, which housed an office of the American Telegraph Company. The gathering at Bull's Head "divided into two or three gangs, with leaders bearing pieces of boards for banners on which were written . . . 'No draft,' etc., and it was unsafe to express a single word in dissent from the proceeding." One of these gangs held up a tantalizing sign that read "Independent."[28] By supper time, one group had set fire to the splendid Colored Orphan Asylum on Fifth Avenue and another began attacking black men and boys in the tenement district along the downtown waterfront. But anti-Republicanism remained the refrain of the violence as crowds returned to Greeley's *Tribune* office in the early evening, stormed the building, and set it afire before police drove them away.[29]

Late Monday we also find the earliest evidence of some rioters abandoning the violence and in many cases joining forces with city authorities to protect property and suppress the uprising. Thomas Fitzsimmons, leader of the committee which closed Jackson's Foundry during the morning hours, had by nightfall organized a patrol to guard property on his block.[30] The other leader of the Twenty-eighth Street patrol was Richard Hennessey, a local soap manufacturer and acquaintance of the Fitzsimmons family.[31] Fitzsimmons and his vigilante police force maintained a twenty-four-hour watch through the week and successfully protected a black man from a lynch mob on Wednesday.[32] Meanwhile Fitzsimmons's assurance that Jackson could reopen his shop Tuesday morning proved false. The iron works was kept closed through the riot week and well into the next by threats of arson from a new committee with intentions far different from those of Fitzsimmons's Monday delegation.[33]

The most obvious repudiation of the violence Monday evening was that of the volunteer fire companies. Firemen were prominent in the anti-draft demonstration of Monday morning but solidly arrayed against the rioters by Tuesday. Engine Company Number Thirty-three, leader of the Monday assault on the Ninth District draft office, was by Tuesday defending its Upper West Side neighborhood against riot and arson.[34] Local residents furnished the Black Joke men with refreshments through the week, and railroad directors and property owners published a letter of gratitude to the fire laddies in the city press.[35] On Tuesday evening, Hook and Ladder Company Number Twelve fought its way through a barricade to douse the flames at the Twenty-second Street Police Station, and then, clearing another set of barricades, rushed to a Fourteenth Street lumberyard where the firemen battled rioters until dawn.[36] Engineer Henry Lewis and his vigilance committee arrested eight "thieves" nearby and recovered $1200 worth of stolen property in the course of their patrol.[37] Still, many fire companies feared their efforts to uphold public order would obscure their ongoing hostility toward the draft. The Forrest Engine Company Number Three printed a notice affirming their opposition to the draft, which it declared to be "unnecessary and illegal." "But in the present exciting times," the resolution continued, "we deem it our duty . . . to protect the property of the citizens of the Eleventh Ward to the best of our abilities."[38] Some rioters, then, continued to denounce the policies of the Republican administration even when they disavowed or campaigned against the unfolding insurrection against property and the social order.

The tensions among Monday's crowds and the abandonment of the revolt by some rioters suggest that from the outset the uprising had different sorts of participants with diverse grievances. At least through early Monday afternoon, the course of violence was much influenced by rioters concerned primarily with the administration of the draft. These workingmen and their families, whose actions call to mind Brooks's "Conscription Riot," denounced Republicans Lincoln and Greeley, carried placards against the draft, and cheered as the Ninth District draft office burned.[39] Like the committee at Jackson's Foundry they used the technique of marching from shop to shop and corraling men into the crowd to increase their numbers and inform the city's workers about their action. This style of protest was familiar to all New York City workers in the 1850s and 1860s as a method of enforcing strikes.[40] These rioters displayed great hostility toward government representatives and property and even condoned the burning of federal buildings and the attacks on Metropolitan Police and United States Army Invalid Corps arriving to investigate the disturbance on Third Avenue. It was even conceivable that a rioter of this sort might have joined in the savage beating of Police Superintendent Kennedy on Monday morning.[41] But these rioters seem to have shunned the looting, the hanging and mutilation of black men, and the attacks on war industries that began Monday and continued through the week; this was the crowd William O. Stoddard (private secretary to President Lincoln and a member of the "volunteer special" police during the bloody week) described as "honest laboring men, of all political parties, thousands of them . . . willing to parade the streets in an 'anti-Draft demonstration,' and to do any required amount of shouting and all that sort of thing, who were at the same time not at all inclined to commit either burglary or arson or murder."[42] Their enemies were the Republican administration and the draft law with its offensive substitution and three-hundred-dollar commutation clauses. Some of these rioters may have regarded a conscription law without the invidious provisions as a legitimate and equitable way to hasten a Northern victory.[43]

Alongside Monday's rioters of limited aims were others with larger designs. These men, women, and children helped to destroy the Provost Marshal's Office, attacked city police and federal officials, and closed factories. But this cohort also severed telegraph lines, tore up railroad tracks and committed the first acts of theft Monday morning. It was probably rioters of this sort who later in the week destroyed the track of the Hudson River Railroad, attacked the Weehawken and Fulton ferries, and attempted to burn the Harlem and McComb's Dam bridges and the Manhattan and Metropolitan gas works.[44] Early Monday, one crowd within earshot of a reporter cut down telegraph poles after discussing the possibility of the authorities summoning troops from Albany.[45] The assaults on telegraph lines, ferry slips, railroad tracks, and gas factories went beyond mere machine breaking to disclose a grander anticipation of—or even plan for—protracted confrontation with the authorities.[46] We do not know how elaborate this plan was or whether such attacks were coordinated, but the appearance of deliberate scheming was compelling enough to convince Republicans and Democrats alike that the destruction of transportation and communication lines and efforts to darken the city were the *coup de main* in a Confederate conspiracy to coordinate Southern military victories with a mob takeover of New York.[47] Contrary to such perceptions, the rioters' attacks seem to have been independent in motive and free from outside manipulation. If anything, the attacks revealed these crowds' keen sense of the dynamics and structure of the city.

Only as the violence entered its second day—the rioters of limited aims now retired and the more ambitious crowds in full force—did it become possible to identify each of the riot's constituencies. Monday's rioters were by no means all Irish and Catholic. Some were German-speaking (one observer thought the early morning procession looked like "some German festival") and some were native-born and Protestant.[48] Artisans in the building trades, who formed the backbone of the uptown fire companies, figured prominently in reports of the Monday morning factory closings and antidraft procession.[49] These were the painters, carpenters, bricklayers, stonecutters, and small building contractors who pushed the city's line of settlement north toward Central Park in the 1850s and 1860s.[50] By the middle of the week, the social complexion of the crowds had changed. Midweek rioters were more predictably Irish and Catholic, and they were more likely to be members of trades restricted to sons of Erin.[51] German-Americans had by Tuesday organized against the insurrection and were commended by the Republican press for their loyal and orderly comportment.[52] Tuesday's and

Wednesday's rioters tended to be industrial workers and common laborers employed in the city's iron foundries, railroad shops, and dock and street construction gangs, especially in the upper wards. The laborer rioters, it should be said, generally did not toil alongside or under the supervision of skilled craftsmen; instead, these quarrymen, street pavers, cartmen, and longshoremen independently organized and managed their own work. Observers frequently commented on the role of "half-grown boys" in the midweek crowds.[53] Of the two dozen identifiable wounded rioters brought into Bellevue Hospital on Thursday of riot week, seven were listed as "minors," six as "laborers," two as "boilermakers," and one as a "cartman." Only one, sixteen-year-old plumber John Ennis, was associated with the building trades. By midweek, artisans in building had largely disappeared from the newspaper accounts and presumably had abandoned the streets.[54]

The lines between constituencies were sometimes blurred. Some Germans appeared among Tuesday's and Wednesday's rioters, some Irish left the streets Monday afternoon. Occasionally artisans rioted through the week, and sometimes industrial workers and laborers dropped out of the crowds early on. But the most clear and abrupt social division was that between journeymen in the older artisan trades, who limited their participation to Monday's demonstrations, and workers in newer industrial occupations and common laboring, who persisted in the midweek revolt. By Tuesday the riot had entered a new phase in which the animus came primarily from Monday's most violent rioters.

As the uprising began its second day, it also became clear that the rioters had altogether failed to lure one important segment of the work force onto the streets. Observers noted that neighborhoods which had witnessed the most violent antebellum strikes and riots were relatively quiet in July 1863. Street violence had time and again punctuated the political life of the "Bloody Sixth" Ward: an anti-abolitionist riot in 1834, a bread riot in 1837, a tailors' strike in 1850 and finally the sanguinary police riot of July 1857.[55] The "Irish Fourteenth" Ward had been home to Michael Walsh and his boisterous "Unterrified Democracy" of the forties and early fifties. During the draft riots, the *Tribune* remarked, "the people of the Sixth and Fourteenth Wards . . . refrained from participating in the outrageous conduct of the mob. . . ."[56] This was meager consolation for resident black families, who found themselves abandoned by white neighbors and easy prey for marauding bands.[57] Notwithstanding such racial violence, the demonstration against the draft in either of its phases never took root in the heartland of antebellum labor protest. Instead Tammany officials including Alderman John Fox, Judge Joseph Dowling, and Comptroller Matthew Brennan presided over an early return to work on July 15.[58] Though this area had long been Democratic, midweek events confirmed it as Tammany territory. Tammany editor John Clancy told his readers on July 18, "Let us hear no more the libellous epithet 'Bloody Sixth.' "[59]

The Bloody Sixth and its northern neighbor the Fourteenth were the heart of the seaport's downtown manufacturing district. The hundreds of sweatshops clustered in the Sixth and Fourteenth turned out clothing and shoes for the trade in consumer-finishing goods and employed the city's most proletarianized poor.[60] These workers toiled in occupations where craft skills had largely been eroded by industrialization. That the employees of the sweatshop district were uninterested in draft rioting—though they were generally faithful Democrats likely to be critical of Republican policies—did not augur well for the laborers and industrial workers who resumed the revolt Tuesday morning.

Tuesday, July 14, to Friday, July 17

The line dividing the riot-week movements of common laborers and industrial workers was more finely drawn than the broad stroke that separated their activities from those of the artisans. The artisan penchant for discipline and rationalism, which may have helped to bring the limited demonstrations to a halt Monday afternoon, was foreign to the thinking of both laborers and industrial workers. These more destructive rioters were ardent foes of the Republican government and all its works, emancipation and federal taxation no less than the draft. They regarded the strike as their most potent weapon against Republican rule and continued to tour the city closing factories and laboring sites. The account of the riot's second and last phase thus begins with Tuesday's vandalism of Republican homes and hostile displays against patriotic symbols, episodes that may well have involved all types of midweek rioters. But increasingly the rift between laborers and industrial workers appeared in vivid relief. The account ends with a discussion of differences in the relations of these two groups with the black community, the "better classes," and the official representatives of Republican authority.

A rage against well-known Republicans and their property remained a dominant theme of the violence. Rumors of an attack planned against the home of abolitionists James Sloane Gibbons and Abby Hopper Gibbons and that of their neighbor James Sinclair (a relative of Greeley's) circulated through the uptown shops Tuesday afternoon.[61] Indeed the Gibbonses' problems with "the Irish mob" had begun the previous winter when they celebrated the Emancipation Proclamation by illuminating their house and draping the windows with red, white, and blue bunting; later that evening they found their front door, steps, and pavement smeared with pitch.[62] The draft rioters' raid on the Gibbonses' house seems to have been one of the more concerted and systematic attacks of the riot week. Late in the afternoon of July 14, two men on horseback waving swords and shouting "Greeley! Gibbons! Greeley! Gibbons!" (the journalist had boarded at the Gibbonses' and was rumored among the rioters to live there) galloped up to the house; a crowd of men and women followed close behind. The horsemen stopped one at each side of the courtyard and allowed about a dozen men with pick-axes into the house while they kept the rest of the mob back. Inside the men "destroyed what they could and threw things out the windows."[63] Finally the advance team was joined by the throng without. The rioters were driven away by a company of soldiers on two occasions but each time returned to the house. Fires were lit and the house was saved only by the intervention of neighbors afraid of spreading flames.

The rioters vented their greatest anger against the Gibbonses' domestic effects:

> The witnesses all agree that a great deal of the furniture was thrown out of the windows, most of it having been previously injured. . . . Many of the books and papers in the library were used to kindle the fires, placed under the furniture collected for that purpose; others were scattered about and trampled upon. . . . The pictures and works of art were mostly defaced or injured in the house. . . . The crockery, etc., was demolished in the house. The carpets and oil cloths were greatly injured, and after having been nearly destroyed, were mostly carried away.[64]

The official who would later examine the Gibbonses' claim for riot damages estimated that only "20 per cent of the articles lost were stolen."[65] James Gibbons recalled that "the piano was actually fired and broken up, and carried off in fragments."[66] The "riot of thieves" à la Brooks was in this instance bent primarily on defacing and destroying property. It was as if demolishing cultural possessions—what one friend of the Gibbonses reverently referred to as the "household gods"—was the rioters' way of destroying the essential attribute of the Republican elite and guaranteeing their departure from the neighborhood.[67]

During the vandalism of the Gibbons home, two drunken Irishmen, Michael O'Brien and John Fitzherbert, led a nearby crowd in the tearing of an American flag to Fitzherbert's chants of "Damn the Flag!"; in a later incident Fitzherbert cheered "Jeff Davis."[68] Midweek rioters saluted a long list of Peace Democrat and Confederate heroes, from Horatio Seymour and General McClellan to the Wood Brothers to Jefferson Davis.[69] By contrast, the Wednesday rioters who marched through Pitt and Broome streets closing factories and machine shops held an American flag aloft.[70] Such diverse and often conflicting demonstrations of allegiance suggest that pro-Confederate statements were a convenient way for some rioters to denounce the Republican Party. More likely than not, these displays and salutes did not represent deep-seated Confederate sympathy. The most striking historical analog to this use of pro-Confederate cheering as a rallying point for anti-Republican Party sentiment was pro-French sloganeering during the violent popular resistance to the Militia Act in Ireland in the summer of 1793. Militia rioters in the late eighteenth-century Irish uprising spoke of their expectations of aid from France.[71] As the draft rioters began to lose hope late in the week, they seized upon any negative point of reference to the Republican authorities. If they only held out a little longer, some whispered, "Baltimore" would come to their assistance. The rioters may have remembered Baltimore as the city in which crowds attacked the Massachusetts Sixth Regiment on its way to the front in April 1861.[72] This heightened and often desperate feeling against the government was characteristic of much of the midweek violence.

After Monday the crowds increasingly turned their attention toward the local black community. Threats of violence, and the occasional attacks on black workingmen which gave such threats their bite, were a regular feature of race relations in the city during this era. But racial tension was high during late spring 1863 because of the shipping companies' decision to employ black labor to break a longshoremen's strike. The week before the riots, Police Superintendent Kennedy reported a growing incidence of physical assaults on black people. Fearing an outbreak, he urged Secretary of War Stanton not to parade the black Fifty-fifth Massachusetts Regiment through the city as planned.[73] On Sunday, July 12, there were a number of arson attempts against houses on Carmine Street, the heart of one of the city's several black enclaves.[74] In a sense, the racial violence of the draft riots was the quickening of an already accelerated tempo of intimidation and assault.

What made the July riots' brand of racial attack new was its sweeping character. Intimations that white working people were going to approach the matter of racial domination with new intensity and thoroughness could be found in the Monday afternoon razing of the Colored Orphan Asylum, followed by the smashing of its furniture and the uprooting of surrounding trees, shrubbery, and fences.[75] The crowds' desire not merely to destroy but to wipe clean the tangible evidence of a black presence surfaced a few hours later when a waterfront lynch mob hanged William Jones, then burned his body.[76]

The riots were an occasion for gangs of white workingmen in certain trades to introduce into the community the "white-only" rule of their work settings. Bands of Irish longshoremen, many of whom lived within blocks of the piers they worked, began the first racial attacks Monday afternoon. Committees of the "Longshoremen's Association" patrolled the piers in the daylight hours insisting that "the colored people must and shall be driven to other parts of industry, and that the work upon the docks, the stevedoring, and the various jobwork therewith connected, shall be attended to solely and absolutely by members of the 'Longshoremen's Association,' and such white laborers as they see fit to permit upon the premises."[77] Irish street pavers, cartmen, and hack drivers followed suit in other parts of the city, though not with the longshoremen's visible organization and proclamations.[78] Any talk of associations ceased at sunset when parties of men and boys abandoned watch over the piers, factories, and laboring sites for a tour of the surrounding tenements. "Dock laborers" were responsible for the Wednesday night beating and near drowning of black workingman Charles Jackson.[79] They were probably also involved in the Monday, Tuesday, and Wednesday evening attacks on waterfront dance houses, brothels, and boarding houses that catered to black laborers and sailors or, as one city official put it, "negroes . . . of the lowest class."[80] Regardless of their own race, boardinghouse keepers and runners known for black clienteles were stripped of their clothes and threatened with the hangman's rope or driven from the area.[81] On Roosevelt Street, tenements that housed black families were torched and the victims' furniture demolished and burned in sidewalk bonfires. The crowds directed their greatest fury against black men, though black women who protected their husbands and sons could become targets through association. Waterfront rioters seized Jeremiah Robinson, a black man trying to escape to Brooklyn wearing his wife's clothing, beat him senseless and threw his body into the East River.[82] By midweek, the rioters had virtually emptied the harbor front of people of color.[83]

The most violent racial purges of tenement districts were the special province of the men and boys of laborer families. Black sailor William Williams was assaulted by longshoreman Edward Canfield and two other laborers at dawn Tuesday when he walked ashore at an Upper West Side pier to ask directions.[84] Like many of the racial murders, this attack developed into an impromptu neighborhood theater with its own horrific routines. Each member of the white gang came up to the prostrate sailor to perform an atrocity—to jump on him, smash his body with a cobblestone, plant a knife in his chest—while the white audience of local proprietors, workmen, women, and boys watched the tragedy with a mixture of shock, fascination, and, in most instances, a measure of approval. A couple of onlookers did slip away to notify the police, but, as was so often the case, no member of the white audience tried to intervene. To the contrary, milkman Daniel Greenleif reported, "after the occurrence there were several cheers given and something said . . . about vengeance on every nigger in New York." The performance over, the assemblage retired to a nearby liquor store. Some minutes later and down the block, a cartman opening his stable to begin the workday was warned by a man, possibly Canfield, "not to put any niggers to work." When the police arrived, they found the street quiet and Williams in a bleeding and insensible condition from which he died soon after.[85]

Elaborate dramas of a similar kind were reenacted during the course of three Upper West Side lynchings. At dawn Wednesday, nineteen-year-old William Mealy spotted black shoemaker James Costello on West Thirty-second Street and gave chase; Costello fired a shot in self-defense.[86] The shot drew the attention of five or six white men, laborer Matthew Zweick maybe among them.[87] The party pulled Costello from the house where he sought refuge, alternately beat, kicked, and stoned him, trampled on his body, and finally hanged him. Before the episode was complete, two of the party dragged him half-dead to a mudhole where, in a variation on the theme of tar and feather, one immersed him in water while the other emptied a barrel of ashes over his head. Finally the rioters plundered and burned down the house where Costello had attempted to hide.[88] Some hours later laborer George Glass yanked crippled black coachman Abraham Franklin and his sister Henrietta from their rooms a few blocks away, roughed up the girl and dragged Franklin through the streets. A lamppost was found and Franklin was hanged. The military arrived, scattered the crowd and cut down Franklin's body, but when the soldiers departed, the corpse was hoisted up again with cheers for Jefferson Davis.[89] Then the crowd pulled down Franklin's body for the final time. In a grisly denouement, sixteen-year-old Irishman Patrick Butler dragged the body through the streets by the genitals as the crowd applauded.[90] After yet another hanging in this neighborhood, rioters cut off their black victim's fingers and toes.[91] The houses of these black residents were often identified, if they needed to be, by bands of small boys who "marked" them by stoning the windows. The boys later returned with their male elders to pull out the black tenants and complete the bloody mission.[92] Through such elaborate routines, these white workingmen and boys cultivated a dehumanized view of their black neighbors.

Sexual mutilation, burning, and drowning of victims call to mind a traditional and highly symbolic strain of popular violence dating at least as far back as the early modern era.[93] Patrick Butler's public display and appropriation of Abraham Franklin's body and one crowd's amputation of a victim's fingers and toes pointed to a need among rioters of this stripe to prove sexual conquest of the black male community through symbolic acts. Startling even to New Yorkers accustomed to the bloody street melee, these acts no doubt served to dehumanize and objectify black men further in the minds of their white attackers.[94] After the manhood of black workingmen had been publicly reified and debased, white laborers seem to have imagined, an objective black male presence could be cleansed from the neighborhoods. We must be careful here not to ascribe too much structure and rationality to such emotional behavior. But it is certainly worth wondering whether bonfire lynch murders and drownings of black victims were the final acts in much improvised dramas of conquest and purification. Fire and water would symbolically render harmless what these rioters perceived as the post-emancipation social power of their black neighbors. In the view of many white Northerners, the political ascendancy of the Republican Party was causing a revolution in race relations in their communities. The demise of slavery as an institution and a national question, almost inevitable by summer 1863, seemed to signal the end of an era in the North as well as the South. But no one set of rules emerged to replace an older *modus vivendi* between the races in Northern cities. Through sexual conquest and purification, many white workingmen may have hoped to erase the threat of a new black dignity at a time when the social and political status of black people was especially unsettled.[95]

The behavior of Patrick Butler requires additional comment. The image of the Irish youth dragging Franklin's corpse through the streets burns itself into the mind even as our sensibilities become dulled from many descriptions of bloody scenes. Butler himself, it should be noted, was listed in the court records as a "butcher": practitioner of an old and violent craft as renowned for knife-flourishing on the streets as within the market stalls. Nonetheless it is hardly sufficient to associate this rioter's brutality with the behavior of an occupational group. The sexual intensity, exaggerated gestures and bravado of Butler's act call to mind, more than anything, the mentality of a sixteen-year-old boy.[96]

Boys often led the most violent and sexually charged attacks on black men. Young and unmarried males were, of course, especially liable to be drafted. But, as Mayor George Opdyke noted in his account of the riots, so many of the rioters were younger than the minimum draft age of twenty that the simple motive of self-protection against the Conscription Act does not explain their actions.[97] If not always vulnerable to the draft law, New York's young immigrant workers were vulnerable in a larger economic and social sense. They were easily the most underemployed members of the white male labor force. The marginality of young male workers was especially acute at a moment when, in many trades, older traditions of apprenticeship were breaking down or altogether gone and a factory work force based in part on youth labor was only beginning to emerge. During this especially bleak interregnum, many poor boys, regardless of their chosen trade, may have wondered whether they would ever become full-fledged participants in the family and job networks that defined social maturity for the mid-nineteenth-century adult workingman. Attacks on both Republican draft offices and the bodies of black workingmen make sense if we imagine white working-class youths to be seeking to restore to the community not just political order but also social and sexual order. If a sense of insecurity helped to inspire the racial attacks of white laborers who competed with black workers, it played an even larger role in the white youth riot that figured as a leitmotif in the midweek violence.[98]

Fear of racial amalgamation, the theme of so many earlier anti-black riots in New York City, was also an element of the violence.[99] Crowds visited and often attacked the homes of racially mixed couples and white women who kept company with black men. Tuesday night, downtown saloonkeeper and small-time Democratic politician William Cruise tried unsuccessfully to instigate a crowd of men and boys to burn the house of Mary Burke, a white prostitute who included black men among her clientele.[100] Cruise had better luck when he gave straw and matches to a gang of boys and led them around the corner to the rooms of William Derrickson, a black laborer at a local loading depot. William escaped but his son Alfred was pulled out on the street, stripped, and beaten. A fire was started under a lamppost, and Alfred would have been lynched and burned to death were it not for the eleventh-hour intercession of grocer Frederick Merrick and a group of local German residents who chased the rioters away.[101] That same evening, laborer Thomas Cumiskie joined physician Thomas Fitzgerald and a crowd in an unsuccessful effort to burn the house of Harlem abolitionist and Internal Revenue Collector Edgar Ketchum.[102] The next night Cumiskie led rioters around East Harlem to solicit money for a round of drinks. The rowdies' destination was the shanty of Ann Martin, where they interviewed Mrs. Martin and debated whether her dwelling should be burned.[103] In these incidents, the women involved—Ann Derrickson and Ann Martin—were white wives of men of color. The sexual policing of black men may have been a motive here much as it was in the murder of Abraham Franklin.

For many crowds in which laborers were prominent, it became important to confirm the loyalty of friends and identify enemies among the neighborhood "chamber of commerce." On Wednesday, Martin Hart, a gaspipe layer recently arrived from Ireland, Adam Schlosshauer, a German gardener, and John Halligan, another laborer, led a band of "at least eight" revellers around Harlem asking storekeepers and employers for liquor money. One man made the request for fifty cents or a dollar while two others serenaded with flute and tin kettles. The group marked the names of stinters on a card and left them with the warning "We'll settle with you tonight."[104] John Piper, a teamster whose name had come up in the lottery, asked Upper West Side slaughterhouse owner Thomas White for "money to treat the boys" and, receiving a few shillings, invited some bystanders into White's office for a drink. Piper announced that he hoped to rally "all the G. D. drafted men to resist the draft" but had come "to inform his friends" what the mob was going to do and "see that his friends would be protected." Promising White immunity, the teamster boasted that a neighboring abbatoir owner would be burned out because "he was a mean G.D.S. of a B.," he had prevented the burning of a black residence, he was German, and "he had informed on the rioters."[105]

In these and other incidents, treating was serious business. The crowds often brandished clubs and revolvers; the threat of arson was carried out against proprietors who refused to pay up.[106] Under such duress, the offer to stand drinks cannot be said to have been freely given. But even then treating carried connotations of friendship and sympathy between donor and recipient, connotations magnified by the black and white moral universe of midweek in which capitulation to the mob furnished grounds for suspicion of treason. While the motives of Hart and company were more purely bacchanalian than those of Piper, the issue of support for the ongoing draft riot never wholly disappeared from these interchanges.

Some rioters were easily appeased: a suspicious party who performed an act of generosity, produced the proper "friendly" sponsorship, or proclaimed anti-draft sentiments could readily undermine the damning evidence of dress, occupation, or political views. In the Piper episode Policeman William McTaggart was among the bystanders ushered into White's establishment after White had offered to stand drinks. Piper asked White if McTaggart was "reliable," and when reassured, proceeded to reveal the rioters' plans of destruction on the Upper West Side.[107] A speech could be as effective as an offer to treat in quelling the rioters' suspicions. Director of the Sixth Avenue Railroad Alfred G. Jones was pulled from his depot on Tuesday by a gang seeking to force those men still working to join their ranks. "I promenaded arm in arm with a drunken Irishman," Jones confided to his diary, "and was let go upon making 'a speech against the draft.' . . . They said I was all right and left me after shaking hands generally."[108] For these rioters—and they were frequently laborers—the project of sweeping the city clean of individuals and institutions responsible for the draft was as much one of social inclusion as it was one of exclusion. A rage against elite offenders went hand-in-hand with chivalry toward elite friends.

As the week progressed, the attention of such crowds began to drift from the Republican government and its works to other community institutions they suspected of hostility to the insurrection, exploitative behavior, or moral reform. On Monday evening, fifty or more boys set fire in the parlor of Republican Postmaster Abram Wakeman's abandoned Yorkville townhouse. Next the crowd burned a police station across the street. Then the rioters moved to the Magdalen Asylum, a home for aging prostitutes on Eighty-eighth Street. The moral reform of prostitution had been a favorite project among the evangelical Protestant middle class since the 1830s, but the Magdalen Asylum hardly qualified as a Republican Party outpost.[109] With laborers Richard Lynch and Nicholas Duffy at the fore (Duffy was a well-known local character), the crowd interviewed the asylum superintendent, departed, and then returned for a follow-up discussion; finally they announced "they were going 'to burn the building the same as they had [the Republican] Wakeman's house.' . . ." They took a desk, fired the building, and released the prostitutes.[110]

In this speculative fashion some rioters explored the political and social allegiances of the Protestant middle and upper classes. Such attacks often surprised the victims themselves who, like the Magdalen superintendent, were not sure why they had become targets. Late Monday, rioters approached Professor John Torrey's house near the grounds of Columbia College "wishing to know if a republican lived there, and what the College building was used for."[111] "They were going to burn Pres. King's house," Torrey wrote, "as he was rich, and a decided republican. They barely desisted when addressed by the Catholic priests. The furious bareheaded and coatless men assembled under our windows and shouted aloud for Jefferson Davis!"[112]

The Catholic priests calmed the rioters' suspicions, as did the family of a Doctor Ward later that evening. The Ward family saved their nearby Fifth Avenue mansion by assuring the crowd that "they were all Brackenridge [*sic*] democrats and opposed to the draft."[113] The entreaties of the superintendent of the Magdalen Asylum were less persuasive. Some rioters lengthened their list of enemies to include an array of middle- and upper-class Protestant individuals and institutions. For these crowds, the midweek revolt had moved beyond an attack on the agencies of the wartime state.

As the Magdalen and Columbia episodes suggest, Catholic loyalties mattered and Catholic antipathies could broaden the swath of the midweek riot. At Tuesday dawn a Harlem temperance and music room was burned, and later in the week two Protestant missions were wrecked.[114] On Wednesday we find the first evidence of rioters proclaiming Irish and Catholic identity as an explanation or legitimation for their attacks. In one incident a Central Park laborer named Doherty incited a gang of Irishmen to warn away and burn the house of Republican lawyer Josiah Porter, whom they called a "black orangeman [and] . . . a black republican at that"; Porter had refused Doherty permission to build a shanty on Porter's land.[115] Earlier that morning Michael McCabe announced his membership in the "Hibernian Society" in the course of extorting forty dollars from a Harlem grocer (whose connection to the Republican Party, if any, remains unknown).[116] Wednesday night, uptown rioters' query to well-dressed gentlemen, "Are you for the Union?" was safely if evasively answered, "I am a Democratic Catholic."[117] While Catholic resentment of Protestant authority smoldered through both phases of the riot, it glowed brightest in the midweek violence.

Another "illegitimate" personality rioters sought to drive away was the waterfront brothelkeeper. The brothel district along the downtown West Side docks was the scene of attacks beginning Tuesday night. Unlike the firing of Roosevelt Street dance houses Monday night and William Cruise's threats to prostitute Mary Burke's rooms on Tuesday, the destruction of the West Side brothels was not necessarily linked to racial animosities. "Bands of rioters" toured the waterfront stores and wharves during the day on Tuesday and rallied laborers for the "procession" that evening.[118] A little after eight o'clock the assemblage divided into two parties and the march began.

Consisting "almost entirely" of boys and led by a man sporting a white hat with a feather, one group headed up Washington Street while another column turned up West Street along the wharves.[119] The first group, now joined by a party of women and men, entered and destroyed the saloon and brothel of Heinrich Strückhausen on Greenwich Street. Wash basins and water pitchers were broken, furniture smashed, liquor and cigars carried off.[120] The crowd continued on to John Smith's brothel at 157 Greenwich, where boys began throwing bricks at the door. Young longshoreman Martin

Haley was about to usher his comrades inside when the proprietor Smith suddenly appeared in the doorway and shot him dead; the marchers scattered.[121]

Prostitutes were never injured in these attacks—the rioters sought only to tear down the offending structure and drive the owner out of business. Musical instruments were thus the first objects to meet with the rioters' axes.[122] German proprietors were frequent victims, but these were not anti-German attacks per se. Though there is no direct evidence on this point, rioters may have regarded brothelkeepers, especially those who catered to German or black workingmen, as petty exploiters who attracted large numbers of migrant laborers to their districts and undermined their ability to control the labor market.[123] The raids on houses of ill-fame Tuesday night and through the week reflected the special social situation and needs of the waterfront laborer families.[124]

Such attacks were at best loosely linked to concurrent goals of repudiating the Republicans and their draft and destroying telegraph lines, ferries, and bridges.[125] Irish rioters now began to attack German and Jewish store owners.[126] Some of these shopkeepers were known Republicans, and occasionally attacks began with a formal discussion (a *Kleindeutschland* bartender managed to save his employer's premises by treating rioters to a round of drinks).[127] More often, though, mobs dispensed with interviews and tests of allegiance and attacked the German shops on sight.[128] One of the earlier reasons for destroying property—to prevent those parties "connected to the draft" from returning to the neighborhood—now disappeared in favor of outright theft.[129]

The most troublesome obstacle Irish rioters faced as they struggled to keep uptown factories and work sites closed through the week was the attitude of their German neighbors and fellow workers. Though many German workers had turned out for Monday's uprising, by Tuesday their enthusiasm for the strike against the draft had flagged. A report came into the *Tribune* office Tuesday night that a gang had approached the German residents of Third Avenue between Fortieth and Forty-first streets "and threatened that if they did not join the rioters, they [the gang] would set the whole block on fire."[130] But what began as an attack on local scab elements had now become an indiscriminate race riot. On Wednesday, downtown rioters included on their list of targets a few defenseless Chinese peddlers suspected of liaisons with white women.[131] As the revolt drew to a close, some crowds seem to have assaulted all groups which on sight could be labeled exploitative, responsible for the presence of a strike-breaking element (real or imagined) in their districts, or merely unsympathetic. Here systematic and concerted attacks or exploratory interviews gave way to impulse.[132]

As these crowds grew less fastidious in their choice of targets, they also became more open to suasion by Democratic Party leaders. What may have been the last collective act of laborer rioters Thursday evening began as a challenge to Republican authority and ended as a dialogue in which Democratic elites convinced workers not to riot. The Seventh Avenue Armory was the great symbol of government military presence in the uptown wards. Rioters had threatened the building all week, and consequently it was heavily guarded. Rumor of a planned attack on the Armory reached Governor Seymour at the Saint Nicholas Hotel dinner time Thursday. Seymour dispatched N. Hill Fowler—Corporation Attorney, Peace Democrat, and longtime resident of the Upper West Side—to address the hundreds of working people who by nightfall were crowding the streets around the Armory. Fowler read a letter from Seymour suspending the draft in New York City and Brooklyn and received loud applause. After some further remarks advising opposition to the draft in the courts and not on the streets, Fowler left and the throng quietly dispersed.[133] By Thursday, laborers seemed to have imprinted their characteristic outlook on the class dialogue of the Upper West Side.[134] If the Democratic upper classes could protect them from the draft, these workers may have speculated, there would be no immediate need to contest the Republicans for power.

As laborers lengthened their list of social enemies and grew more tentative in their challenges to the federal government, industrial workers focused their choice of targets. Eighteen-year-old blacksmith Peter Dolan led a gang in a Tuesday attempt to destroy Republican Mayor Opdyke's house.[135] On Thursday, boilermaker Edward Clary assaulted and may have tried to shoot point-blank policeman William H. Carman (Clary denied only the final pulling of the trigger).[136] Such direct confrontation with the authorities became the special cachet of industrial worker riot activity by the middle of the week. While laborers and industrial workers agreed that their common political enemy was "Black Republicanism," laborers were more likely to emphasize "black," while industrial workers put their stress upon "Republicanism." Industrial workers wished less to coax upper-class loyalty and good will than to separate their world from that of the Republican elite and its authority.

The racial attacks of industrial workers, while often quite violent, were different in style from those of laborers. Blacksmith John Leavy and his son led a Tuesday night assault on West Indian broker Jeremiah Hamilton's Upper East Side residence. The Leavys worked together at a coach factory a few doors away.[137] Though one of the boys in the raiding party announced with some bravado, "There is a nigger living here with two white women, and we are going to bring him out and hang him on the lamppost," little violence occurred after it was obvious Hamilton was not home. "Sentinels" were posted at the corner of each nearby avenue. The elder Leavy and his gang asked Mrs. Hamilton for liquor and cigars, the boy Leavy requested a suit of old clothes, and the band departed. No attempt was made to burn the house.[138] In fact, few firings of black residences or bonfire murders of black men occurred in districts with large numbers of industrial worker families.[139] The uptown Eleventh Ward, known for its massive machine shops and its bitter hostility toward blacks, Republican politicians, and the draft, witnessed no bonfire lynchings and no arson against black homes, though there were more than a few black dwellings in the area from which to choose.[140] Of course racist sentiment in this neighborhood was intense. But industrial workers connected their racism to their primary targets, the Republicans and the draft. "It would have been far from safe for a negro to have made his appearance in that locality," wrote one reporter of the Eleventh Ward, "for the laboring classes there appear to be of the opinion that the negroes are the sole cause of all their trouble, and many even say that were it not for the negroes there would be no war and no necessity for a draft."[141] Like laborers, industrial workers hoped to drive black families from their districts. But sexually charged purges of black men did not occur in industrial worker neighborhoods.

In those Upper East Side neighborhoods, crowds continued to assault Republican homes and factories through the week (as late as Thursday morning, mobs were threatening the stately Republican residences along Gramercy Park). However, the Republicans who suffered the largest share of violence on the Upper East Side were draft officers, soldiers, and policemen. Women led the rioters' week-long crusade to drive from the factory districts all armed representatives of the Republican government. Women of the industrial Eleventh Ward, an observer noted, "vow vengeance on all enrolling officers and provost marshals and regret that they did not annihilate the officers when they first called to procure the names for the draft."[142] Most draft officials wisely stayed off the streets during the uprising (though in one instance rioters broke into the home of enrollment officer Joseph Hecht, called him "Mr. Lincoln," and beat him in his own parlor).[143] Government troops and police were left to bear the full brunt of popular wrath. On Tuesday an Upper East Side mob attacked, beat, and killed Colonel Henry O'Brien of the Eleventh New York Volunteers. The day before, O'Brien had used a howitzer to clear Second Avenue of rioters and killed a woman bystander and her child. On Monday night, O'Brien's house was gutted.[144] The next morning, neighborhood people spotted O'Brien returning to inspect his property. One man approached O'Brien from behind and clubbed him to the ground. The murder of Colonel O'Brien that afternoon lasted six hours.[145] Women dominated the crowds that first beat his face to a pulp, later pulled him through the streets and into his own backyard, stripped him of his uniform, and finally "committed the most atrocious violence on the body" before he died.[146] The crowds turned on a local druggist who offered the half-dead soldier a drink of water and wrecked the druggist's store. A girl who protested the violence was beaten and the house where she boarded destroyed. Only Father Clowry of nearby Saint Gabriel's managed to calm the crowd sufficiently to administer O'Brien the extreme unction.[147]

The assault on O'Brien was not special treatment reserved for Irish supporters of the government or the murderers of women and children, though O'Brien's deeds no doubt distinguished him as a target. A similar pattern of violence characterized the dozens of assaults on policemen and soldiers through the week.[148] Fierce beatings were administered to these officials, but the women and men in the O'Brien crowd and others like it were not concerned with purification as were the racist lynch mobs. There were no burnings or drownings of police or soldiers. The participation of working-class wives suggests these events were not merely the outgrowth of the male workplace experience and may have relied as well on the neighborhood networks of poor Irish women. No rioter announced that his or her attacks on policemen and soldiers were inspired by hostility to the Republican Party. Yet there is reason to suspect that the city's immigrant poor did associate such armed authority with Republicanism. It was not that all soldiers and policemen were Republicans (many were not). Rather, the Union Army and Metropolitan Police were institutions that had emerged under the auspices of the Republican Party. The Army's relation to the Republican national government was self-evident; the Metropolitan Police was a creation of the local Republican elite, which had wrested control of city law enforcement from the Democratic Party in July 1857 only after suppressing a riot opposing the change.[149] Under the zealous leadership of Superintendent John A. Kennedy, whom the rioters singled out for vengeance early on, the Metropolitan Police practically became an arm of the Republican government in Washington. Secretary of War Stanton appointed Kennedy a special provost marshal and the police a provost marshal's guard in August 1862, as the city prepared for the state draft. Kennedy's police now had the power to apprehend any who interfered with the war effort, and they exercised that power to the utmost. They arrested four thousand deserters in little over a year, defined disloyalty broadly enough to include harmless statements against the Republican Party, detained suspects on meager evidence, and on Election Day 1862 used information procured during state draft enrollment to challenge the legal status of immigrant (and presumably Democratic) voters.[150] This was the context in which Kennedy's police were popularly identified with Republican authority.

The O'Brien incident and others like it set the tone for the titanic struggles between the crowds and the police and military in the Upper East Side factory district. Monday's and Tuesday's battles for control of the Union Steam Works on East Twenty-second Street presaged the entrenched style of conflict soon to characterize the fighting in this neighborhood. Working-class families from the blocks surrounding the firearms factory hoped to seize the hundreds of carbines rumored to be inside. A wire factory converted to wartime arms production, the Steam Works was owned by Mayor

Opdyke's son-in-law George Farlee. "Mr. Opdyke's Armory" had become a popular symbol of Republican control of the uptown industrial landscape.[151]

Rioters broke into the factory on Monday afternoon. In some of the most bitter hand-to-hand encounters of the week, Inspector George W. Dilks and two hundred policemen drove off with clubs the crowds of men and women streaming into the factory. Through the day neighborhood working people stormed the armory and several times seized the building. Police wrested the factory back only after a stair-by-stair struggle.[152] Tuesday afternoon, crowds captured the factory again and repulsed an attempt of the Eighteenth Ward police to dislodge them.[153] Inspector Dilks and Captain John C. Helme then returned to the scene with a company of the Twelfth United States Infantry.[154] In this next encounter, the women were "very desperate," barring the policemen's path and assaulting them with stones and clubs.[155] Only repeated volleys of gunfire into the dense mass of working people enabled the combined forces of police and military to rescue the remaining boxes of weapons from the building.[156] Tuesday night, the rioters finally did assert their claim to the Union Steam Works, burning the building to the ground. The Eighteenth Ward Police Station on the same block was also burned that evening.[157]

What Brooks called the "collisions of the military and the mob" were now centered on the streets surrounding the large uptown factories, as the crowds battled to seize and hold the industrial terrain. On Tuesday barricades went up east of First Avenue in the Seventeenth and Eleventh wards, east of Third Avenue in the Eighteenth Ward, and along Ninth Avenue in the Twentieth Ward.[158] These boundaries cordoned off the waterfront residential and work world of the heavily Irish Catholic industrial working class from the center island that was the domain of a more native-born and Protestant middle and upper class.[159] Eighth and Third avenues, thoroughfares of retail stores and artisan shops, were both decidedly on the far side of the barricades from this working-class perspective. By Wednesday, Second Avenue had become the critical boulevard for workers to control: "Crowds of excited men occupied the corners of the streets and no one was allowed to cross the Second Avenue without first being placed under a rigid and scrutinizing examination."[160] East of that avenue, as one reporter so examined put it, any "stranger" was suspected of being an agent of the Republican authorities, either a "special officer or a spy."[161] It was here, under the shadow of the massive factories of the Upper East Side riverfront, more than on the West Side, that industrial workers were concentrated and their involvement in the riot was most evident.[162] Through the middle of the week laborers remained highly mobile: touring the city, closing factories, and demanding obeisance from proprietors in the uptown neighborhoods. By contrast industrial workers and their families now became intensely local in their thinking. They hoped to establish zones of the city free from a Republican presence. Irish Catholicism was an important part of the identity of many East Side rioters, and by excluding Republicans, they may have believed that their territory of justice would also be free from the usurpations of a nativistic and homogenizing Protestant rule.[163] When the wives of Upper East Side workingmen enjoined their husbands to "die at home" on Wednesday and Thursday, "home" meant the dozen or so square blocks in which these families lived and worked.[164]

One of the large factories behind this working class *cordon sanitaire* was Jackson's Foundry, scene of Thomas Fitzsimmons' peaceable Monday demonstration against the draft. By Wednesday morning, the mob outside the Twenty-eighth Street factory grew so dense that two companies of the Sixty-fifth New York National Guard and a detachment from the Eighth Regiment, along with some police reinforcements, were sent in to occupy and protect the factory.[165] A committee of rioters demanded that the soldiers give up the policemen to the crowd. The committee promised to disperse the gathering outside the foundry if the policemen were "delivered up."[166] Otherwise the rioters threatened to storm the factory at all hazards. The police officers donned workmen's garments and escaped the building in disguise. Hitchcock and Jackson, owners of the iron works, begged Brigadier General Harvey Brown to withdraw the troops from the building "as their presence only exasperated the people."[167] State Senator John Bradley and Judge Michael Connolly repeated the request on Friday, informing Commissioner of Police Thomas Acton that "the presence of the military . . . incited the mob to acts of violence."[168] Brown and Acton both refused to remove any troops, while Acton wrote back, "The Eighteenth Ward is a plague spot and must be wiped out."[169] Meanwhile the committee besieged the foundry and nearly flushed out the starving soldiers before military reinforcements arrived.

The rioters' committees had better success keeping the tenements free of military and police. Committees moved block-by-block through the Upper East Side riverfront district, posting sentinels at street corners and searching each house for soldiers wounded in the fighting and secreted away by sympathetic neighbors. Exacting tests of loyalty were administered by the rioters to those suspected of aiding military men, and the officials were treated ruthlessly when caught. On Wednesday night, a crowd beat and nearly hanged an army surgeon found hiding in an East Nineteenth Street cellar.[170]

By Thursday, regiments returning to the city from Gettysburg seized and occupied the streets and key factories of the uptown wards. An Eighth Regiment artillery troop and a mountain howitzer surveyed the streets around Gramercy Park.[171] One company of the Seventh Regiment occupied Day's India Rubber Factory on East Thirty-fifth Street and took control of that block.[172] Other troops from the Seventh were stationed as pickets from Third Avenue and Thirty-second Street east to the river and north to Fortieth Street. "Not more than one citizen at a time is allowed to enter the picket line, by permission of the officer of the guard," observed one reporter, "and even then he is not permitted to stand still or look around, but must briskly walk to his destination, under penalty of being shot by the sentries. . . ."[173] By Friday, many of the six thousand active troops now in the city were stationed in the uptown districts.[174] Industrial workers' hopes of liberating their neighborhoods from the policies and personnel of the Republican government had to be deferred.

Now the battle on the Upper East Side moved indoors. Late afternoon Thursday, Colonel Thaddeus Mott led a company of volunteers down East Twenty-second Street between Second and Third avenues when a crowd attempted to blockade the street and prevent the soldiers' passage.[175] The residents of the block fired bricks down on the military from the rooftops, and a sniper shot killed Mott's company sergeant. The company beat a quick retreat but returned with police and citizen deputy reinforcements and orders from General Brown to recover the officer's body. General Putnam, commanding the returning troops, retrieved the body and ordered his soldiers to clear the houses of rioters, first on Twenty-second Street and then on Thirty-first Street. The neighborhood barricaded its doors against the soldiers but finally the troops prevailed. Breaking down doors, bayoneting all who interfered, the soldiers drove the crowds to the tenement roofs. "A large number of the crowd" jumped to certain death below.[176] Colonel Mott marched sixteen male prisoners from the two blocks back to General Brown's headquarters. Two of the five traceable Twenty-second Street prisoners were metal workers, one an iron molder and one a blacksmith.[177] On the Upper East Side, the homes of industrial workers became the final redoubts of the draft riots.[178]

A true calendar of the events of July 13–17 reveals that the draft riots unfolded in two phases, each with its own characteristic participants, motives, and dynamics. Through early afternoon Monday, much of the violence bore the stamp of rioters who were conducting a one-day demonstration against the administration and inequitable provisions of the Conscription Act. This "big show against the draft" was hostile to Republican leaders and officials and conceived the draft as one of several obnoxious Republican wartime policies. Such rioters left the streets by late Monday and in many instances joined the organized effort to restore order. Alongside these rioters were others who from the outset were willing to employ far more violent means to put down the draft and were opposed to conscription on any terms. By Tuesday they dominated the action, and the riot entered a new, more murderous and destructive phase. The midweek rioters proceeded to connect the draft to many of its social bases in the community. Only the offensive behavior of an entire cast of characters, these rioters felt, could account for the subversion of just authority and the unprecedented power and arrogations of the Republican Party. A problem of this magnitude called for major surgery. Midweek crowds aimed to cut out the tumor whole, to isolate and remove all manifestations of the Republican social presence. As they watched a centralizing national government become increasingly identified with the prerogatives of a local elite they associated with exploitative and interventionist authority, rioters of all persuasions sought to reclaim the polity in the name of the community.

Would the draft rioters realize their ambition of establishing the political authority of the community in New York? By the end, the riots had revealed a popular opposition to Republican rule broad enough to astonish even the most optimistic Copperhead. The rioters were in most cases not the economically marginal or criminal poor, though vagrants and thieves, ubiquitous in the unruly port city, certainly did join the mob. Nor were the rioters the most proletarianized and degraded workers. The revolt was primarily the doing of wage earners accustomed to considerable control over the conduct of their jobs. Judging from the aggressive tone and wide-ranging scope of the riots, these workers also had a sense of their own political importance. Yet this popular opposition was fragmented, drew upon diverse constituencies, and deployed many, often conflicting, ethnic, religious, racial, class, and political strategies. The theme of varied strategies was clear both in the different ways artisans, laborers, and industrial workers rioted as well as in the decision of workers in the sweatshop district not to participate. The fate of the draft rioters' assertiveness would depend upon what social and political leaders could make of these disparate materials.

One of the questions left open to speculation in mid-July was the political future of those workers who had expressed their grievances independently of Democratic Party leaders. While the employees of the sweatshop district allowed Democratic politicians to speak for them, the artisans, industrial workers, and laborers who rioted did not. These wage earners were doubtless interested in what Democratic orators of all persuasions had to say, but had a capacity for self-directed and sustained collective activity that would make their participation in any future political movement something to be watched closely, not taken for granted. Laborers, who deferred to Democratic leaders at the end of the riot, were hardly less independent-minded than their artisan and industrial worker associates. From the perspective of a Democratic boss, these rioting workers were potentially a fractious lot.

Two other groups opposed to the draft—organized workers in the building trades and German-Americans—expressed their views apart from the party machinery. The strikes and riots of 1863 began a decade of unprecedented organization among New York City workers, but during the war most local trade unions were still new and fragile.[179] Few unions were strong enough to influence the behavior of many wage earners during the July riots. Even when sympathetic to the strike against conscription, most unions preferred not to associate themselves with treasonous and violent acts (the exception here, of course, was the Longshoremen's Association). The one journeymen's society to condemn the riots publicly and warn its members against treason was Patrick Keady's Practical House Painters' Association, which included some uptown Irishmen. Known for his criticism of Republican measures and his insistence that unions be kept free of party politics, Keady sent his membership a letter that began, "I do not for a moment suppose that any of you took part in the late riots. You are too well aware of your own interest to do that; but you can in many ways exercise your influence to prevent recurrence of such disgraceful scenes as were then enacted. . . ."[180] Keady and the organized few for whom he spoke hoped to establish the trade union movement as an independent arena of opposition to both treasonous violence and Republican policies.

Similarly, German-Americans sought to restore peace to their home districts at arm's length from the personnel of the major political parties. When most Germans left the crowds late Monday and organized to protect property and battle rioters, they necessarily placed themselves in league with Republican authorities. But like some of the fire companies, Germans made clear that they acted independently of Republican allies. "We believe," one *Kleindeutschland* group announced on the 16th, "that the draft is unconstitutional and uncalled for, [and] said draft now being stopped we organize ourselves as a body to protect the lives and the property of the people in this district. . . ."[181] While such a stance chilled German relations with the Republican Party, it was by no means a declaration of allegiance to Tammany Hall. Germans tended to take to the streets not as self-proclaimed "Democrats" but as "Germans" (as in the "Germans of Division Street" who patrolled the woodworking district on the 16th) or as squads of *Turnverein* or *Schützenverein*.[182] When the Democrats of *Kleindeutschland* did organize a citizens' brigade to suppress the violence, it was an anti-Tammany or Peace Democrat outfit.[183] The prospects of popular anti-Republicanism also depended on the political identities that the building trades unions and German-American associations would choose for themselves.

Finally, the draft rioters' success or failure rested on the responses of the many groups which comprised the local middle and upper classes. An internally polarized and embattled elite confronted the political challenge of the rioters. The eventual outcome of the Civil War and the fate of the Republican experiment in nation-building were highly uncertain during the second week of July 1863. Unresolved, too, was the outcome of a dramatic contest for authority among New York City's "better classes."

CHAPTER 2

The Two Tempers of Draco

For New York's middle and upper classes, no less than for its wage laborers, the riot week was a time of evaluating commitments and choosing sides. The riots provoked sharp disagreement among local elites over whether to declare federal martial law and allow a Republican standing army to enforce the draft in New York City. The debate over federal power also became a dispute over relations with the poor and between the races. By midweek there were two well-defined elite positions on the violence with no neutral ground in between. The polarization of the middle and upper classes turned the draft riots from a working-class challenge of unprecedented scope into something larger—a political crisis.

In what sense political? As one would expect, the plan to put New York under military supervision was generally a party issue, urged by Republicans and opposed by Democrats. But in 1863, as New York Senator Edwin D. Morgan observed, political divisions were determined less by party "antecedents" than by whether "men *were now right*" on the questions of the day. New Yorkers often chose the labels "radical" and "conservative" to describe the supporters and opponents of Lincoln's wartime expansion of federal powers. The same political categories were applied in the debate over martial law. "Certain Republicans of the radical sort," one writer recalled, "were busily engaged . . . in making efforts to get General Benjamin F. Butler sent to command in New York . . . [so as] to see a few hundred 'copperhead' corpses." Democratic General John A. Dix, who hoped to keep federal forces out of New York City, was regarded by such Republicans as "too mild and conservative."[1] In this context, Democrats were almost always "conservative," but not all Republicans were "radical." Senator Morgan, wealthy city merchant, founding member of the Republican Party, and longtime Republican National Chairman, was a conservative critical of the Lincoln administration and opposed to martial law.[2] The riot-week political crisis was a contest between Republicans and Democrats, but more accurately and broadly, between radicals and conservatives who argued the martial law question much as they had other administration policies.

The debate that prompted the crisis was articulated by four groups. As the riot unfolded, it revealed economic and cultural fractures among the middle and upper classes. One of the groups against martial law was composed of Democratic businessmen such as August Belmont, Samuel L. M. Barlow, and Samuel J. Tilden, who had come to the city between the mid-thirties and early fifties. By the time of the war they had attained vast political and economic power as directors of the national Democratic Party and as investment bankers and lawyers prominent in the Wall Street market in railroad securities. These wealthy Democrats favored negotiation with white rioters, neglected black victims, and sought to have the draft law declared unconstitutional by the courts. Another source of opposition to military rule came from men of more modest means, the contractors and small proprietors whose political fortunes were so often tied to Tammany Hall. The Tammany men shared the Democratic merchants' racial views but did not care to overturn the draft law. Instead they sought to block its obnoxious provisions through public philanthropy—the government would pay the commutation fees of poor drafted men. Martial law was championed by the Union League Club, an ultranationalist organization composed of patrician merchants and members of the learned professions who could trace their metropolitan roots back to the 1820s, and their would-be-patrician allies, often intellectuals of recent arrival. The Union Leaguers excoriated Irish rioters, lavished their charitable attentions on the black poor, and demanded that conscription be enforced at all costs. Also supporting a no-holds-barred execution of the draft were the evangelical industrialists whose machine-building and metal-working shops provided a backdrop for much of the midweek violence. Speaking through their reform organ, the Association for Improving the Condition of the Poor, the industrialists condemned the draft rioters as an "unworthy poor" undeserving of philanthropy and better persuaded by force than moral ministration. Instead industrialists sought to cultivate a white "respectable poor" more receptive to their message of industry and uplift. A profound division of elite interests and values was exposed by the riots.

The crisis unfolded in several stages. On Monday afternoon, July 13, diverging elite responses to the violence were already visible. The crisis was revealed Monday night as city leaders debated whether to declare martial law and was elaborated through the week as the four elite perspectives took shape. The crisis eased late Thursday and Friday when troops pacified the last remaining areas of resistance and Lincoln decided against martial law by appointing Democratic General John A. Dix as Commander of the Department of the East; the draft was carried out in late August under the auspices of Dix and the conservatives. Finally, counteroffensives by the radical Union League Club and Association for Improving the Condition of the Poor in the months following the insurrection demonstrated that a long-term problem of political rule still remained.

Diverging Responses

With all their forebodings of violence, well-to-do New Yorkers were not really prepared for trouble in the second week of July 1863. Most of the state militia were engaged with Lee's forces at the seat of war north of the Potomac, and the city was defended only by a few troops at the Navy Yard and harbor forts, the untested Invalid Corps and Provost Guard, and the Metropolitan Police.[3] In an era when urban authority relied much on the personal appearance and intervention of local gentry during social upheavals, the 13th found many notables summering at country retreats: August Belmont at Newport, Hamilton Fish at Garrison, Horatio Seymour at Long Branch. Many "substantial and weighty and influential men," the lawyer and keen civic observer George Templeton Strong noted, were thus absented from Monday's emergency meetings.[4]

Before ten o'clock Monday morning, reports of the large concourse of working people moving toward the Ninth District Provost Marshal's Office reached Police Superintendent Kennedy and Mayor Opdyke.[5] Kennedy ordered two ten-man squads to report to the uptown draft offices and directed reserves on the Upper West Side to the Seventh Avenue Arsenal.[6] Opdyke wrote Police Commissioner Thomas C. Acton, Major General Charles W. Sandford, Commander of the State Militia, and Major General John E. Wool, Commander of the Department of the East, to request permission to use force against the rioters and, if necessary, call out United States troops.[7] In his letter to federal officer Wool, Opdyke distinguished between the "present demonstrations against the U.S. draft," which he felt did not yet merit military assistance, and the possibility of a "serious riot," which would require military action.[8] Provost Marshal Nugent had already asked Kennedy for a police guard at each draft office in the city and ordered the Invalid Corps to assist the police in their work.[9] By early afternoon, Provost Marshal General James Fry in Washington had been apprised of the "riot in progress in Third Avenue." Fry, in turn, authorized Wool to supply Nugent with available federal troops.[10] Not long afterward, Opdyke acknowledged that "a serious riot" did exist and, with Wool's help, contacted the governors of New Jersey, Connecticut, and Rhode Island as well as officials in Newark, West Point, Albany, Utica, and Rochester in a frantic search for military aid.[11]

In New York City, the resort to federal forces was new and required justification. In May 1849, city leaders had called out the state militia to quell the Astor Place riot. Though the militia had been summoned several times before, Astor Place was the first instance of troops opening direct fire on a crowd.[12] Cooperation between local and federal authorities developed later. The November 1857 bread riots prompted the city government to defend downtown federal buildings with state forces.[13] During the June 1863 longshoremen's strike, convicted deserters and convalescent soldiers were employed as strikebreakers and loaded army transports under police and militia protection.[14] The 1863 Conscription Act, investing unprecedented powers in the federal Provost Marshal Bureau, sought both to routinize and to expand well beyond these modest beginnings of a centralized response to urban disorder. But as yet there had been no instances of large-scale federal intervention into the governing of New York City.

Through Monday, Opdyke chose not to invoke new federal powers and held government forces well within the bounds of local precedent. His one o'clock telegram to Secretary of War Edwin M. Stanton did not solicit troops or other federal assistance.[15] An early afternoon communication with Governor Seymour limited troop requests to "whatever force you can command from adjoining counties" and made no mention of the state militia Seymour had sent to the front.[16] On Monday police were armed with locust-wood clubs, and early police requests to shoulder muskets were denied.[17] In consultation with General Wool, Opdyke concentrated his available forces around federal installations and local arsenals in the manner of 1857: a detachment of the Tenth Regiment at the White Street Arsenal and lines of sentries surrounding the Customs House, the Post Office, and the Seventh Avenue Armory.[18] During the afternoon Opdyke issued a proclamation calling on "loyal citizens" to return to work or, if possible, to report to the Mulberry Street Police Station to be deputized as special police.[19] While Opdyke's meetings with federal authorities pointed toward new departures from precedent, his defense of government property and proclamation drew upon an existing repertoire of responses to disorder.

The mayor's strategy of dispatching small bands of police and military to the uptown working-class wards enraged rather than pacified the crowds.[20] The appearance of men in uniform often precipitated Monday's violence. Fire Department Chief Decker had obtained permission from the rioters outside the Ninth District draft office to save adjacent homes when the Metropolitan Police arrived. Immediately the crowd united to drive both police and firemen from the scene.[21] Next to arrive were the forty men of the Provost Guard who again escalated the violence. Awed by the size of the crowd, the Provost Guard fired into the throng, then turned and fled. Soldiers unable to escape were pursued into side streets, "hunted like dogs," and beaten to the point of death.[22] The first decisive victory for the police Monday afternoon occurred not amidst the shanties and open lots of the uptown wards but on Broadway, in the heart of the middle-island commercial and manufacturing district. Holding a "No Draft" banner aloft, the rioters on Broadway headed toward Lafarge House where they planned to drive off black servants rumored to be employed there. Before the rioters reached the hotel they met Inspector Daniel Carpenter and a company of police, armed only with clubs but heady with orders "to take no prisoners."[23]

Carpenter's force of two hundred scattered the crowd with little difficulty. In contrast to the earlier uptown conflicts, the Broadway Metropolitans fought to the applause of a local population warmly sympathetic to the police.[24] But the Broadway triumph of the police was a footnote to the emerging rule of the day. When not backed by superior numbers and weaponry, displays of government power in the poor uptown districts would lead to massive and successful neighborhood retaliation.

While some New Yorkers saw such armed confrontation as the only acceptable response to the violence, others explored more temperate methods. Democratic Judge George G. Barnard was present during the first appearance of the crowd at Mayor Opdyke's Fifth Avenue home (Barnard probably feared for his own house a few doors away).[25] Recognized by the rioters, the judge was cheered and asked to give a speech. He proclaimed the conscription law an unconstitutional and despotic act—"the administration had gone too far." But he urged the crowd not to destroy the home of its "legally elected chief officer" and "sully the reputation of our city for its obedience to the law." "The courts," Barnard announced, "would protect . . . the exercise . . . of just and legal rights."[26] Coming from a high official of the court, this assurance had the ring of authority. A voice in the crowd cried out, "You're about right, Judge," and the threatening assembly disbanded.[27] Later in the day the Peace Democrat C. Chauncey Burr dispersed a Harlem crowd by reiterating Barnard's theme of legal opposition to the draft, though now with an anti-abolitionist variation. The courts, predicted Burr, would insure that a fanatical abolitionist minority would not use conscription to deprive white workingmen of their liberties.[28] Less typical were the speeches of Aldermen Peter Masterson and Terence Farley, who jettisoned the cautious legalism of their fellow Democrats. Masterson and Farley praised the courage of the rioters who destroyed the Ninth District Provost Marshal's Office but assumed that the violence would cease once the draft had been suspended.[29] These elite spokesmen all openly opposed the Conscription Act and, by negotiating with the crowds, legitimated Monday's anti-draft demonstrations by implication if not by declaration.

During Monday other differences in physical and mental posture toward the riots became evident. A dazed Mattie Griffith sat awaiting the mob at the window of her Upper East Side apartment above the American Freedmen's Commission. The abolitionist wrote an English friend of a "confused sense of half-being" as she watched the city overtaken by "the strange wretched abandoned creatures that flocked out from their dens and lairs."[30] Republicans like Griffith soon found themselves limited to a navigation of safe blocks or stranded at home, cut off from the city and vulnerable to social invasion. We have seen how, by contrast, anti-draft Democrats hovered close to Monday's demonstrations and addressed the crowds to approve and caution. Monday also witnessed the first hardening of attitudes toward the riots' growing racial violence. Morgan Dix, rector of St. John's Episcopal Church, dismissed the story of a servant who flew into his library and with pale face reported that "while passing through Clarkson Street . . . she had seen a colored man hanging on a tree, and men and women setting him on fire as he dangled from the branches." Soon Dix came to understand the full scope of the tragedy and learned further that his church had been chosen as a target by the rioters because it housed a religious school for black children. He requested and received a detachment of troops to patrol the chapel grounds.[31] In another Monday episode, a black laborer named George Spriggs was evicted by his white landlord with the explanation that if he "didn't get out of the house the mob would burn the house."[32] Sprigg's landlord was typical of hundreds of proprietors who drove blacks from their premises—while they often disapproved of the violence they collaborated with the rioters' racist purges in the name of self-preservation. Dix also wanted to safeguard property but deeply sympathized with black victims and commanded enough prestige and influence to secure military protection. By late Monday, elite New Yorkers were already beginning to react in different ways toward both the white rioters and the threatened black community.

The Crisis Revealed and Elaborated

On Monday evening the problem of quelling the riot became at once easier and much more difficult. For the time being the draft was over. The rioters had succeeded in crippling the draft administration and postponing (or possibly even preventing) conscription in New York City. Democratic elites that had to this point supported or sympathized with the anti-draft violence now confronted an insurrection that had become much more than a "big show" against conscription. No matter how benign the riot had seemed to some Democratic leaders early on, Democrats and Republicans now agreed that the violence had gotten out of control and had to be put down. This perception brought notables from both parties to the Saint Nicholas Hotel Monday night to seek a means of restoring order. Now, it seemed, the leading men of New York would sit around a parlor table and find a solution.

But the mood at the Saint Nicholas was more one of point-counterpoint than calm collaboration. When George Templeton Strong and Oliver Wolcott Gibbs arrived at the hotel, they found fellow Republicans and Union League Club members Frank E. Howe, John Jay, George W. Blunt, and John A. Stevens, Jr., urging "strong measures" upon Mayor Opdyke and General Wool.[33] Strong and Gibbs forced the issue by demanding an instant declaration of martial law. The Union Leaguers believed all anti-draft demonstrations, no matter how orderly, should be ruthlessly suppressed. A federal military occupation with sweeping powers and of indefinite term would enforce the draft, guarantee against future outbreaks, and improve the loyal climate of the city and the nation. Opposing military rule were New York State's Judge Advocate General Nelson J. Waterbury, Captain Isaiah Rynders, and other Democratic officials. They argued that "in the present excited state of public feeling" such a draconian act would result in conflicts more severe than those already witnessed.[34] Some Democratic leaders, in the manner of Masterson and Farley, approved of Monday's more peaceable demonstrations against the draft; others, like August Belmont, were sympathetic to conscription.[35] Regardless of their views of the draft, the Democrats all still hoped to finesse or cajole the riots to a halt. This conciliatory style required a frequent resort to armed force to be effective but it stopped well short of martial law. By late Monday, elite differences over the uses of federal power had crystallized. Instead of producing consensus and concerted action, the Saint Nicholas conference had brought Republican and Democratic leaders to a political impasse.

Opdyke refused to declare martial law Monday evening. By temperament the Republican mayor was not given to extreme pronouncements and measures. Further, he believed the riot had run its course. He held out to an impatient Union League Club delegation the possibility of military rule later in the week if the riot continued and postponed the decision until that time. But ultimately the Mayor was more convinced by the arguments of Waterbury and Rynders. Martial law, Opdyke later wrote, "would have exasperated the rioters, increased their numbers, and those in sympathy with them, for the Democratic party were, to a man, opposed to the measure. The probable result would have been the sacking and burning of the city, and the massacre of many [more] of its inhabitants. . . ."[36] The Mayor still hoped to mend the growing schism between elite camps Monday evening and find a political rather than a purely military solution to the violence. Meanwhile, a disgruntled George Strong and several comrades left the hotel doubting that Opdyke and Wool were "equal to this crisis" and telegraphed President Lincoln "begging that troops be sent on and stringent measures taken."[37]

As Strong observed, a crisis was at hand. Two poles of opinion were established at the St. Nicholas, and the dispute over federal intervention was left unresolved. The question was not merely whether Republicans or Democrats would control New York but whether Lincoln's expanded federal regime would survive and the momentum of the war be sustained. When Navy Secretary Gideon Welles imagined an "understanding between the mob conspirators, the Rebels, and our own officers" or John Jay warned War Secretary Edwin M. Stanton that the riot was "the last great card . . . of the rebellion," they were acknowledging that national and local issues were fused in a multilevel problem of political rule.[38] The riot became a struggle between conservatives who opposed martial law and radicals who advocated it. Moderates such as Mayor Opdyke were forced to choose sides.

What were the dimensions of this crisis? The full articulation of elite responses to the insurrection would require the remainder of the week, and in some instances subsequent weeks and months. Yet by the middle of the week, Monday night's positions had been expanded and refined into coherent outlooks. Tensions appeared within the conservative and radical camps, as four middle- and upper-class groups elaborated their views on the problems of federal power, relations with the poor, and between the races. At the center of the controversy stood the Governor of New York State, Horatio Seymour, who arrived in the city Tuesday morning, July 14. More than any leader, Seymour had the power to determine which combination of these groups would direct the course of metropolitan social relations over the next few days. The Governor would do all he could to help conservatives retain control of the city and to make conciliation of the white rioters and neglect of black victims the dominant themes of the post-riot settlement.

Historians have long debated the merits of Horatio Seymour's riot-week behavior without considering it in the context of his Jacksonian education in the canal towns of upstate New York.[39] The building and commercial development of the Erie Canal was the formative event in Seymour's life: it occasioned his father's rise to wealth as a Utica merchant and Canal Commissioner for the Albany Regency, defined the upstate conflicts which accompanied Seymour's own rise to prominence in the Democratic Party in the 1840s and 1850s, and provided the model for his career as a capitalist. Seymour invested heavily in schemes to forge canal links between the major waterways of the Midwest in the 1850s and 1860s.[40] He imagined the United States as a great national market system, canals and railways working in harmony, delivering to both Midwestern towns and Eastern cities the same fruits of enterprise brought by the Erie Canal to the Mohawk Valley in the second quarter of the century. Industrial capitalism figured lightly in this vision, if at all. Seymour's New York City was the entrepôt of an agrarian nation rather than the center of factory production it was fast becoming. When the Governor arrived in New York City he brought an agrarian sense of the importance of local independence and a decentralized polity. These values would shape Seymour's treatment of the rioters and his position on the martial law question. Indeed, Horatio often admitted privately that visiting New York City made him uncomfortable. He was more at ease speaking to farmers on the Mohawk Valley political circuit than to an urban throng outside City Hall.[41]

It was at City Hall, however, that Seymour addressed "a large crowd of men and boys" late Tuesday morning in the famous "My Friends" speech.[42] Flanked by Tammany leaders—District Attorney A. Oakey Hall, Supervisor William M. Tweed, and Street Commissioner Charles G. Cornell—Seymour promised the peaceable gathering he would uphold their rights so long as they "refrained from further riotous acts."[43] He pledged to enforce federal conscription if the courts declared it constitutional but noted that he had urged Washington authorities to postpone the draft until the courts decided its fate. If the courts ordered the law carried out, he would have money allotted to relieve "those who are unable to protect their own interests."[44] Finishing the speech to applause, Seymour retired to a conference with city and military officials, whom he advised not to declare martial law until all other means of pacification had failed.[45] After announcing rendezvous points for citizen deputies, Seymour wrote Archbishop John Hughes asking the Catholic leader to exert his "powerful influence" to stop the violence.[46] Finally Seymour toured the city with Sheriff James Lynch, Judge Connolly, Supervisor Tweed, and Street Commissioner Cornell at his side. The Governor repeated his City Hall address to a gathering on Wall Street and to another in an Upper West Side working-class district. Both speeches were well-received, and the entourage was cheered along its route.[47]

Seymour's conciliatory speeches to the crowds reinforced the emerging conservative response to the riots. Like Judge Barnard, Chauncey Burr, and other Monday orators, Seymour viewed the Monday demonstration against the administration of the draft as a rightful undertaking. A proclamation Seymour issued from City Hall on Tuesday read, "I know that many of those who have participated in these proceedings would not have allowed themselves to be carried to such extremes of violence and of wrong except under an apprehension of injustice. . . ."[48] Seymour and the other orators repudiated the murder, theft, and arson of the midweek riot but saw face-to-face negotiation with Tuesday and Wednesday's crowds as a legitimate way to restore order. The Governor's headquarters at the Saint Nicholas came to resemble a conservative speakers' bureau as Seymour, with the counsel of Samuel Barlow and General George McClellan, dispatched local gentry around the city to address threatening gatherings.[49] Seymour believed that once rioters heard from trusted leaders that the draft had been suspended, the violence would cease.[50] Finally on Tuesday he declared the city to be in "a state of insurrection" and invoked a New York State statute regarded by many Democrats as an alternative to federal martial law.[51] In so doing, he defined the restoration of order as a state rather than a federal responsibility and affirmed the efforts of the conservative middle and upper classes to put down the riots without federal intervention. As the pro-Seymour *Evening Express* phrased it Tuesday night, "The civil authorities, state and city, are now amply able to preserve the peace of the city, even without federal military cooperation, provided we have here the ten thousand State Militia we have so cheerfully sent from the city to defend Pennsylvania from invasion."[52]

Alongside the gentry whom Seymour sent out among the crowds, uptown small employers, proprietors, and clergy mobilized a midweek campaign to restore order. Tuesday afternoon, a crowd destroyed the Harlem Railroad track and threatened to burn the house of Provost Marshal William Dunning in the Upper East Side hamlet of Yorkville.[53] A Catholic priest named Father Quarters convinced the rioters to leave Dunning's property and attend a street-corner meeting that afternoon to discuss the draft. Carefully orchestrated by Quarters, that gathering heard speeches from Yorkville lawyer John Keynton, a Methodist minister Doctor Osborn, and Quarters himself. Each speaker called for opposition to the inequitable three-hundred-dollar clause but demanded a halt to the destruction of property. City clerk William Hitchman read a petition asking Mayor Opdyke to request a stay of the draft in New York. Should conscription eventually be enforced, the document continued, it would be the duty of the Common Council to pay the commutation fees of poor workingmen. A committee of twenty-nine with a sizable contingent of local saloonkeepers was selected to present the petition to the Mayor and Common Council. The assembly cheered Quarters and Keynton and agreed to reconvene at the end of the week.[54]

The assurances offered the working classes by Judge Barnard, Chauncey Burr, and N. Hill Fowler differed from those of Quarters, Keynton, and Hitchman. Judge Barnard dispersed Monday's crowd outside Mayor Opdyke's house by promising to protect the community against the unjust acts of Republican "outsiders" in Washington. He relied solely on an assumption of shared interest with his working-class listeners and the respect his presence commanded. Father Quarters cut a less imposing social figure than Barnard, but he offered something more tangible to his audience—a proposal for the city to purchase draft exemption for the poor. A municipal draft exemption fund would not merely deflect the class bias of the Conscription Act but also grease the wheels of deference. Rioters demanded that small proprietors prove their allegiance to the crowds by treating and token "gifts" of money. The display of municipal philanthropy proposed by Quarters would satisfy these demands on a grand scale. By midweek, two styles were emerging among the conservatives. One reflected a confident upper-class paternalism, the other the more fragile prerogatives of an uptown *petite bourgeoisie*. This latter group was represented on the Board of Aldermen by the uptown contractors Terence Farley and Jacob M. Long. The draft exemption fund approved by the Board of Aldermen and Common Council Wednesday morning well reflected the outlook of men of their social stamp.[55]

Drafted by Tammany Democrats, the ordinance "to relieve the City of New York from unequal operation of conscription and to encourage volunteering" appropriated $2,500,000 to pay the three-hundred-dollar exemption for every New York conscript. The fund for this municipal expenditure was to be raised through the sale of "Conscription Exemption Bonds" at up to 7 percent interest, redeemable in the year 1880. The Board of Aldermen also directed the Corporation Counsel to test the constitutionality of the Conscription Act and asked the Comptroller to activate the exemption fund only in the event that the courts upheld the draft.[56] The nature and extent of the proposed expenditure were without precedent. But the most controversial aspect of the July 15 ordinance was the attitude toward philanthropy which informed it. Elites on both sides of the Atlantic viewed gifts to the poor as a major form of social influence weighted with consequences for both donor and recipient.[57] With their multimillion dollar handout to the poor, the Democrats of the Common Council established themselves as the preeminent donors in the city. They hoped that their indiscriminate style of giving—without detailed investigation of the recipient or personal tie between recipient and donor—would become the city's dominant form of charity. Upper-class conservatives might have preferred to see the draft nullified in the courts rather than through Tammany's great giveaway.[58] Nonetheless these wealthy men cooperated with small proprietors in the attempt to find a local solution to the riot. Both social groups sought to confirm conservative political control of the city and validate a conciliatory style of class relations.

Though conservative attitudes toward federal power and relations with the poor crystallized early, responses toward the racial violence emerged more gradually. Even the most shrill organs of white supremacy condemned the attacks on black workingmen as excessive and misdirected.[59] "Do Not Harm the Negro," advised the *Daily News,* "he is not to blame for the misfortune that has befallen us. The poor negro is beginning to taste of the bitter food prepared for him by Abolition philanthropy."[60] The Tammany-run *Leader* also saw the attacks on blacks as the byproduct of Greeley-style agitation against slavery: "The negro in his best aspect is repulsive to the white man's instincts; but when he is mixed up as a reason for civil war he becomes odious. . . . Doubtless 'H.G.' and his negro allies when hunted through the city last week by disgraceful throngs may have remembered his [abolitionist] articles. . . ."[61] Such criticism of the lynch mobs was seldom accompanied by sympathy for the devastated black community.

Well-to-do conservatives spoke of estrangement from their friends prompted by discussions of race. Maria L. Daly, wife of Democratic Judge Charles Patrick Daly, confided to her diary:

> I saw Susanna Brady, who talked in the most violent manner against the Irish and in favor of the blacks. I feel quite differently, although very sorry and much outraged at the cruelties inflicted. I hope it will give the Negroes a lesson, for since the war commenced, they have been so insolent as to be unbearable. I cannot endure free blacks. They are immoral with all their piety.[62]

At a social gathering on the 19th, the Dalys discovered that their friends James and Susanna Brady sought stiff punishment for the Irish rioters and relief for suffering blacks. By contrast, Maria Daly empathized with the motives of the crowds—it was "exceedingly unjust" that all who could not "beg, borrow, or steal" the commutation fee had to go to war—and saw some justice in the distress of black victims. This intertwining of dispositions toward the Irish poor and the black community had doubtless begun well before the draft riots. But the insurrection forced upper-class conservatives to articulate feelings previously half-formed or unspoken, and quickened a fusion of ethnic and racial perceptions. What middle-class conservative journals such as the *Leader* condemned as riot-week "abolitionism," upper-class conservatives such as Charles and Maria Daly confronted as a highly personal dilemma: a sudden feeling of alienation from close friends.

By Wednesday, July 15, conservatives seemed well situated to carry out their solution to the uprising. Governor Seymour had publicly endorsed a style of social control that combined force against the rioters with a measure of appeasement. So decisively had Seymour entered the conflict on the side of the conservatives that some began to fear for his secret arrest by federal officials.[63] General Wool named State Militia Commander Charles Sandford head of "all the troops called out for the protection of the city."[64] A veteran of the 1849 military action against the Astor Place rioters, Sandford was nonetheless an extremely cautious soldier responsible by law to Governor Seymour.[65] At noon Tuesday, Mayor Opdyke finally asked Secretary of War Stanton to send New York City what military force he could spare.[66] Stanton ordered north five regiments from the Pennsylvania and Maryland front—now possible as Lee's army had retreated south of the Potomac.[67] It was unclear how these troops would be employed when they arrived. General Wool was a War of 1812 veteran who many thought too old and infirm to take aggressive action.[68] Though he had called federal troops into the conflict, Opdyke had now made clear that he preferred a conservative solution.[69] Even the loyalty of the troops sent to New York was open to some conjecture. General Halleck was advised that among the New York militia only the native-born and Protestant Seventh Regiment could be trusted with riot duty.[70] Finally the unanimous passage of a draft exemption ordinance by both chambers of city government suggested that peace, when it came, would be purchased dearly. Tammany legislators had committed themselves to a de facto nullification of conscription through an unprecedented exercise of municipal philanthropy.

As conservatives positioned themselves to restore order in the city, radicals predicted chaos on a grand scale. "It is the beginning of a new era," wrote William Helfenstein to Secretary Stanton on Tuesday, "of violence, resistance to law, contempt of the government, and disregard of all public and private good. . . . I am satisfied the civil authorities here will but smoulder and strengthen the fire and not subdue it."[71] A business visitor to the city, Helfenstein captured the mood of a widening elite circle. These men and women now saw themselves fighting a two-front battle: against the crowds who by Tuesday had beaten back the police and military and against conservatives of their own class who, it seemed, would compromise with treason. The belief that the insurrection was "not simply a riot but the commencement of a revolution, organized by sympathizers in the North with the Southern Rebellion" had by Tuesday become a commonplace among radicals.[72] These businessmen looked to Washington for drastic remedies—

against both the rioters and wealthy neighbors who contented themselves with halfway measures.

By Tuesday the demand for "an immediate and terrible" display of federal power in New York City had spread from the Union League Club leadership to a broader group of merchants and financiers.[73] The widening appeal of martial law could be seen at a Tuesday meeting at the Merchants' Exchange on Wall Street. Not far from where Governor Seymour had addressed an earlier gathering, merchants William E. Dodge, Jr., and Prosper Wetmore and bankers John A. Stevens, Jr., and William M. Vermilye called a meeting to raise volunteer companies to suppress the riots. After unanimously supporting the proposal to form a merchants' volunteer brigade, the meeting split on the question of martial law. A resolution in favor of federal intervention was finally passed against vocal opposition. The martial law vote at the Merchants' Exchange revealed both the degree of division in the commercial community and the growing numbers of capitalists willing to consider placing the city under federal control.[74]

The influential Republican lawyer David Dudley Field spoke for many radicals when on Wednesday he asked President Lincoln to appoint General Benjamin Butler military governor of New York City.[75] William Hall, shoe contractor for the army and investor in Greeley's *Tribune,* seconded Field's request in a letter to Postmaster General Montgomery Blair: "Enforce the draft and give us General Butler and all will be well with us."[76] For Field and Hall, Butler's brutal suppression of treason during his 1862 term as military governor of New Orleans made him the preferred replacement for the irresolute General Wool.[77] The call for Butler represented the extreme radical alternative during the riot week. It is worth noting that Greeley himself endorsed martial law as early as Monday but never advocated a Butler regime for the city.[78]

Social fissures soon appeared among the radicals much as they had among the conservatives. The elites who urged federal intervention displayed two very distinct styles of relating to the working classes. One of these styles arose from the merchant and professional ranks of the Union League Club. The other belonged to industrial employers in the metal trades, whose employees were in many cases the most intransigent midweek rioters.[79]

No one better articulated the Union League Club view than Frederick Law Olmsted. The designer of Central Park first promoted the idea of a wartime "club of true American aristocracy."[80] Such an institution, originating in New York but spreading to Washington and other Northern cities, would bring the prestige and social influence of a national business and cultural elite to the work of cultivating loyal opinion among the middle and upper classes. The Club membership list began with men of ancient New York lineage such as John Jay, James Beekman, and Robert Stuyvesant. But it deliberately included, as well, scientist Oliver Wolcott Gibbs, literati Henry T. Tuckerman and George Bancroft, respected professionals such as physicians William Van Buren and Cornelius Agnew and lawyer George Templeton Strong, loyal capitalists Robert Minturn and James Brown, and clergyman Henry W. Bellows.[81] Olmsted's pet project during July 1863 was the creation of a national journal, to be titled "The Statesman" or, alternatively, "The Loyalist" or "The Gentlemen's Review." This publication would bring the loyal cultural influence of the Union League Club to bear on a middle-class reading public.[82] Both the Club and its journal were designed to undermine the social and political leadership of New York's foremost conservatives. Olmsted's solution to the draft riots was simple: "Let Barlow and Brooks and Belmont and Barnard and the Woods and Andrews and Clancy be hung if that be possible. Stir the govt up to it. I did not mean to omit Seymour."[83]

Olmsted was not indulging in hyperbole. For the Union League Club, the riot was an intensification of an ongoing struggle with wealthy conservatives. No surprise that Olmsted's comrades were the earliest and most vocal advocates of martial law during the riots. The Club hoped to preside over a federal inquisition into disloyalty and so expose, discredit, and (if Olmsted had his way) execute elite opponents as traitors.[84] Club businessmen understood upper-class "semi-loyalty" as a root cause of working-class disaffection. They expected that once loyal statesmanship had been vindicated in New York, wage earners would no longer be led into violent opposition to the government by false leaders.[85] A unified nationalist elite could then create an orderly society with new urban spaces (Central Park was one), new institutions (opera and theater were the appropriate agencies), and an active state.

The movement to aid black riot victims began not with these larger intentions but at the urging of *Times* editor Henry Raymond and through the informal charity of the Protestant clergy and merchant families who protected black domestic servants.[86] At the end of the riot week, a committee was formed under Union League Club officer Jonathan Sturges to open a subscription list to aid the devastated black community.[87] Over one hundred merchants met to begin distributing relief to black victims, assisting blacks in making claims against the city, and protecting those driven from their homes and jobs.[88] Over the next month and a half the "Committee of Merchants for the Relief of Colored People" paid out almost $40,000 to nearly 2500 black claimants. Most of the money was administered under the supervision of black clergymen, with some assistance from the Association for Improving the Condition of the Poor.[89] While the neediest cases were handled at the Fourth Street office, most claims were evaluated by black "missionaries" such as the Reverend Henry Highland Garnet, who visited black victims at home and based his determinations on his "extensive acquaintance among the colored people."[90] The Committee's efforts represented an unprecedented liberality to the black poor.

Committee merchants used the riots as an opportunity to champion what they characterized as the "pure" loyalty of the black working class. It was no accident that the merchants' campaign in behalf of the black community came at a time when white wage earners seemed inattentive if not deaf to their cultural ministrations. A shared Protestant language and culture encouraged the protective attitude of these philanthropists toward the black poor. Jonathan Sturges told one merchants' meeting, "Those who know our colored people of this city, can testify to their being peaceable, industrious people, having their own churches, Sunday-schools, and charitable societies and that as a class they seldom depend upon charity. . . ."[91] Surely the nativist merchants needed little prodding to distance themselves further from a heavily immigrant and Catholic white working class.[92] Committee leader Jackson Schultz recalled at an 1886 dinner to commemorate Union League Club efforts in behalf of the black community during the Civil War, "By all it was known as a riot instigated and carried on by the Irish Catholics, although by too many writers erroneously designated as 'Negro Riots.' . . . In no sense, then, was this riot of 1863 a Negro riot. The colored people of this city did not participate in that riot, except as defendants, and most inhumanly were they treated by their assailants."[93] Merchants could uphold the city's blacks as a model deferential working class because of their indisputable status as victims. In a riot-week atmosphere where any public connection with the black community rendered person and property liable to attack, such attention to the needs of black people, paternalistic though it was, required some courage.

Alongside this humanitarian concern was a larger symbolic project. Through their open association with the black poor, merchants sought to counteract working-class disloyalty publicly by exhibiting an ideal relationship between classes. Consequently Union League Club members resolved during the riot week to raise a black Union Army regiment in New York City.[94] From the Club perspective, this seemingly modest undertaking had sweeping consequences for class relations in the riot-torn metropolis. "Its greater significance," LeGrand B. Cannon recalled two decades later, "was the raising from their degradation and bringing into the service of the nation the colored citizens of the State of New York, and by that act purging the City from the taint of that wicked, infamous and inhuman riot of July."[95] Purifying the community became the aim of both the most violent midweek rioters and the Union League intelligentsia.

Industrial employers' response to the riots was different from that of the merchants and professional men of the Union League Club. Unlike Olmsted and most of his colleagues, employers in metallurgy and machine building had factories to defend and needed to sustain a dialogue with their employees to keep their shops running during the strike against the draft. William Boardman feted loyal workers at his Neptune Iron Works after the riots. Other metal trades employers such as Horatio Allen, president of the mammoth Novelty Iron Works, were content with the quieter camaraderie of shoulder-to-shoulder riot duty with loyal workingmen at the factory gates.[96] Compared to the Union Leaguers, who dismissed the immigrant poor (and particularly the Irish) as a faceless canaille, industrialists were much more interested in and personally familiar with the white working class and did what they could to secure its allegiance.

John Roach, owner of the large Etna Iron Works, sought to redress the problem of the inequitable draft law by a shop floor agreement with his employees. Roach asked Etna workers to sign a statement allowing the withholding of one week's pay toward a fund "for the purpose of procuring substitutes or paying the exemption fee for those who [might] be drafted" out of the establishment.[97] As consideration, Roach himself donated five hundred dollars. After the draft, any surplus money would be divided proportionally among the subscribers. Drafted employees inclined to serve could keep their three-hundred-dollar "share" or make it over to their families.[98] It is difficult to gauge the popularity of the "draft fund" with employees or, for that matter, employers. But by the following week, the idea had spread to John C. Parker's Upper East Side carriage manufactory.[99] We know little about Parker; the wealthy Roach, a staunch Republican, had been schooled as a molder in the machine shops of the East Side and was known as a paternalistic employer who set his men "an example of what a saving, resolute workman could make of himself."[100] Both Roach and Parker presided over factories whose employees may have been sympathetic to the riot in at least one of its phases.

The industrialists' draft insurance plans were distinguishable from the conservatives' riot-week philanthropy. The Roach and Parker funds required monetary "consideration" and signed agreements from participating wage earners, in contrast to the largely indiscriminate and impersonal giveaway of the Tammany draft ordinance. It was more than coincidence that the metal trades bosses who devised draft insurance funds also established profit-sharing plans in the sixties—whereby certain "good" workers became limited partners in the firm. It also made sense that such industrialists were active participants in the Association for Improving the Condition of the Poor, a moral reform organization that employed a highly personal and closely supervised charity to cultivate a worthy poor. Industrialists and Tammany men agreed that philanthropy was the key to restoring social peace; they disputed which style of giving, personal and supervised or indiscriminate, would bring about that peace.

There were also differences in outlook between the industrialists and their fellow radicals at the Union League Club. Both groups condemned the conservatives' conciliatory approach to the violence and tried to distinguish a loyal clientele among the poor. But the industrialists sought to isolate respectable white workingmen from the treasonous crowds, while the Club businessmen championed the cause of the pious black poor. While Roach and Parker made their aid conditional upon monetary consideration and written agreements, Sturges and Schultz allowed the black community surprising latitude in the distribution of funds. Even if these radical groups succeeded in importing military rule, they would use that power to promote very different styles of social leadership.

The high tension between conservatives and radicals made the restoration of order problematic even as hundreds of federal troops poured into the city late Wednesday and Thursday. On the Upper East Side, employers and proprietors who had received threats asked Judge Connolly and State Senator Bradley to petition the authorities for the removal of the military from the Eighteenth Ward, as the troops only "incited the mob to acts of violence."[101] Connolly and Bradley presented the request to General Harvey Brown and were refused. On the West Side, General Brown ordered one Colonel Mayer to disperse a lynch mob attacking black residents. Mayer's company was fired on from the housetops and returned volley. At that moment City Judge and Peace Democrat John H. McCunn appeared and "informing the Colonel that he [McCunn] was acting under the authority of Gov. Seymour, requested him not to fire upon the mob and he would talk to them."[102] Mayer repeated to McCunn his orders "to put down the rioters by force of arms" and "proceeded to do so, cutting down a negro who was hanging upon a tree with his own sword."[103] McCunn departed after briefly speaking to the crowd, and Colonel Mayer fought rioters threatening the black community through the afternoon and evening. Two days later, Judge McCunn irritated the Republican authorities once again by issuing a writ of habeas corpus releasing prisoners arrested during the riots. The writ was ignored by police and prison officials.[104] The encounters among Connolly, Bradley, and Brown and between McCunn and Mayer dramatized conflicting elite social styles and highlighted the obstacles federal authorities faced in New York City.

Across the North, political and business leaders anxiously awaited Washington's full response to the New York upheaval. Provost marshals in Des Moines wired the President on Wednesday the 15th, "Suspension of the draft in New York as suggested by Gov. Seymour will result disastrously in Iowa."[105] A Wilkes-Barre Republican warned Secretary Seward, "Depend on it, sir, the draft cannot be enforced in this county if the Administration compromises with the rebels in New York. The loyal men here will not sustain the draft unless it is enforced in New York."[106] Writing from Philadelphia, Robert A. Maxwell urged upon Lincoln the same theme: "The whole country is observing with interest the course of the Administration in dealing with the New York Conscription. If not proceeded with, say, by an officer of known determination such as General Butler with military and naval forces to support him, the Union goes up in a blaze of States Rights. An exhibition of resolution will insure Seymour's submission, the execution of the draft elsewhere and avoidance of foreign intervention." Maxwell was hardly exaggerating when he pinned Republican political fortunes, the fate of the draft throughout the North, and the post-emancipation war effort itself on the administration's actions in New York City: "If you enforce . . . [the draft] in New York the Democrats have nothing left to stand on. If you suspend it there and execute it elsewhere, dissatisfaction will destroy the Administration Party."[107]

The Crisis Eases

High as these stakes were, Lincoln and other federal officials rejected a ham-fisted solution to the insurrection in New York City. While Republican governments imposed military rule on other draft riot districts, various Southern-sympathizing cities, and whole regions of the South during the sixties, martial law was never declared in New York.[108] Nor did New Yorkers witness the impromptu sentencing and execution of draft rioters by military street tribunals and firing squads. Compare this with the mass-production killings of Parisian workers by the *gardes mobiles* in June 1848, or the executions of Communards in 1871 which left stacks of corpses three-deep and hundreds abreast near the Trocadero and outside the Ecole Polytechnique.[109] That firing squad sovereignty was never attempted in draft riot New York was in large measure a consequence of the brittle elite politics of the nation's economic capital. Federal military rule of New York raised the grim prospect of a withdrawal of conservative support for the restoration of order, followed by desertions from the army and an escalation of the midweek violence.[110]

The decision of Republican authorities not to declare martial law denied city officials certain military alternatives but created others. The concern of George Opdyke, August Belmont, and others—that martial law would alienate the conservative middle and upper classes from the restoration of order—could be put to rest. One major issue of the riots had been settled: New York would for the time being remain under Democratic control, and radicals would fail in their campaign to import Republican rule. Local Republicans could rely on a measure of cooperation from Democratic leaders and, no less important, were freed from the necessity of occupying an entire city. Now that the threat of outside intervention had been allayed, conservative elites quickly restored order to many neighborhoods. These quieter districts more often experienced the insurrection as a reverberation from uptown or the waterfront and not as a direct challenge by local wage earners. In addition, authorities could now expect some cooperation from two groups—organized workers in the building trades and German-Americans—that did not gravitate toward either pole of elite opinion. The failure to declare martial law created the possibility of a range of neighborhood political solutions to the violence and allowed authorities to concentrate troops in the so-called "infected" districts.

The plan adopted Wednesday the 15th by General Brown and Police Commissioner Acton abandoned the earlier tactic of clustering soldiers and police in defensive positions around City Hall, federal property, and the armories. Now authorities divided the city into unequal quadrants and gathered forces at strategic points (two on the Upper East Side, one downtown at City Hall, and one at the northern end of the island).[111] This scheme was better suited to exploit the political geography of the midweek revolt. The rioting of laborers had taken on a guerrilla aspect as bands of men and boys struck targets in different neighborhoods with little or no warning. "No sooner . . . had the conflict ended in part of the city then it began in another," wrote Upper West Side Police Captain George Walling of these itinerant struggles.[112] Industrial workers' barricaded defense of territory required a more concentrated assault against factory-district streets and tenements. The authorities' task was complicated by the barricading of three uptown areas, two of them non-contiguous. One massive onslaught against a working-class stronghold in the manner of Paris 1848 would not suffice.[113] The Brown and Acton plan allowed the massing of troops against what Brown called "the plague spots" of the Upper East Side and took advantage of the successes of elites in quieter districts by leaving those areas to the care of citizen militia and police.

This military isolation of the factory districts mirrored the industrial workers' own attempts to cordon off that world from neighboring middle- and upper-class quarters. The Seventh Regiment ran its pickets up Third Avenue, one block west of the draft rioters' barricades, presumably to take advantage of the loyalty of Third Avenue shopkeepers. As the leaders of the midweek uprising stationed sentinels at street corners and searched homes for soldiers, policemen, and weapons, so too did the military move from house to house inside *their* cordon seeking stolen goods, weapons, and pockets of resistance. The final battle on the East Side turned on a military attempt to recover the body of a slain soldier—in a district where workers had invested the physical presence of soldiers and policemen with great political meaning.[114] On the West Side, late week battles centered on the retrieval of the corpses of lynched black workingmen.[115] In the tragic finales to these racial attacks, authorities ordered the military to cut down the bodies of lynch mob victims. Soldiers who attempted to lower the corpses met fierce opposition from rioters. Often the military succeeded, only to have the crowds hoist the bodies up again hours later. Such military actions were an obvious way to restore order and an appearance of respect for the government. But it is worth observing that after rioters set the symbolic program of the revolt by focusing struggle on industrial territory and the bodies of uniformed Republican representatives and black men, the military kept its actions essentially within the limits of that program. The willingness of the Republican authorities to battle draft rioters on the rioters' own symbolic ground may help explain the restraint of national troops in New York relative to their Parisian counterparts.

Restraint also characterized the Republican response to the last event of the riot week, the speech of Archbishop John Hughes "to the men of New York who are now called in many of the papers rioters" on July 17.[116] Many East Side workers came to the Archbishop's house in response to an invitation printed in handbills and the morning press. Flanked by a dozen Catholic priests, Hughes counseled his audience to change the government through constitutional and not revolutionary means.[117] Anticipating trouble, Major General Wool instructed Brigadier General E. R. S. Canby to station a regiment at Madison Square, thirteen blocks from the Archbishop's house and well removed from the view of Hughes's audience. Wool decided to keep federal troops outside the factory district unless absolutely necessary for the preservation of order.[118] Though the crowd interrupted the speech with cries of "Stop the Draft," it remained peaceable and retired quietly when Hughes finished. Only a few blocks away, the Seventh Regiment surveyed the insurrectionary riverfront slums with its howitzers. Still, in the Hughes episode, Republican authorities had begun to act with some circumspection in their relations with conservative leaders. The conciliatory style was confirmed as the dominant mode of social relations in the city.

The new circumspection was best symbolized by Lincoln's appointment of Democratic General John Adams Dix to replace General Wool as Commander of the Department of the East at the end of the week. It has not been unusual for presidents to reach out to the opposition in time of war—Woodrow Wilson "adjourning" partisan politics during World War I, Franklin D. Roosevelt appointing Republicans to crucial posts in World War II. Even if we regard the appointment of Dix as wartime politics as usual, we should credit Lincoln with having made an extremely shrewd choice. Dix's political roots reached back to the Albany Regency of the twenties (like Seymour's) and the Free Soil Party of 1848 (like Tilden's). His credentials as a conservative lawyer and financier were impeccable. During the secession winter of 1860–61, New York City capitalists had demanded Dix's appointment as Secretary of the Treasury when the Buchanan administration had begun to falter. As Southern states were leaving the Union, Dix obtained $5 million in loans to the federal government from Eastern bankers and financiers.[119] But it was General Dix's wartime reputation as a relentless opponent of treason as Commander of the Maryland Department that won him the job of preserving order in post-riot New York. President Lincoln was committed to enforcing the New York draft—this was the message that Samuel Barlow received from T. J. Barnett, his Hoosier informant inside the White House.[120] Lincoln and Stanton knew that General Dix would carry out the draft in New York at all costs. They no doubt also realized that Dix commanded the respect of metropolitan elites of all political stripes and would succeed, if anyone could, in mobilizing conservatives behind conscription and the post-emancipation war effort.[121]

General Dix hoped to conduct the draft with as little reliance on federal power as possible. Minimizing the federal military presence in the city, Dix understood, would remove a major cause of the insurrection. This strategy required the cooperation of local Democratic leaders. Tammany Hall's pre-riot record as a patriotic defender of the war made it the direct beneficiary of post-riot indignation against Southern-sympathizing rioters and politicians. Tammany City Recorder John T. Hoffman and District Attorney A. Oakey Hall presided over the prosecution of draft rioters beginning the last week of July and continuing through the months of August and September. This Tammany tribunal took the place of a proposed federal inquiry into the insurrection.[122] Though few draft rioters were convicted and fewer sentenced to long terms, Tammany's aggressive prosecution of rioters won applause from both Republicans and Democrats and launched the political careers of Hoffman and Hall.[123] General Dix could rely on the support of local Tammany Democrats in enforcing the draft.

Governor Seymour, who commanded the state's military forces, proved less cooperative. During late July and the early weeks of August, Seymour tried to convince President Lincoln that the state's draft quotas discriminated against Democratic counties, and that even if the quotas were adjusted, the draft measure was unconstitutional and unenforceable in New York City.[124] In a bipartisan effort, Samuel Tilden and Edwin D. Morgan went to Washington to press Seymour's arguments upon the President; if anyone could persuade Lincoln of the draft's illegality, it was Tilden, the nervous but brilliantly articulate corporate lawyer. Lincoln rejected all Seymour's appeals. The Governor wrote dejectedly to Tilden on August 6 that the anti-draft campaign would "do no good—except making up a record—I look for nothing but hostility but I shall do my duty, demand my rights, and let consequences take care of themselves."[125] Meanwhile, Dix's letters to Seymour—seeking cooperation in enforcing an August draft exclusively with state militia—were answered evasively or not at all.[126] An angry Dix wrote Seymour on August 18:

> It was my earnest wish that the federal arm should neither be seen nor felt in the execution of the law for enrolling and calling out the national forces but that it might be carried out under the aegis of the State which has so often been interposed between the government and its enemies. . . . Not having received an answer from you I applied to the Secretary of War on the 14th inst. for a force adequate to the object. The call was promptly responded to and I shall be ready to meet all opposition to the draft.[127]

Stanton ordered to New York ten thousand infantry and three batteries of artillery from federal forces in Virginia.[128] Few of the regiments sent to New York City that week were of New York State origin.[129] Along with these troops, Dix armed himself with a presidential proclamation empowering him to call out the state militia in the event Seymour refused to suppress a second draft riot or commanded state regiments to resist the federal government.[130] While Tammany Democrats accepted the eventuality of the draft, Seymour, Tilden, Waterbury, and many upper-class conservatives clung to the hope that federal conscription would be nullified in the courts.

Federal troops were only just arriving in the city on the morning of August 19 when Dix resumed the draft. While the first July lotteries had been held in the poorer uptown wards, the August draft began in the affluent and Republican Greenwich Village and was concluded without incident. Stationed in Washington Square, Mason Whiting Tyler of the Thirty-seventh Regiment Massachusetts Volunteers later recalled, "Everything was as quiet and orderly as a New England Sabbath. No disturbance occurred anywhere, and in the evening the various details were gathered again in the park and we slept with our guards and our pickets thrown out as if we were in the enemy's country. . . ."[131] The first August drawings were staged in quiet Republican districts, and troops bivouacked in neighborhoods where their presence would be least offensive. This new Republican discretion helped explain the August calm.

The draft lottery in the industrial and Democratic Eleventh Ward also proceeded quietly on August 24. Republicans Edwin D. Morgan, Orison Blunt, and Mayor Opdyke made appearances at the Provost Marshal's Office.[132] More important, Tammany officials Francis I. A. Boole, Smith Ely, Jr., William M. Tweed, and Charles G. Cornell vouched for the fairness of the enrollment, counted the name ballots, and stood watch over the drawing through the day.[133] Unlike the July draft, August conscription was clearly endorsed by leaders of both parties and by a broad segment of the local middle and upper classes.

Why then was the August draft enforceable while the July draft was not? Of course, the federal government's show of force in New York promised bloody consequences to any repetition of July's violence (authorities in Philadelphia were alerted by New York's experience and avoided draft violence altogether by posting ample military force in the city while the lotteries were being carried out).[134] By the last week in August, the momentum of the war was undeniably in favor of the North, and European intervention was unlikely. The mood of July—that sense that the direction of the war was uncertain and could even be changed by popular intervention—had passed. Most important to the August calm was the new position of Tammany leaders. Tammany's public participation in the August draft helped account for its widespread acceptance. Tammany Supervisor Elijah Purdy and many fellow conservatives staked their political futures to a quiet and equitable enforcement of the draft.

Before a draft lottery audience in *Kleindeutschland,* Purdy pledged that "those men drafted who require and [can] not procure substitutes [will] have them procured for them."[135] Tammany guaranteed an equitable draft through its "county loan" ordinance, which was finally passed over successive mayoral vetoes and became law in September.[136] By September 20, New Yorkers had contributed $885,000 to the draft fund.[137] This money was distributed to conscripts by the "County Substitute and Relief Committee" composed of Republican Chairman Blunt, Tammany Supervisors Purdy and Tweed, and Tammany Comptroller Brennan.[138] The exemptions process required the cooperation of the provost marshals' offices, the Republican chairman, and the Tammany officials who numerically controlled the committee. Under the ordinance, each conscript accepted by the provost marshal as a soldier could claim relief from the Committee if he were a "policeman, fireman, or a poor man with a family dependent on him."[139] The last category gave Tammany an opportunity for much latitude in applying the ordinance. Before the Supervisors' Committee paid the poor conscript's commutation fee, it required that the man attempt to find a substitute. The Committee encouraged draftees to visit the agents and brokers who supplied substitutes for a fee, if indeed the Tammany Supervisors did not hire the services of brokers directly.[140] "Had it not been for the policy thus put in practice," the *Sun* reported, ". . . the Government might have obtained a large amount of money but few soldiers. The probability is that not 1000 of the originally conscripted will enter the army from this city."[141] The *Sun* may have exaggerated the number of New York City draftees who actually reported for duty. The County Substitute and Relief Committee announced on September 28 that 1,042 of the city's conscripts had furnished substitutes, 49 had paid the $300 commutation themselves, and two had gone to war. The total cost of this arrangement under the ordinance, through September 28, was $327,000.[142] The Tammany legislation virtually guaranteed conscripts who desired not to serve that they would not be compelled to do so and solved the immediate problem of the July riots, that of the politically explosive draft.[143]

A Lingering Problem of Rule

While a conservative middle and upper class had disposed of the draft as a political question and set some limits to the degree and kind of political centralization Republican leaders could attempt, it still had to negotiate a host of other issues raised by the July insurrection. Intense conflicts among the elite over the racial and class legacy of the riots had to be mediated before conservative political hegemony became secure in New York City. Two post-riot events, the presentation of the colors to the black Twentieth New York Regiment in March 1864 and the publication of the 1863 Annual Report of the Association for Improving the Condition of the Poor, revealed the vast dimensions of the problem of political rule posed by the July insurrection.

Many black men, women, and children fled the city during the third and fourth weeks of July. Some sought refuge in the New Jersey hills, others in the Brooklyn countryside. Only in the vicinity of Republican Greenwich Village did some black families arm themselves and respond collectively to the assault on their persons, property, and livelihood.[144] For months white longshoremen kept black laborers off the waterfront. Black workers in most other occupations had difficulty reclaiming their old jobs despite the feeble movement among upper-class families to employ black domestic servants in place of the Irish.[145] In the weeks after the riots, white omnibus drivers and passengers prevented blacks from using public transportation.[146] Beatings of blacks by gangs of white boys continued through the summer.[147] Left to the uncertain protection of the Metropolitan Police, some blacks who remained in the city moved from the uptown or Five Points to safer Republican or German districts.[148] The post-riot decline in the black population continued a long-term trend.[149] "A large number of colored families have left the city," the *Tribune* reported on August 3, "with the intention of never returning to it again."[150] For months after the riots the public life of the city became a more noticeably white domain.

This contraction and withdrawal of the black community occurred with the blessing of most conservatives. The Tammany Common Council moved quickly to protect poor white families from the inequitable draft but took no action on behalf of black victims. Tammany's version of white supremacy was not the extreme variety of Tilden and his associates George T. Curtis and Samuel F. B. Morse, who in spring and summer 1863 penned pro-slavery leaflets for the upper-crust Society for the Diffusion of Political Knowledge.[151] As loyalty to the Union cause became increasingly identified with support for black rights in 1863, Tammany leaders may have feared that a too-shrill racism would compromise their post-riot position as patriotic defenders of the social order. These differences in tone acknowledged, both Tammany and the pro-slavery SDPK tolerated or encouraged the prevalent racist mood.

In this atmosphere, the Union League Club's plan to raise a black regiment and parade it through the city became a bold attempt to reclaim urban public spaces in the name of an alternative racial and class relations. The Club's black military recruiting campaign emerged from two sources: the Committee of Merchants for the Relief of Colored People and John Jay's scheme to shift some of the weight of the state draft quota to blacks in order to allow an equal number of white workingmen "to remain quietly at home."[152] But by the time War Secretary Stanton gave the Club permission to raise a black regiment in December 1863, the recruiting campaign had become more than an attempt to transfer the burden of the draft away from the white working class. The planning of a spectacular presentation of flags to the new regiment followed by a march down Broadway to the debarkation point on the Hudson suggested a far more ambitious project.

On March 5, 1864, an audience estimated at one hundred thousand crowded Union Square to witness the ceremony.[153] The "Mothers, Wives and Sisters of the members of the New York Union League Club" appeared in front of the Club House, addressed the black soldiers, and presented them with a stand of colors.[154] Witnesses connected the event directly to the July riots. Club member Francis Lieber wrote his friend Charles Sumner on the 6th, "yesterday I omitted mentioning the historic fact—the, to me, great symbolic fact—of the presentation of colors to the regiment of blacks in Union Square by our club. There were drawn up in a line over a thousand armed negroes, where but yesterday they were literally hunted down like rats. . . ."[155] In the wake of an insurrection which linked community disapproval of amalgamation to a range of conservative middle- and upper-class prerogatives, the Union League Club sought symbolically to associate black men with the most loyal element of the white upper classes.

It was significant that the Club "Mothers, Wives and Sisters" occupied center stage. In their address to the Twentieth, the women pledged the black soldiers their patriotic devotion: "[We] will anxiously watch your career, glorying in your heroism, ministering to you when wounded and ill, and honoring your martyrdom with benedictions and tears."[156] LeGrand B. Cannon described the relation of the Club women to the black regiment in different terms at a banquet commemorating the ceremony twenty-two years later:

> The three hundred founders of this Club who braved social ostracism and contempt, by marching as an escort for the first colored regiment down Broadway, are worthy of commemoration and honor; but the ladies who gave the regiment its flag must share in its glory. . . . The potent influence of the women of position and power in our New York world stamped out prejudice, turned hisses into applause, exalted the humble and despised to places of honor, and in giving the black man not only the right but the invitation to fight for his liberty, created the force which emancipated the slaves and saved the Union.[157]

The Club women invoked the nineteenth-century view of the women's sphere as a preserve of pure republican virtue and an inspiration to patriotic self-sacrifice during war. Cannon fused this republican reading of the ceremony with a decidedly elite view of women as keepers of the gates of upper-class social life. The draft riots had confirmed anti-abolition and anti-amalgamation as the twin pillars of a conservative middle- and upper-class outlook. The March 5 ceremony symbolically stood the anti-amalgamationist motive of the riots on its head by bringing elite women and black men together on an august public occasion. The conservative and Democratic *Herald* acknowledged this theme of March 5 in mocking fashion, calling the ceremony "a pretty fair start for miscegenation."[158] For conservatives, the message of the ceremony and the parade down Broadway was clear: entry into the upper reaches of loyal "society" would now be predicated on willingness to embrace emancipation and the attendant changes in the status of free black people.

The Union League Club thus attempted to characterize the July uprising as a race riot which could be remedied with a racial response such as the March 5 spectacle. The ceremony honoring the black Twentieth Regiment did not directly address the insurrection's class dimension or its challenge to conscription and the wartime state. In its Annual Report for 1863, the Association for Improving the Condition of the Poor more freely acknowledged class antagonism as a cause of the insurrection. The AICP regarded the alienation of the poor from the middle and upper classes as the problem most urgently raised by the draft riots.

The revolt posed an especially grave challenge to AICP members. Most were Republicans, and some were singled out by the rioters for attack. The AICP had long believed in the essential moderation of the American working classes. During the secession crisis of 1861, the Board of Managers wrote, "Our mechanics and laborers form not only the broad granite base of society, but they are a great conservative power. We have too few of the dangerous classes, for any well-founded apprehension of violence and social disorders."[159] The draft riots jolted this confidence: how might the Association resurrect a respectable working class receptive to its reform message of industry and sobriety from the ashes of the July insurrection?

The 1863 Report addressed itself to this post-riot task of rehabilitating a deserving poor. During the insurrection, the Report emphasized, the "industrious poor" had been misled by "false men" such as Peace Democrat Fernando Wood. To AICP members, Wood was notorious for his endorsement of public welfare schemes to pacify bread riots during the Panic of 1857.[160] The Association sought to battle the demagogic influence of such indiscriminate philanthropy with its own "scientific charity," which relieved the poor on a private and individual basis so as to stimulate incentives for saving and industry. The draft riots would not have occurred, the Report insisted, had it not been for the blandishments of conservatives like Wood and the susceptibility of the many "ignorant poor."

But the grand scale of the riots and the involvement of so many different kinds of workers mocked the reformers' narrow focus on the worthy poor as a solution for class conflict. A more comprehensive strategy was required to reach the "semi-brutalized masses" who were deaf to the AICP message.

The AICP remedy for the riots amounted to an intensified campaign of moral education and health and housing reform. The centerpiece of the effort was "the training and discipline of the children of the poor."[161] Juvenile rioters had been "among the most active to burn, to rob and to torture."[162] The reformers placed great faith in the public education system and planned to increase their agitation for strict enforcement of the state truancy law. By diminishing the "neglected juvenile population," the Report concluded, "so essentially conservative and law-abiding . . . would the mass of people become, that future riots would be impossible."[163] The AICP also sought to intensify its campaign against overcrowding and health nuisances in slum tenement districts. The Board believed that such conditions encouraged barbaric behavior among the working classes. "The debasing effect of tenement-house life on its inmates," the AICP would later argue, was a direct cause of the riots of 1863.[164] This increased AICP interest in working-class living conditions after the riots laid the groundwork for the sweeping municipal reforms of the Citizens' Association of the mid-sixties.[165]

Such broad-gauged reforms verged on the sort of indiscriminate aid to the poor that the AICP elite abhorred. The Association was forced to justify the liberality of the post-riot reform campaign in its 1863 Report:

> Would it not be more politic, not to say more noble, humane, and Christian, to minister discreetly to the necessities of the unfortunate, even at the risk that some ignorant, deceived, perhaps guilty persons may share in such bounty, rather than doom the great mass of the deserving poor to suffer unrelieved.[166]

The expansive and optimistic tone of the Report was undercut, finally, by a realization that a large cultural chasm divided the immigrant rioters from the Protestant and native-born reformers: "It is unreasonable to expect that the classes of persons in question, being almost exclusively Irish Catholics, can to any considerable extent be reached and influenced by Protestants, whom they regard as schismatics and heretics."[167] The AICP was forced by the July insurrection to widen its definition of the deserving poor and direct a new campaign of moral education at the broader masses. But these moral reformers saw limits to the extent they could cultivate the morality of the "intelligent classes" among a volatile metropolitan working class. The reformers' moral influence would necessarily have to be accompanied by "a more strict and rigorous surveillance" of the poor.[168]

Most of all, the AICP middle and upper classes sought to establish themselves as the true and legitimate donors in the city. The Tammany draft exemption ordinance, indiscriminately rewarding deserving and unworthy poor alike, violated the Association's principles of scientific philanthropy. Such undisciplined giving, the reformers believed, discouraged the development of habits of industrial discipline among the poor and consequently made wage earners vulnerable to the appeals of false leaders. Though not truly a radical, Mayor Opdyke was a committed AICP member who expressed Association fears of the Tammany ordinance in one of his veto messages:

> Stopping for a moment to consider the composition of the [County Substitute and Relief] Committee, I cannot but observe that it would afford great opportunities for abuse. Too numerous for individual responsibility, too unwieldy for the investigation of particular cases . . . the applicants for relief

> would be many . . . this would be in effect an appropriation of three millions of dollars for such objects as should afterwards be determined. . . .[169]

Opdyke and other reformers preferred a private and closely supervised relief of cases of "special hardship" among the conscripts.[170] Only such a monitored aid would bind a volatile working class to its proper elite patrons and allow donors to educate recipients. Perhaps even more than their Union League Club comrades, the poor reformers wanted to make the working classes over in their own image. In the wake of the draft riots, this was no small project.

Among the middle and upper classes, the draft riots disclosed contrary notions of legitimate federal power, relations with the poor, and between the races. Conservatives hoped to keep federal martial law armies at bay, tolerated or even supported the early phase of the revolt, and approached the more insurrectionary rioters with a mixture of force and conciliation. While they negotiated with white rioters, they ignored black victims. Radicals were quick to summon the full power of the federal government against treason and social disorder and were hostile toward the riot in all its phases. They advocated draconian measures against the white rioters and their sympathizers and campaigned for the relief of the black community. In his poem "The House-top: A Night Piece," New Yorker Herman Melville wrote approvingly of the riot's bloody finish: "Wise Draco comes, deep in the midnight roll / Of black artillery. . . ." But as Melville's Draco attempted to restore order, he displayed two very different tempers to the city.[171]

By late Monday, July 13, the draft riots had exposed a deep fault in the metropolitan political order, as conservative and radical wings of the middle and upper classes debated how best to bring peace to the city. At this moment of extreme political and social polarization, it became necessary to resolve some of the questions posed by the insurrection. All leaders were interested in the protection of private property, and most did not want to endanger the momentum of the Northern war effort. The Lincoln administration along with local allies such as Mayor Opdyke decided to forgo a declaration of martial law in order to secure a measure of cooperation from city conservatives. Lincoln and Stanton decided that a moderate solution in New York City, though it conceded the ambitions of local Republicans, would most quickly restore quiet to the nation's economic capital at a critical juncture in the war. The draft was eventually enforced in New York, but as part of the conservatives' program.

What were the prospects of this post-riot conservative ascendancy? The pro-war Democrats of Tammany Hall were its direct beneficiaries, while the Peace Democracy, now tainted with treason, became its first political casualty.[172] But Tammany's victory after the riots—and it would not be until after the war that Boss Tweed solidified his position—was incomplete and much compromised from the outset. That victory relied upon Lincoln and his successors' continuing reluctance to intervene militarily in New York. Another outbreak of violence might bring General Butler to the city after all—the Beast of New Orleans stood poised in New York harbor on Election Day 1864 (fighting between supporters of Lincoln and McClellan was feared), but did not have to be called ashore. Nor could Tammany take for granted the good will of the wealthy Democratic investment banking interest. While Tammany proprietors sought a draft tempered by public charity for poor conscripts, wealthy Democrats preferred to void conscription in the courts. Even after Tammany disposed of the draft issue, tension between conservative groups remained, never far beneath the surface of Tammany rule under Tweed and quick to explode after Tweed's fall. Tammany also faced two radical challenges: one from the Union League Club and the other from the Association for Improving the Condition of the Poor. The Club merchants challenged Tammany's racist style by publicly associating themselves with the black poor. By raising the black Twentieth and marching with it down Broadway, the Club exhibited to insurrectionary white workingmen the deferential loyalty of a "model" urban poor and to conservative elites the loyal nationalism of a "model" upper class. AICP industrialists countered Tammany's indiscriminate giveaway with a program of private and personalized charity, volunteer brigades, company-based draft insurance plans, housing improvement, and education to cultivate loyal white workers; the AICP recommended truancy laws and heightened surveillance for an "ignorant" pauper class. Tammany also faced an independent challenge from German-Americans, who on Election Day 1863 sent to City Hall one of their own, independent Democrat C. Godfrey Gunther. Nor could Tammany lay full claim to the allegiance of a labor movement which began endorsing workingmen's candidates on an eight-hour platform in the fall of 1864. Most important, Tweed rule depended on the favor of the sorts of workers who took to the streets in July 1863. These wage earners—so quick to intervene when they perceived a wrong the politicians could not redress—bore the closest watching.

The crisis of July 1863 revealed a fractured and unstable political order. An assertive poor confronted an elite deeply divided. Argument over the relative justice of the Conscription Act spilled over into and merged with a debate over the definition of justice in New York—a broad-ranging and politicized debate that had been going on for some time. To appreciate the many voices in this debate we have to put aside the disillusioned Herman Melville of the sixties, who celebrated the military suppression of the draft riots as a simple victory for the forces of order.[173] Order in mid-century New York was no simple thing; it was a contrivance of considerable complexity. We have to cultivate instead the vision of the democratic Melville of the early fifties, the Melville of *Moby-Dick,* who was willing to entertain, as few of his contemporaries were, the multiple claims to authority in this era. Just as Melville scrutinized his great white whale from every angle, we have to be willing to move around our subject slowly and view it from all perspectives. The crisis of the riots originated in the many conflicting claims to authority during the pre-war decade. It was then that workers became so assertive and the "better classes" so divided.

PART II

Origins of the Crisis, 1850s and 1860s

CHAPTER 3

Workers and Consolidation

During the draft riots workers pushed aside the middle- and upper-class reformers of the Republican Party and sought to intervene in politics on their own terms. To explain how workers came to behave so assertively we must look for earlier evidences of this politicized style of class relations. Historians have provided scant clues. While many scholars have stressed the leading role played by the Republican middle classes in the national transformation of mid-nineteenth-century America, few have located wage earners within the context of that middle-class initiative.[1] Further, one finds little mention of politicized struggles between workers and reformers in the literature on the 1850s labor movement. Instead the fifties have frequently been characterized as the seedtime of an apolitical and wage-conscious craft unionism. John R. Commons and his associates, Norman Ware, and other students of the mid-century labor movement pointed to the New York City strike wave of 1853–54 as the first expression of a "modern trade unionism" preoccupied with "pure and simple" economic issues and permanent unions. Wages and organization, and not politics, contended Commons and his followers, were the concern of the 1850s labor agitation in New York and other American cities.[2]

One need only begin to read the passionate working-class debates over organization and politics in the fifties to sense that the problem of working-class political consciousness in those years was far more complicated than the Commons school would have it. Those debates do bear out Commons's threshold premise: that many workers became disenchanted with the usual practices of the major political parties and began to search for other structures and institutions. But many New York City workingmen would have felt uneasy with Commons's characterization of their activities as a retreat from politics and reform. They would have instead portrayed the fifties as the dawning of a new age of social and political experimentation.

When Horace Greeley told the spring 1850 meeting of the American Union of Associationists that "so long as society remains in its present incoherent and warring state, the contests of its political parties must partake of the prevailing antagonisms" and that "the measures of our political parties can have only an inconsiderable and temporary effect for good," he spoke for a vast cross section of Northern urban and industrial society. Beginning during the fifties in the large Northern cities, reformers of many different social and political stripes cried out against chaotic individualism and competition and for "Organization" and "Association." New national professional societies such as the American Medical Association and the American Society of Civil Engineers denounced what in 1851 one leading physician called the "dishonorable competition, which an uninterrupted individualism is so apt to engender." By 1853, Frederick Law Olmsted was complaining to his friend Charles Loring Brace of an anarchic "materialism" tied only to the prerogatives of the marketplace and of the need for new urban institutions to bind together a society torn by unbridled individualism. Beginning in the late 1840s, Unitarian minister Henry W. Bellows expounded this same theme to his merchant parishioners. Ira B. Davis, the middle-class land reformer and business agent for New York's Protective Union Labor Association, told the city's Industrial Congress in 1850 that they lived "in an epoch fraught with circumstances of momentous importance to the present and future generations of humanity" because they now had the opportunity to "destroy avarice and competition" through "cooperative association." In the early fifties, an urgent sense that the major political parties were unequal to the task of binding a society ravaged by individualism and competition and that new structures were needed swept the urban and industrial North. John Higham has argued that Americans then began "exploring new possibilities for organizing and disciplining a culture of rampant individualism" and has characterized this new cultural movement as "the emergence of a pattern of consolidation."[3]

Consolidation is a useful way to describe what certain mid-nineteenth-century groups were after, but the term must be used with care. First, these mid-century critics were embarked on something more than a generalized search for order, which presumably most members of any society desire. They shared a very particular sense that their overly competitive society had corrupted the "old parties" and that the social and political systems together required overhaul. Second, many different urban groups—each for its own reasons—became troubled by excessive individualism and the inadequacy of party rule. Individualism and party rule, it hardly need be said, had distinct meanings for different social classes. It is most fruitful, then, to view the highly charged urban social, cultural, and political conflicts of the 1850s as a debate over consolidation among antagonistic, shifting, and sometimes overlapping groups. In New York City, the result was not a new pattern of consensus but the beginning of an epoch of relative disequilibrium and recurring challenge to political order.

While many different groups of New Yorkers began to advance their own solutions for an unrestrained individualism in the fifties and sixties, workers were at the fore. For no other group was imposing coherence on a new industrial environment of rampant competition, topsy-turvy markets, persistent unemployment, and squalid living conditions more urgent. In a period of unprecedented organizational creativity, city workers devised scores of producer, consumer, and building cooperatives; presided over a citywide Industrial Congress in 1850–51, an "Amalgamated Trades Convention" in 1853, and a Workingmen's Congress during the war; sent their own representatives to scores of "independent" reform meetings; and, finally, created the permanent trade unions celebrated by Commons. These new institutions were conceived at remove from the party oligarchies and were political in the many instances where they were part of an effort to complement or replace an effete party system unresponsive to the problems of urban industrial life.

In another way, the term consolidation captures the aspirations of New York City workers in the 1850s. To bind a competitive industrial order, some workers sought to "consolidate" their own realm within that society. Some of the institutional experiments of the fifties drew the boundaries of a separate working-class social domain and invested it with new political potentialities. For some New York City wage earners, then, the era's spirit of consolidation led to the formal proscription and exclusion of certain groups from working-class political activity. Nothing more dramatically signaled the end of Higham's Jacksonian era of "boundlessness" than the proclamation by several New York City trades in 1850 that certain white, middle-class, Protestant groups were barred from workingmen's institutions.

In these efforts at consolidation, in both senses of the term, lay the working-class origins of the draft riots. New tendencies among workers in the early fifties—the blurring of distinctions between economic and political action and the proscribing of middle-class politicians and reformers—presaged the politicized conflict of July 1863. Beginning with the Industrial Congress of 1850, artisans in the draft riot trades began to elaborate a plan for an authentic popular rule, a way of converting their prerogatives as craftsmen into political influence apart from the major parties. The Republican centralization of government and economy during the war had the effect of directing such pre-war discontent against Lincoln's regime. For these artisans the Conscription Act, with its class-biased commutation clause and sanctioning of new kinds of political intervention, was the quintessence of exploitative authority. By the fifties, other artisans had lost much of their status as craftsmen as a result of industrialization. These most proletarianized workers sharply distinguished between economic and political spheres of action, shunned the political experiments of the fifties' labor movement, and did not riot in July 1863.

Industrial workers and common laborers also sought to translate their trade prerogatives into political power. Among these poor New Yorkers—and they were usually Irish and Catholic—workplace and neighborhood life were most intimately connected. For many industrial workers, the integration of work and community was part of both their trade culture and their relations with employers. The Irish Catholic boilermaker or blacksmith was often employed by a Republican and Protestant boss who sought to reform him in the shop, at home, and in the neighborhood. Common laborers had a less direct relationship with the Republican Party. They regulated the labor market in their neighborhoods through ethnic networks of family and friends, and Republicans were one of several groups suspected of hostility or indifference to their kinship and friendship circle. For all their managerial discretion over the work process, many industrial workers and laborers came to feel especially vulnerable to displacement in the urban economy as the war progressed. By spring 1863 many could believe the Copperhead prediction that emancipation would bring north hordes of low-wage black freedman to compete for employment. As Irish Catholics, these industrial workers and laborers felt still more vulnerable and saw other defects in the Republican Party. They understood conscription as one of a train of wartime Republican assaults on their livelihood and culture.

Differences in economic location and experience are the most appropriate starting point for explaining the dynamics of the riot and analyzing the origins of the rioters' challenge to the Republican Party. Not all poor immigrants, Irish, or Catholics became draft rioters, and those that did disagreed about what draft rioting meant. Artisan craftsmen, industrial workers, and common laborers traveled distinctive paths to arrive at the steps of the Ninth District Provost Marshal's Office on July 13, 1863. Because they understood the interconnectedness of economy and polity and acted in ways that challenged authority in both realms, these New Yorkers came to riot against the Republican draft and command surprising power in the political order of the mid-nineteenth-century city.

Artisans

Artisans either abstained from all phases of draft rioting (as did most in shoemaking and tailoring) or helped lead the limited anti-draft demonstrations of July 13, 1863 (as did many in building and woodworking). In either event, they were guided by styles of political thinking and acting conceived during the 1850s. Each group of artisans developed its style in response to economic changes in the crafts during the second quarter of the century.

Adam Gamble was the presiding officer of the United Society of Journeymen Cordwainers. The shoemakers' union was one of the few New York City trade associations that had endured the hard times that followed the Panic of 1837; most of the city's unions were disbanded in the forties and then reconstituted in the economic upturn of early 1850. We know little about Gamble except that one observer at an April 1850 Cordwainers' meeting described him as "a venerable old man" and noted in passing that he had been "a member of the Society since 1803." Gamble was not, it seems, an active leader of the shoemakers' association: he was silent through most of their debates that spring and would not serve, as so many other trade union leaders did, as a delegate to the citywide Industrial Congress. Instead, if we can exercise our imagination, Gamble's leadership of the Cordwainers' Society had symbolic value. The old shoemaker was a valued relic from the early years of the century. He was living evidence of a time when journeymen shoemakers and their organizations occupied a station of prestige and influence among the city's artisan trades.[4]

When journeymen in shoemaking and tailoring observed the disarray of their trades in 1850, they may well have required the presence of an Adam Gamble as a tonic remembrance of better days. A journeyman shoemaker or tailor of middle age who in 1850 reminisced about the state of his trade in 1825 would have recalled a drastically different work world. In New York's consumer-finishing trades, the second quarter of the century brought major changes in the organization of labor: multiple divisions in the work process converting the skilled journeyman artisan into a producer of only a piece of the finished article; the rise of contracting and "outwork" arrangements that transformed the tailor's custom of bringing work home to his family and the shoemaker's traditional home work into the exploitative "sweatshop" system; and a growing separation between a coterie of custom workers and a sea of journeymen in the "ready-made" or "slop" trade. In 1845, the *Tribune* observed that "the system of apprenticing to the Shoemaking trade is now pretty much done away with in the city"; by that year, formal apprenticeship had also disappeared from tailoring. This altered organization of labor furnished the context for the entrance of thousands of Irish and German immigrant workers into the consumer-finishing trades. While mechanization in these trades would not begin until the mid-fifties, the factory system was prevalent by mid-century. Highly subdivided and seasonal manual production, performed either in large factories or in secluded garrets, became the rule in the manufacture of consumer-finishing goods during the pre-Civil War decade. By then women workers, often toiling under the supervision of a journeyman paterfamilias, predominated in the most deskilled and subordinate precincts of clothing and shoe manufacture. The new prominence of women in these trades further signaled the full erosion of an older, male-dominated craft world. With all of these changes it made sense for artisan communist Wilhelm Weitling to place shoemaking and tailoring at the bottom of his 1855 pecking order of the crafts.[5]

Most male wage earners in these trades occupied an ill-defined intermediate ground—between a mass of proletarianized women and child workers, whose labor the journeymen often supervised in home and garret settings, and the large clothing and shoe merchants and contractors who presided over the ebb and flow of production and credit. Particularly in the clothing industry, this middling ground was vast. By 1850, there was little to distinguish the experience of the subcontracting journeyman who rented a garret and supervised the labor of a dozen hands from that of the small master tailor. Years later tailor Conrad Carl called attention to the power of those journeymen tailors who had access to "a house or a pile of money": by employing "hands besides their own families, ten or twelve or fifty hands," such journeymen could produce as many as a thousand coats a week. By the fifties, many journeymen in shoemaking and tailoring—either as heads of households or as garret masters—also exercised some of the prerogatives of the small employer.[6]

But the limited managerial prerogatives of such journeymen owed little to craft traditions. Many of the fixtures of the artisan craft experience had by the fifties disappeared from the consumer-finishing trades. The traditional cobblers' or tailors' shop had been reputed for its bonhomie among workmen. In the village shops of both trades, the institution of the "reader" survived into the nineteenth century—men would take turns reading books and journals aloud or the youngest boy would fetch and read the news. Throughout late eighteenth and early nineteenth century Europe, shoemakers were known as artisan "intellectuals," "philosophers," and often, by extension, political radicals. The advent of the family system of labor of mid-century New York undermined what one historian has called "shoemakers' culture" and brought in its stead physical isolation and the drudgery of unending toil. Sequestered in home shops, shoemakers labored, as one worker put it, "sixteen, eighteen and sometimes nineteen hours" in an "unwholesome trade" with "no opportunity of getting fresh air nor any exercise unless what they got by the movement of their arms." The insistent demand at the 1853 meeting of the ladies' branch of the trade, then, was merely that "the men working for each shop should know each other and . . . shop meetings . . . be held." In the fifties and sixties, shop-floor actions in shoemaking and tailoring were common but rarely supported by an effectual central union; centralized organization was made especially difficult by the isolation and dispersal of home work. While mid-century journeymen shoemakers and tailors frequently managed the work of others, they lacked the sense of status and *esprit de corps* that belonged to the craft worker. With craft traditions no longer intact, there was even less to differentiate the experience of the subcontracting journeyman from that of the small boss.[7]

These social circumstances gave rise to a distinctive class relations. In July 1850, New Yorkers turned their attention to the bleak conditions in the sweatshop district when nine hundred tailors halted production to demand an increased pay scale and a formal bill of prices signed by each employer in the trade. From the outset, employers sympathetic to the strikers' grievances and willing to sign the bill of prices were regarded as full-fledged members of the tailors' movement. On July 26, the German Central Committee of the United Trades observed that the object of "the Tailors' movement" was "to unite together all the Journeymen Tailors, as well as their Employers who adhere to this bill, into one great Protective Union, which should guard the rights and promote the interests of all." The *Tribune,* whose editor Horace Greeley was a favorite speaker at labor meetings that summer, observed as early as the 16th that boss tailors had "very generally given in their adhesion to the Society, and in many instances compel[led] their Journeymen to join"; the *Herald* even reported a rumor that the strike originated not with the journeymen but "with the sub-bosses, who have tailors' shops, and employ journeymen to work for the bosses." Tailors relied on the sympathy and even the leadership of small bosses in their struggle to impose the price scale on the larger shops involved in the Southern market; journeymen did not unilaterally legislate wages for the trade but consulted with friendly employers through the strike. The tailors' strike demonstrated, above all, that the "honorable" employer or subcontracting journeyman was a welcome participant in the tailors' social enterprise.[8]

All told, industrialization in the second quarter of the century overwhelmed traditional arrangements of craft production in tailoring and shoemaking. Because the contract system grew up in an environment in which craft traditions had largely been eroded, the small boss and the subcontracting journeyman by mid-century shared much of the same social experience. Journeymen tailors consequently saw no need to exclude subcontractors and small masters from their organizations.

The local building industry underwent many of the same transformations as the consumer-finishing trades before 1850. Like clothing and shoe manufacture, building witnessed the rise and spread of the contract system during the 1820s. The new building contractors were less successful than the clothing manufacturers in subdividing the work process but found other ways to undermine the skilled journeyman's control of craft. In 1834, contractors led by the notorious Elisha Bloomer, the builder of New York University, sought to undercut the wages of local stonecutters during a building boom by employing upstate prison labor. Many employers and subcontracting journeymen introduced cheaper "Carpenters' Gothic" and "Carpenters' Doric" styles of work turned out by low-wage men. By the 1840s the building trades had spawned a class of wealthy contractors and master artisans who speculated in uptown construction ventures and invested their profits in metropolitan real estate.[9]

But the middle-aged carpenter or bricklayer who in early 1850 reflected on the events of the last quarter-century observed many more continuities than his tailor or shoemaker neighbor. The basic social structure of the building trades had been little altered. In 1850 most journeymen carpenters, bricklayers, plasterers, painters, and stonecutters still served a full apprenticeship. Subdivided in many other crafts, the work process here remained largely intact. Through the fifties house construction continued to employ considerable numbers of native-born workers. Despite the rise of the contract system of house building and some attempts to dilute the crafts in the thirties and early forties, journeymen retained control over a wide range of issues relating to the organization and management of work. Regulation of the hiring of apprentices and helpers, the length and pace of the work day, and subcontracting among journeymen and small masters were still assumed to be the prerogatives of journeymen through the mid-1840s. These journeymen's practices were not successfully contested by most employers. Accordingly, journeymen had no reason to promulgate or even write down such work rules. The use of machinery was just beginning to affect carpentering in the late 1840s but hardly touched the other building trades. The journeyman carpenter who in early 1850 mused on the industrial changes of the last generation saw little or no change in a broad realm of craft control. The reorganization of work and resulting social conflicts in building in the first four and a half decades of the century concluded with a set of practices and relationships that secured considerable power to the journeyman.[10]

The journeyman in building, like his counterpart in shoemaking and tailoring, exercised many of the prerogatives of the small employer. Indeed he still had a reasonable expectation of attaining "independence" as a small master when this hope was fast disappearing in the consumer-finishing trades. Such "independence" was usually not a secure new identity but a revolving door that might lead into the ranks of the employing classes one building season and out the following. While during strikes the division between journeymen and small employers hardened, sometimes the two groups found themselves confronting together common enemies like the large speculative builder or banker. The Master Marble Manufacturers' Association was accordingly an outspoken supporter of the journeymen stonecutters' 1834 campaign against some builders' use of convict labor. Further, journeymen carpenters, bricklayers, and plasterers frequently paid their helpers and laborers out of their own wages or independently took out subcontract work.[11]

Yet, for all these similarities with their employers, journeymen in building had one special quality which set them apart and gave them a heightened sense of their own status. Because their prerogatives were grounded in a control of craft, these journeymen believed that they, and not their employers, were primarily responsible for maintaining the welfare of their trades. As the pace of industrialization in building quickened in the late 1840s, these craftsmen's sense of their own social importance would become especially evident.

The late 1840s witnessed four changes in metropolitan construction. First, between 1846 and 1850 there was a large increase in the number of buildings erected annually.[12] Among these new buildings were the resplendent commercial emporiums of Broadway above Chambers Street, which by the fifties made that thoroughfare the equivalent of London's Regent Street or Naples's Strada Toledo. It may also have been the case that as the physical scale of metropolitan building grew, the tendency for building sites to serve as arenas of association for different groups of artisans likewise increased. The geographic concentration of construction work at mid-century contrasted with the growing dispersal of production in shoe and clothing manufacture. Next, real estate auctioneering became a "business," with some auctioneers now specializing in brokerage and commission transactions.[13] This emerging real estate business created new financial instruments like the "buy-back" building lease which facilitated investment in construction by speculative capitalists.[14] Third, with the new scale of building and the new instruments of credit came a new class of builders and investors, many of them immigrants. These boss carpenters and masons frequently became sublandlords and agents in the burgeoning tenement-house districts on the city periphery.[15] Finally, this expansion of the scale and personnel of the speculative building industry brought on a new wave of subcontracting and assaults on journeymen's work rules, as builders on the make sought to capitalize on the booming construction market. This spurt in urban growth would disrupt established usages in building and prompt an assertive response by craftsmen accustomed to guiding the pace of expansion in their corner of New York's economy.

Journeymen bricklayers in the late forties and early fifties saw a new incidence of "unmanly" or acquisitive behavior among master masons and many journeymen. Bosses now seemed to flaunt longstanding journeymen's work practices. "Some employers," reported one unnamed bricklayer at a May 1850 meeting, "now employed as many as 20 'scabs,' and were principally engaged on subcontracts, erecting poor buildings with sand and a little lime." When this bricklayer alluded to "scabs," he did not mean non-union workers, for in May 1850 a bricklayers' society had not yet been formed. The speaker referred to employees who violated customary work arrangements and lacked a proper journeyman's training. Another speaker at the meeting attacked " 'Hoosiers,' a class of workmen who could not, as a general thing, 'strike a line.' " For his part, he "would never work alongside one of them, nor with outsiders either, who worked for under wages." Such criticism of speculative builders who violated longstanding practices by employing "Hoosiers" and "outsiders" was in another instance directed against a new class of subcontracting journeymen. At another May gathering, bricklayers condemned the recently established practice "to give out the building of cellars, basements, etc., by sub-contract" to "men who are not mechanics and therefore [are in] no way competent to carry out such responsibilities safely and properly." These unscrupulous "lumpers," frequently journeymen bricklayers themselves, took subcontracts at a low rate, worked on the jobs themselves to increase the margin of profit, and hired low-wage men and unapprenticed boys in violation of time-honored work rules. It seemed to many journeymen bricklayers that such inconsiderate bosses and exploitative journeymen had begun to proliferate in the late forties.[16]

To combat the sudden rash of rule-breaking among boss masons and journeymen, the bricklayers and plasterers wrote down and promulgated an array of customary rules when they reconstituted their trade society in July 1850; other workers in building and wood also felt the need to codify work customs that spring and summer. The new Bricklayers' and Plasterers' constitution announced that

> from the 1st of March to the 30th of November, inclusive, the wages shall not be less that $1.75, and no three-quarter days, except caused by weather or other justifiable purposes. No member is permitted to work for any employer who is not a regular and legitimate boss-mason or plasterer, who shall be privileged to membership, to speak and vote on any subject, but not to hold office or serve on Committees. (Excellent regulations are also made with regard to apprentices.) No member shall work on any building erected by the owner, except such building is in the charge of some competent man acknowledged by the Society.

The journeymen bricklayers and plasterers legislated appropriate behavior for both journeymen and employers. "Lumpers" were outlawed, and while "legitimate" boss masons and plasterers were "privileged to membership," they were prevented from overly influencing the actual decisions of the organization. In their new constitutions, the House Carpenters and the Coach Painters had fewer reservations about the inclusion of bosses and subcontractors in their associations but still subjected such employers to special scrutiny and rules. The Carpenters "discouraged" subcontracting; the Coach Painters required that all subcontractors who joined their society employ fully trained journeymen and adhere to the society's ratio of apprentices to journeymen on all jobs. For the first time, then, wage earners in several building and woodworking trades formalized sanctions against acquisitive behavior. Such rules, and the centralized trades unions that enforced them, arrogated for these journeymen wide-ranging powers of regulation over the conduct of their trades.[17]

Informing the language of the constitutions of the early fifties was a preoccupation with craft. The references, on the one hand, to "Hoosiers," "scabs," and "the botching system of subcontracting," and, on the other, to "competent men," "regular and legitimate boss-masons," and "practical journeymen," associated the journeymen's condemnation of the new system of subcontracting with the prerogatives of skill. The new constitutions denounced the acquisitive mentality that induced some journeymen to assume subcontracts employing half-trained journeymen at low wages—but they denounced that mentality as a violation of craft. The great pariahs of the age, announced the Bricklayers' 1850 constitution, were the "butchers and tinkers who have never learned the trade, usurping the places of legitimate mechanics, manufacturing annually herds of unfinished workmen, piling brick together unsightly to the eye and disgraceful to the trade. . . ." At its most punitive, the craftsman's critique of the acquisitive subcontractor led to full exclusion of all employers from the journeymen's councils. In 1851 the Bricklayers' Association amended its constitution to restrict membership to competent journeymen "of good moral character" and exclude any who drifted into employer status during their Association tenure. What, these workers continually asked, were the appropriate social boundaries of their crafts, which groups fell beyond the pale, and what would a newly defined working-class realm look like? Craft workers always viewed the subcontractor and the employer as controversial figures.[18]

The rise of subcontracting, then, was the most visible and immediate way that the problem of rapid and disruptive urban growth presented itself to journeymen artisans. Workers in both the sweatshops and the craft shops sought to set limits to the accumulation of capital as a means of harnessing that growth. They did not mind an incremental and orderly expansion; rather it was the enlargement of capital by leaps and bounds, so frequently accompanied by exploitation of the wage earner, to which they objected. The situation of the subcontractor here posed a keen dilemma. The ability of the "competent" or "legitimate" journeyman to manage the labor of others independently—which often meant the license to take out subcontracts—enhanced his economic and social authority. At the same time, the situation of the subcontractor made him especially vulnerable to the acquisitive inducements of the expanding urban economy and gave him special power to debase the value of labor. Workers in the sweated consumer-finishing trades preferred to set limits to exploitative growth in tandem with the sympathetic subcontractor and employer—they seemed to need the prestige and authority of honorable employers to make their case. By contrast, workers in building and wood censured the subcontractor. They did so in the interest of a more potent authority, the cooperative and esthetic standards embedded in the experience of craftsmanship. These craft workers believed that the maintenance of such standards and the supervision of bosses and subcontractors prone to violating them could be entrusted only to journeymen like themselves. Disagreements about the role of the subcontractor were not some fuss over administrative niceties—they exposed workers' most basic understandings of the structure of society and their status within it.

How these two groups of artisans interpreted the rise of a middle class in their trades helped to define their respective political styles in July 1863. In the working-class movements of consolidation, beginning with the Industrial Congress of 1850, craft workers began to explore the possibilities for an authentic popular rule—a way of translating their substantial authority in the trades into political influence. Of course, the political controversies of the Industrial Congress and subsequent reform movements had their own dynamic, apart from the economic debates on middle-class participation occurring in the journeymen's associations. In the Congress the controversial personality was the meddling politician or middle-class reformer more than it was the acquisitive contractor. But the craft delegates who provided the Congress with much of its creative energy did not draw an acute distinction between their activities on economic and political fronts. The sharpest distinction, craft workers thought, was not between the affairs of their trade unions and those of quasi-political trade conventions like the Industrial Congress, but between the activities of those two institutions and the ways and means of the major political parties. They believed that the competitive urban economy could be consolidated only if wage earners, and not party politicians or middle-class reformers, took a leading political role. Workers in the consumer-finishing trades occasionally shared the craftsmen's hostility to party politicians but readily deferred to the political leadership of middle-class reformers or so-called "workingmen's friends." As a result of Republican wartime policies, both groups of artisans came to focus their critique of exploitative rule on the Republican regime. But craft workers, who insisted throughout this era that wage earners themselves had to intervene politically when the parties failed to remedy injustice, were the only artisans to join the anti-draft demonstrations in July 1863.

It becomes essential to our purpose, then, to look closely at the debates of the Industrial Congress in order to pinpoint the origins of the political styles evident among artisans during the riots thirteen years later. Founded in June 1850, the Congress became a forum for virtually every labor, land, sanitary and political reform in vogue on either side of the Atlantic. Proposals for producers' and consumers' cooperatives that would replace the wages system, trade-run financial institutions or *gewerbetauschbanken, Conseils des Prudhommes* (similar to what Horace Greeley called "mutual councils of delegates from both employers and journeymen"), a craft-controlled hiring hall, land reform, municipal reform—indeed, the whole 1848 associational program—received serious hearing in the weekly meetings of the Congress. By July the citywide assembly had moved into a wing of City Hall, where it began directing the reform activity of the trades in the manner of Louis Blanc's spring 1848 Luxembourg Commission. French, German, or British class politics were not carried over into the Industrial Congress in any formal way, nor did European events serve as an explicit point of reference for the Industrials. Instead, the legacy of 1848 for New York was an institutional grammar from which the Industrials selectively chose. Though it would continue functioning for nearly two years as a switchboard for labor reform affairs in the metropolis, the Industrial Congress was in its first summer and fall a powerful generator of ideas about class relations and politics.[19]

It should first be said that the Industrial Congress revealed a remarkable new unity and cooperative energy in the metropolitan labor movement after a decade of bitter ethnic and religious strife. New York never experienced the nativistic rioting which devastated the immigrant Catholic suburbs of Philadelphia in 1844. Still, in that year New Yorkers saw the rise of a popular nativistic movement which elected a mayor, hired the city's first professional police force, and heightened tensions between Protestants and Catholics in lower-class neighborhoods. With memories of the divisive nativism of the mid-forties still vivid, wage earners explicitly sought to forge unity among diverse ethnic groups when they reconstituted their trade unions in spring 1850. Indeed, with the massive immigrations from Ireland and Germany in the late forties, cooperation among workers of different nationalities had become a prerequisite for any trade-wide organization. By the mid-fifties, three-quarters of the city's work force would be foreign-born. The mood of ethnic cooperation that pervaded the artisan trade unions of 1850 was reflected in the deliberations of the Industrial Congress. From the outset, the issue of greatest dissension in the Congress was not ethnicity or religion but class boundaries.[20]

There was no mistaking the Congress's working-class coloration. "Small masters and middle-class labor advocates like Greeley assumed important roles," one historian has written, "but they retained their influence only so long as they were able to enlist and maintain working-class support." In a telling acknowledgment of the newness of the Industrial Congress and its working-class character, Greeley wrote in June 1850, "The New York workingmen have the right sort of men for leaders, and we see no such disgraceful scenes enacted at their meetings as shown at the gatherings of the old political parties." For Greeley and many of the other reformers who gravitated to the new movement, the working-class temperament of the Congress was in part what helped to distinguish it from "the old political parties" and gave it promise as a new instrument of social and cultural cohesion.[21]

But the question of whether or on what terms non-workingmen could participate in the Congress was present at the outset when the delegates drew up their constitution. Representatives of the journeymen craft unions proposed that every delegate be a "practical laboring man, elected by any chartered or unchartered association." The language of exclusion used by the craft delegates was remarkably uniform. Representing the Silversmiths' Protective and Benevolent Association, John Lowe proclaimed on June 14, ". . . we will have no Doctors (there's nobody sick)—no Capitalists (we want no office and can't give you any)—and no Lawyers (we'll call at your office as soon as we have a law suit)." Two weeks later, Cornelius McCloskey, a new delegate from the Bricklayers' and Plasterers' Association, elaborated Lowe's position:

> I came here because it was rung in my ears that this body were [*sic*] humbugging the mechanics of New York. When the roll was read over tonight, I heard the name of Daniel B. Taylor read as a delegate from the laborers. Is he a laborer or did he ever carry a hod? What right has he to represent the laborers of New York? What right have bosses whose names are also on the list, to represent the working man? We want working men to represent working men, and not men who represent political interests, and want to turn this Congress to their own account. . . . This man and that get into Congress on the shoulders of the working men. What did they do for us when they got there? They were the greatest demagogues that ever lived. To effect our object we do not want members of benevolent societies. We want no teachers, no orators. . . . We want no doctors, or lawyers or men of that kind. We hope they will leave us with a good grace, for their own sake, for if they do not, we shall be under the painful necessity of putting them out.[22]

The accused Daniel B. Taylor was a hod carrier turned attorney and politician. In 1843 he had helped found the Laborers' Union Benevolent Association, which by 1850 counted six thousand members and was the largest labor organization in any American city. By that time Taylor was a rising figure in Tammany Hall. From the perspective of craft workers like Lowe and McCloskey, Taylor and the Laborers' Union embodied several different threats to the Congress.[23]

Lowe and McCloskey imagined the Industrial Congress as a separate and pure working-class institution, free of the presence and intervention of both politicians and middle-class reformers. It was not sufficient that a prospective delegate be independent of the machinery of the "old parties": the taint of what they perceived as middle- and upper-class callings—"lawyers," "doctors," and "capitalists"—was just cause for exclusion. Lawyers were doubly suspect as the henchmen of both the political parties, and often enough, of capitalists. Only "practical laboring men"—and to McCloskey that seems to have meant wage earners without any claims to employer status—were privileged to participate in the working-class enterprise. Benevolent associations such as LUBA, designed to administer funds to needy members of the trade, were particularly dangerous. Nearly always they included small master artisans, reformers, and professionals. Usually they were far less militant than "protective unions" such as the Bricklayers' (though LUBA did protest working conditions and even supported its members in a strike in spring 1850). Eliminating the participation of benevolent associations, even at the cost of denying thousands of hod carriers a voice, would be one way to help insure the Industrial Congress remained under the control of wage earners. Disgusted with the hybrid social character of the Congress, Bricklayers' delegate P. J. Downey told an approving gathering of his trade "to oppose the word 'benevolent,' as many would be smuggled in as delegates who were no workingmen, but had only their personal speculations with this Congress." After several weeks of impassioned debate the

Congress refused to bar Taylor and the benevolent unions from its ranks. Influential at first, Lowe and Downey now vanished from the deliberations.[24]

McCloskey and two representatives from the Carpenters' Union, Benjamin Price and James Bassett, retained their authority in the Congress by articulating a second outlook on class and polity, distinct from that of Lowe and Downey. These three believed that the Congress was not a perfect reproduction of an ideal working-class domain, but as a reasonable likeness it had many uses. Though the Congress met each week in City Hall, its exact political status and its projected political influence—its relation to government at all levels—remained unclear; McCloskey, Price, and Bassett began to outline a political program. Price in particular was one of the more imaginative and enduring leaders in a labor movement that recast its leading players almost every year. In late July he announced himself in favor of "the people turning their own lawyers, exercising their rights on all proper occasions, especially the inestimable right of suffrage which would sweep from the statute books all laws injurious to the workingmen. . . ." He went on to redefine the exclusionary social formula of Lowe and McCloskey—no lawyers, no doctors—as a politicized call to action. A mechanic, the carpenter proclaimed, "can do anything, if he will try. He can change the law by political action, and he can make new laws as well as any lawyer. A good mechanic is a good lawyer, a good doctor, a good everything. . . ." While they shunned the "old parties," Price and his fellow craft delegates would let middle-class reform politicians such as Horace Greeley and Daniel B. Taylor remain in the Congress on the condition that they allowed wage earners to direct its political energies.[25]

By the end of the summer, Price, Bassett, and others had given content to that tantalizing phrase, "the people turning their own lawyers." Price urged the Industrials to remove the laws against combination from the state code (a standing threat against trade unions since the notorious 1836 conspiracy trial of New York City tailors) and petition the state legislature to create elective "Inspectors of Rents" empowered to halt payment of rent to landlords unable to keep their tenements fit for occupation. Bassett proposed a bill creating "District Surveyors" to inspect all city buildings and prohibit occupation of any that endangered public health. A special committee of Price, McCloskey, and Congress President K. Arthur Bailey offered a memorial to the Common Council reiterating an earlier Bassett and McCloskey proposal—abolition of the contract system on public works (the city government would become the employer instead)—joined with a provision for a minimum wage for laborers on public employment and a requirement that all public job superintendents have apprenticeship or training in the trades they were hired to oversee. In September a new measure advocating "legislative enactments making eight hours a legal day's work" was added to the list. The goal of such proposals, in the words of the Price Committee, was to counter "the principle (so loudly asserted by many) that every and any individual is justified in accumulating *immense* wealth, by any and all means within the pale of the *Law* [*sic*]—that he has a perfect right to depress the price of labor to the lowest ebb—regardless of the misery and distress thereby entailed upon hundreds and thousands of his fellow beings." The Industrial Congress under Price and Bassett's guidance had become a legislative drafting council dedicated to curbing some of the economic injustices associated with New York's headlong growth.[26]

Even as the fate of the nation hung on the grueling negotiations in the United States Congress over the admission of California as a free state and the attendant "Compromise of 1850," Greeley and other leading journalists saw the Industrials' political future as an issue of great moment and inhaled the logic of their arguments. "My own judgment," Greeley confided to a friend, "is that the Working Men, if they nominate at all, should nominate from their own body men who live by their own labor without deriving any profit from the labor of others—not sudden converts but men who heartily engage in the movement. . . ." *Herald* editor James Gordon Bennett warned that if the Congress was not restricted to members of the trades, it would be left vulnerable to "sinister influences," "wire-pullers," and "needy or ambitious politicians." Of course, many of the city's political organs were less sympathetic, but all felt obliged to respond to the Industrials' social analysis. During summer and fall 1850 the Industrial Congress set the terms of political and social debate in New York City.[27]

The approach of the fall elections rapidly forced Price and his colleagues to adopt a posture toward the political parties and their candidates. The Industrials agreed that they needed to keep their distance from the major parties. Most delegates would have applauded land reformer William V. Barr's denunciation of the Whigs and the Loco-Foco Democrats as "controlled by political tricksters" and his plea that "a barricade of ballots . . . be erected against these politicians, and if necessary they should be cast to the Devil!" How then to proceed? A proposal for launching "a separate organization of workingmen," a political party of wage earners, was ardently debated on September 17 but dismissed. Instead, the land reformers led by Barr passed a resolution to submit nine of the Congress's political measures to the approval of all candidates for office. According to Barr, this technique of candidate interrogation had allowed the land reformers "great influence . . . in the old parties."[28]

By mid-September 1850, then, McCloskey, Price, and Bassett's notion of the Industrial Congress as a new structure with a new kind of political influence had evolved into Barr's idea of reformers wielding power "in the old political parties." The Industrial Congress had become committed to a political strategy oriented not toward "the people turning their own lawyers" but toward endorsement of major party candidates so long as they pledged themselves to the reform program. By the early fall, Price found himself in league with men entirely comfortable with the idea that middle-class reformers should formulate the social program and direct the political energies of the Congress. Ira B. Davis was one such reformer. Davis aimed to reorganize the primary nominating system of the Democratic Party to prevent it from remaining "the mere instrument of designing men by which they are to obtain authority and wealth, which have generally been used to the injury of that class who needed and simply believed they would be protected." At a September 16 meeting of the "Democratic Union of the Sixteenth Ward," Davis chaired a meeting which resolved to investigate the claims of all registered Democratic voters of the locality so that "impostors" might be cast out. In their wards William V. Barr and John Stevens led similar "Assembly Committees" chosen to insure that "honest men" independent of the political cliques that controlled the Democratic Party would be nominated for the state legislature. It is not surprising that the one forum where Davis, Barr, and other "independent" Democratic reformers received unquestioned acceptance through the summer was that of the striking tailors. With their ultimate goal of purifying rather than repudiating the Democratic Party, the Assembly Committees of 1850 represented a third outlook on class and polity in the Industrial Congress.[29]

The movement to reform the machinery of the major parties through candidate interrogation and workingmen's Assembly Committees won considerable support among voters in 1850 (the Industrials eventually took credit for helping elect nine candidates to the State Assembly.) But with middle-class reformers of the Davis and Barr stamp at the helm of the Industrial Congress after fall 1850, it was only a matter of time before Tammany leaders found a way to usurp control of the workingmen's body. Barr's own independence was at the very least suspect: his brother Thomas was a Tammany alderman (no surprise, then, that William would himself become a Tammany candidate for alderman in 1852). At a June 3, 1851, meeting of the Congress, Barr was joined at the podium by eminent Tammany politicos Lorenzo B. Shephard, John Cochrane, and Mike Walsh—the boisterous Democratic politician who helped invent the persona of the "workingman's friend" during the previous decade. The Congress was reorganized in the manner of the "old political parties," ward by ward, and the crucial injunction against the admission of "politicians" relaxed; Tammany took over. Barr's electric image of workingmen erecting "a barricade of ballots" against politicians was for the time being discarded and Benjamin Price's vision of the "people turning their own lawyers" deferred.[30]

How did artisans get from the rostrum of the Industrial Congress in 1850 to the steps of the Ninth District Provost Marshal's Office in July 1863? Artisans of all stripes participated in the Industrials' experiment in 1850. Tensions appeared between craftsmen who hoped wage earners would control the new structure and use it to achieve a distinctively working-class influence, and proletarianized artisans who shared the craftsmen's distaste for the clique rule of the major parties but preferred the authority of middle-class reformers within the Congress. These positions and constituencies were not always distinct, and such tensions were not severe enough to destroy the Congress, though a group of craftsmen who preferred an exclusively working-class institution were voted into silence early on. Rather, delegates from the crafts and the sweatshops shared power and guided the course of the assembly together. By the 1863 riots, the divisions of 1850 had become more sharply defined. Craft workers conducted the July 13 strike against the draft, while employees in the sweatshop district shunned the crowds and deferred to the leadership of Democratic notables.

But the draft riots involved more than a crystallization of the artisan views on class and polity explored in 1850—they also represented a redirection of those impulses. It was one thing to criticize the schemery of "politicians," "doctors," "capitalists," and "lawyers" and to urge workers "to turn their own lawyers." It was quite another to take to the streets against the draft and the Republican Party. The Industrials censured an assortment of middle-class enemies, while the draft rioters focused on one obnoxious political party. Further, in many of their proposals for consolidation, the Industrials contemplated an active government setting limits to the exploitative accumulation of capital—what the Price Committee referred to as the amassing of "immense wealth regardless of the misery and distress thereby entailed upon hundreds and thousands . . . [of] fellow beings." The appointment of an "Inspector of Rents" to police slum landlords and "District Surveyors" to inspect city buildings and shut down unsafe structures, or the legislative enactment of an eight-hour work day, conjured up an image of an aggressive government intervening in the economy.[31] In comparison with the political economy of the Industrial Congress, the popular anti-Republicanism of July 1863 was more a defensive effort to keep an intrusive and exploitative government from interfering with the workers' domain. Between 1850 and 1863 artisans' definition of consolidation changed—from a vision of government actively regulating urban growth to a revolt against Republican Party rule. The Amalgamated Trades Convention and the German *Arbeiterbund* of 1853–54 gave the first intimation of that shift in artisan thinking.

Though it might not appear so at first, the opportunities for labor reform narrowed in 1853. The political problem of 1850, the corruption and irrelevance of the old parties, had now become the urgent concern of a broader and more diverse palette of social groups. The "politics of impatience," as one historian has characterized the mood of anti-partyism in the mid-fifties, was rampant in New York City. A New York grand jury revealed that the officials who directed the explosive physical expansion of New York City in the early fifties had routinely taken bribes in the letting of contracts. The Common Council of 1852–53 was nicknamed the "Forty Thieves," and Whigs and Democrats were implicated alike. Middle-class reformers launched a series of overlapping initiatives, all of which emphasized to a degree unprecedented in antebellum reform that the political system was the locus of redemption. One of these projects was "City Reform," led by the "millionaire mechanic" Peter Cooper and devoted to creating a legitimate popular governance through revisions in the municipal charter. Another was the movement to pass and enforce a New York "Maine Law" modeled after that state's 1851 prohibition of the manufacture and sale of alcoholic beverages; now the government and not the individual would be given the final responsibility for monitoring temptation. In short, the middle classes aggressively advanced their own definitions of justice in 1853, each of which centered on political transformation. Gone was the scenario of 1850 that featured workingmen speaking for reformers of all classes from the august rostrum of City Hall. When the new Amalgamated Trades Convention was constituted in early September 1853, it convened in an assortment of meeting halls on the East Side—as it were, on wage earners' own territory. Horace Greeley would not have been a welcome guest at the Amalgamated, showed relatively little interest in it, and, on the night of the Amalgamated's first meeting, could be found several blocks away dining at the festival of the "New York Vegetarian Society." New Yorkers were again talking of consolidation in 1853, but workers did not preside over the discussion as they had in 1850. Further, the very terms workers used to conceive political action were changing.[32]

For the first half of its nearly two-month existence, the Amalgamated Trades Convention was dominated by craft leaders, Benjamin Price prominent among them, who kept the problem of popular political power at the center of focus. The problem for the Convention was how to restore political equality in America. In the 1850s that equality had been subverted through "cunning, or perversion of the law, or the management of legislative acts." As a result, "the few were stealing the power from the many," and "all of the offices of the General Government and those of each State were disposed of to a specific class." Workingmen who "had lost that political equality which they were primarily presumed to possess" now had to "elevate themselves to such a position that they would be themselves the rulers." The issue, these craftsmen proclaimed now as they had in 1850, was how to create an authentic majority rule in the United States.[33]

The solution they offered was a synthesis of two ideas that had remained half-coherent in the Industrial Congress: McCloskey, Lowe, and Downey's vision of a working-class domain wholly separate from the business, professional, and employing classes, and Price's notion of the "people turning their own lawyers" through a politicized trades institution. Begin, bookbinder Thomas Doyle advised, by avoiding the mistake of the Industrial Congress of 1850, which had failed because it was open to the public and operated "on the minds of the public without rather than on those within. . . ." In the new Trades Convention all delegates would be, in silver-knife-maker David G. Croly's words, "men of their standing": "they wanted no lawyers, no doctors, no clergymen, no men in business to act as their representatives." Reformulating Price's idea, Doyle recommended

> that the [Trades' Convention] delegates in question should act as representatives of the various Trades-Unions, and should exercise legislative authority; that they should be to the trades of New-York what the General Committees of the two great parties had been to those parties.

By "legislative authority" Doyle and the other craft leaders seemed to mean the power to propose new state laws in the name of the trades. The Convention would serve as a workingmen's lobby in the state legislature—even "two thousand earnest mechanics so confederated," Croly imagined, "could exercise immense influence and would be able to effect the passing of any law at Albany, which would be necessary to their interests." There was no explicit discussion of creating a political party of wage earners here, except by way of Thomas Doyle's elusive metaphor that compared the structure and influence of the Convention to that of the "two great parties."[34]

Despite the imaginative rhetoric of its early meetings, the Amalgamated Trades Convention rapidly fell apart for reasons that reveal much about the road to the Ninth District draft office in 1863. The division between the crafts and the degraded consumer-finishing trades was clearer now than in 1850. By the third week in September the original coterie of craft leaders had lost influence, and the Convention came under the guidance of the tailors, shoemakers (men's and ladies' branches), and painters. These new men maneuvered passage of a resolution "discarding all pretensions to political action" and defining the Convention as "solely a Social organization, for the protection and elevation of Manual Labor"; soon after Croly resigned as president. No longer, it seemed, could artisans in the crafts and the degraded trades successfully collaborate.[35]

Nor were the inventive economic proposals of the Industrial Congress taken up and elaborated by the Convention of 1853. Here was consolidation without its heart, that is, without creative legislative proposals to limit the exploitative accumulation of capital. No doubt Croly envisioned that his workingmen qua lawyers would enact such economic regulations, but the Convention was noteworthy for its absorption with questions of political structure at the expense of economic program. Croly and his associates were obsessed with the dominant theme of the mid-fifties, fear of powerful political minorities. Croly himself thought that keeping proceedings secret and basing representation in the shops rather than the trade unions would eliminate shadowy political "wirepullers" from the Convention and establish a true majority rule that reached out among the city's poor. Secrecy provisions were the favorite device of another political movement frightened of powerful minorities—the nativist Order of Know Nothings just then appearing. Cooper's "City Reform" also sought to depose political cliques and institutionalize government by the majority. For all its novel and potentially radical thinking about how workers could exercise political power independently of other social classes, the Amalgamated found it difficult to define its own space in the mid-fifties' welter of movements and may have lost much of its constituency to "City Reform" as the campaign of 1853 began. The Trades Convention reflected a changing urban political culture increasingly preoccupied with the misuse of political power. The emphasis of consolidation had subtly shifted. In 1853 craft workers were less concerned with wielding the power of government to regulate the quality of urban growth than with protecting their domain against the political interference of the parties.[36]

There was a defensive tone, as well, to the political thinking of the *Amerikanische Arbeiterbund* founded at a mass meeting in spring 1853. Many of New York's German artisans in the 1850s were emigres from the revolutionary Germany of 1848. The rifts of that revolution were deep, dividing conservatives who supported free trade and the commercial reforms of the *Zollverein* and radicals who opposed free trade and saw the *Verbrüderung*—a national alliance of urban craft organizations—as a foundation for the social republic. These conflicts were both reenacted and redefined in New York, as a conservative *Kleindeutschland* business elite allied with the immigrant Catholic and Lutheran hierarchies, German Democratic politicians, and the Democratic German-language press contested an array of German trade and communal organizations dominated by craft workers. It was among these trade and communal associations, the influential *Vereine,* that the *Arbeiterbund* first appeared. It was as if the spirit of the *Verbrüderung* still lingered in the sawdust of the Attorney Street cabinet shops.

Like Doyle and Croly, the leaders of the *Arbeiterbund* saw political equality ebbing away. "Social relations," the German workingmen observed, "are no longer as they were when the republic was founded. The introduction and development of large industry has brought on a new revolution, dissolved the old classes, and above all created our class, the class of the propertyless workers." The *Arbeiterbund* attacked capitalists who sanctioned the privileged labor market status of "lawyers and doctors" while workers were forced "to compete among themselves" and were "endlessly exploited." If the capitalists promoted economic competition, they also controlled the political legislation which the *Arbeiterbund* called "the beginning of our ruin." Thomas Doyle's indictment of capitalists' "cunning, or perversion of the law, or the management of legislative acts" would have comfortably fit the preamble of the *Arbeiterbund* constitution. The solution the *Arbeiterbund* prescribed was that workers form a political party. The raw material for such an effort, the authentic working-class institution that Doyle and Croly labored to create, was already in place in the form of the powerful *Vereine.* Now the task was to coordinate and politicize the *Vereine,* to extend them into the city.[37]

While the *Arbeiterbund* called upon workers to "seize the political authority" themselves, its actual platform demanded not so much that workers capture the citadels of government but that government not interfere with the freedom of workers in their own domain. The German workingmen sought rapid and accessible naturalization procedures, uniform federal administration of all laws affecting workers, equal access to the court system, elimination of all laws "which encroach upon the worker in the enjoyment of his freedom as for example, sabbatarian laws, temperance laws and the like," enforcement of a mechanics' lien, and the inalienability of the public lands. The notion of freedom from the social intervention of government reappeared in the *Arbeiterbund*'s resolutions against the Nebraska Act six months later: ". . . we desire to consider and shape our own welfare, free from the dictation of lawmakers, wire-pullers and the hireling masses" (this last a reference to all workers who, through ties to the political machine, had lost their republican "independence"). In 1854, German artisans, like their English-speaking comrades, believed that the final goal of working-class political intervention was not the aggressive use of government to limit the social damage of reckless urban growth but the protection of workers from exploitative authority. Almost a decade before the draft riots this apprehension of intrusive political power had begun to dominate artisan thinking.[38]

Consolidation thus defined continued to inspire craft workers to seek ways to intervene in politics on their own terms, and this search kept many at a distance from the Republican Party when it was first created in 1854. A political tempest swept the North when Stephen A. Douglas introduced the Nebraska Act in Congress at the beginning of that year. The bill proposed to repeal the Missouri Compromise, the *sanctus sanctorum* of antebellum politics, and allow slavery north of the 36°30′ latitude. Pro- and anti-Nebraska camps formed in New York and other cities, and the Republican Party emerged out of the anti-Nebraska coalition. If craft workers had heretofore found it hard to translate their authority at the workplace into political power, they had far more trouble once the all-absorbing Nebraska controversy began. But while their trade unions were still intact they retained some hope of political action of their own. The independent tenor of craft workers' anti-Republicanism in 1863 could already be detected in these mid-fifties' developments.

The English-speaking labor movement had made it clear during the debate over the admission of California in 1850 that it regarded the slavery dispute, in the words of one resolution, as a "contest between capitalists of the South and the North." The abolition of wage slavery took precedence over that of black chattel slavery, and workers would take no special interest in the conflict over chattel slavery "except insofar as the American Union may be endangered thereby. . . ."[39] It was difficult to sustain such thinking in 1854 as the question of slavery's status in the territories took precedence over all others. Nonetheless, workers in the building trades tried to define their own political enterprise. At a fall 1854 forum sponsored by the Carpenters' Union, William Masterson proclaimed that "labor is now asserting its right. Nebraska and Anti-Nebraska are nothing compared to it." Plasterers' leader Daniel Walford, the most important intellectual link between the fifties' and sixties' labor movements, suggested the formation of a "National Reform Association" that spoke for mechanics since "such meetings as the Nebraska Workingmen's Meetings, which were managed by lawyers and parsons, were not the thing—not the sitting on rum-casks, and talking about Hard and Soft Shellism." Masterson and Walford were groping for an effective project—"something," in Walford's words, "to excite the interest and attention of the working people." This search for a program that would allow craftsmen to set the terms of political debate apart from the Nebraska controversy had limited success. But it was clear that English-speaking building-trades unions would not provide fertile ground for the Republican Party just then forming. They associated the anti-Nebraska movement with the machinations of "lawyers" and "parsons"—the personnel of the "old political parties." They would not recognize in the Republican coalition of temperance reformers, nativists, and free soilers that new structure to impose cooperative ideals on a chaotic and individualistic industrial order.[40]

The Republicans would also find it difficult to sink roots in *Kleindeustchland.* In contrast to the English-speaking building-trades unions, the *Arbeiterbund* recognized a close relation between the status of waged workers and black chattel slaves and formally opened its political ranks to every worker of the nation "without discrimination of occupation, language, color or gender." German workingmen streamed into the anti-Nebraska movement of late winter 1854. On March 1, the *Arbeiterbund* sponsored a mass rally against the "hunker clique of the *Staats Zeitung*"—the pro-Douglas German Democrats—at which it condemned "the heavy capitalists" and the "land jobbers of the nation" (the Nebraska Act would inhibit land reform), protested "all Slavery, whether white or black," and announced itself independent of the machinery of the major political parties. The German free-soil movement in New York City would not serve as a staging ground for the Republican Party as it did in many Northern communities. The *Arbeiterbund*'s anti-Nebraska sympathies were suffused with a hatred of temperance and nativist reformers and a deep suspicion of interference by "wire-pullers and lawgivers" in workers' affairs. New York City's comparatively patrician and nativistic Republican Party, so intent upon the moral reformation of the immigrant poor, would have difficulty courting the *Arbeiterbund* free-soilers.[41]

The depression winter of 1854–55 was a quiet turning point in the political history of artisan New York. The metropolitan building market collapsed and many trade organizations dissolved. The few unions that remained no longer had the centralized authority that made the political intervention of organized craftsmen a possibility in the first half of the decade. The *Arbei-*

terbund also faded as straitened German workers devoted what energies they had to fending off attacks from the anti-immigrant Know Nothing Party, now enjoying its brief season of popularity.

Middle-class reformers who had struggled to find an audience at the time of the Amalgamated Trades Convention now came to the fore. Ira B. Davis, William Arbuthnot, William West, and a versatile Benjamin Price finally found a working-class audience eager for the intervention of reformers—the jobless laborers who huddled in front of City Hall in search of work. On December 28, 1854, one of these gatherings named Davis chairman of a new "Central Committee of Associated Workingmen" empowered to investigate the state of destitution and "in behalf of their distressed fellow-citizens, to organize an efficient centralized action, whereby to realize present and permanent relief." In memorials to the Common Council, Davis, West, and John Commerford took up two of the cries of workers at the outdoor rallies: a demand for the city government to hire unemployed men on public works and another to provide uptown lots for affordable working-class housing. Veteran land reformers like Davis, Commerford, and West were intrigued by this "urbanized" version of a national Homestead Act. Economic imagination had by no means vanished from labor circles. But now economic program surfaced only during bouts of recession and depicted the government as a provider of public welfare rather than as a regulator of exploitative social relations. Daniel Walford appeared at the winter 1855 meetings but looked on as Davis and his circle increasingly represented the sentiment of the labor movement. Walford told a "workingmen's mass meeting" on January 30 that a high tariff to protect home manufactures was the only solution for unemployment and stressed that such a question of "labor" was "entirely unconnected with politics and religion"; he was interrupted with the cry, "We have heard enough about English labor, now let us hear the American rate of labor." It was Davis's campaign for public hiring of the unemployed, and not Walford's vague-sounding cries for a "labor" program, that captured the attention and allegiance of the workingmen who shivered at the outdoor rallies. By mid-January, Mayor Fernando Wood had put some of these men to work rebuilding City Hall. The independent Democracy, and not Walford and fellow critics of middle-class interference in artisan reform, emerged from the workingmen's meetings of winter 1855 as the dominant influence in the labor movement and, *mutatis mutandis,* would so remain until the eve of the draft riots.[42]

By the late fifties Wood, the dashing publican and merchant shipper turned Democratic popular politician, had become the darling of the middle-class labor reformers. In the wake of the bread riots of early November 1857, Ira B. Davis defended Mayor Wood's public works programs and the actions of the unemployed at a meeting supporting Wood's reelection. Davis's speech, one *Times* reporter observed, was a "vindication of Mayor Wood as the poor man's friend" and an endorsement of the main tenets of the Democracy: the Independent Treasury System, divorce of bank and state, and "the liberal course of Mr. Buchanan," which likely included the President's expansionist foreign policy and efforts to preserve the Western territories as an all-white zone. By 1857, Wood had successfully identified himself as a political independent, a "poor man's friend" uncontrolled by the corrupt machinery of Tammany, even though he had affiliated with Tammany for a time. As a result the Independent Democracy of 1857 included Wood within the working-class domain and brought Davis and labor reform virtually into the ideological mainstream of the Democratic Party. William V. Barr's 1850 call to action—to erect "a barricade of ballots" against the politicians of the major parties—had now been transmogrified and appropriated by Wood and his Democratic allies.[43]

The familiar fifties' complaint against manipulative politicians was now directed against the new Republican Party. By spring 1860, many New York City workers were convinced that the slavery controversy and the Republican Party they believed responsible for it would endanger the nation. In March and April New York's English- and German-speaking shoemakers gathered to express sympathy with striking shoeworkers in Lynn, Massachusetts. Discussion drifted from the need to avoid strikes to the sectional conflict; the meeting quickly assumed an anti-Republican tone. A lawyer condemned the North's "assassin-like attack on the institutions of the South," proposed sectional peace and the protection of Southern markets as a solution for the Lynn conflict and urged shoeworkers "to rebuke that party—he need not name it—which would destroy that comity between States which was the guarantee not only of our national but of our social and commercial prosperity." Another speaker announced that he was "a Southern man" but was also "morally opposed to any interference with this glorious Union":

> It would not take a dozen of us to select men to go to the Chicago Republican or the Baltimore Union Convention, and say,—"Give us Presidents who will respect the laboring classes of this nation, or we will not vote for your candidates," and so force them to take our men. . . . Let this be done and . . . the codfish aristocracy and the pettifogging politicians will retire from our march as rats retire from a burning barn.

The notion of workers intervening to end a sectional conflict aggravated by contentious elites and self-seeking politicians gained some currency in the labor movements of Northern cities on the eve of the war. At the same time that a "Peace Convention" of business and political leaders met in Washington in February 1861, a Workingmen's Committee of Thirty-four met in Philadelphia to condemn the major parties for dragging the nation into an unwanted war. The Committee of Thirty-four endorsed the Crittenden Compromise that proposed the reestablishment of the 36°30′ boundary of the Missouri Compromise as a northern limit to slavery. New York City workers had no representatives in Philadelphia and instead offered their complaint against "pettifogging politicians" at an "Anti-Coercion Mass Meeting" managed not by wage earners but by Fernando Wood's Democratic lieutenants Isaiah Rynders and Levi S. Chatfield. The New York meeting supported Senator Crittenden's proposal, but with a strong anti-Republican, Southern-sympathizing qualification, noting that Crittenden did not "grant the South her full, just and equal rights under the constitution." The Industrials' studied indifference to the slavery controversy in 1850 now gave way to a caustic racism and anti-abolitionism, as the workingmen accused the Republicans of conspiring to "reduce white men to a forbidden level with negroes." On the eve of the Civil War, New York wage earners allied to the middle-class reformers of the independent Democracy recast the defensive critique of manipulative political minorities of the mid-fifties as an attack against the Republican Party. Now, too, workers began to shift their attention from the city and state legislatures—where they had directed some of their most creative reform schemes of the fifties—to the national political arena.[44]

The Civil War brought an outburst of patriotism among city wage earners and new economic hardships which tried that patriotism. A Great Union Meeting was held on April 20, 1861, to proclaim New York City's support for the war effort, and workers turned out *en masse,* some marching to Union Square with their shopmates. Workers also hung flags over factories and volunteered for military service in large numbers, sometimes along with friends in the fire companies and *Vereine*. But as had been feared, the withdrawal of Southern trade triggered a wave of business failures and threw hundreds out of work. As early as July 1861, two thousand German workers met to demand a municipal public works program for the unemployed and an end to a clique-controlled public contract system. Even with the revival that wartime government contracts brought to the clothing and shoe trades, metal and machine manufacture, and ship building, wartime inflation made survival an issue for the most skilled artisans. By May 1862, German cabinetmakers argued that the three to five dollars they received each week failed to feed their families. That fall, ship joiners and caulkers went on strike for an increase in wages, shortly joined by coppersmiths and hat finishers; in early 1863, carpenters, pianomakers, machinists, and tailors at one of the large Broadway shops followed suit. By the time of the draft riots, dealers had raised the price of coal to a high of ten dollars per ton, nearly five dollars a ton more than the average over the previous fifteen years. One of the most vivid images of the riots was one of destitute women and children combing the burned-out buildings for firewood. It was clear to the reformer Albert Brisbane that wartime inflation had political meaning to the city's working classes when he reported in 1864 that "among the mass of people a strong reaction is setting in in favor of the Democrats & against the war. I have been among the mechanics, and the high prices of provisions are driving them to wish a change."[45]

Who else to blame for such troubles but the Republicans, whose wartime innovations in government had indeed altered the Northern economy and placed a severe burden on the poor? When in early spring 1863 the artisan trade unions revived and a citywide Workingmen's Congress was formed, the Republican Party and its policies were the universal target. Delegates from most of the reconstituted artisan trade unions attended the Congress's inaugural meeting on March 24, including the shoemakers, tailors, coachmakers, bakers, plasterers, shipwrights, hatters, bookbinders, painters, and coopers. The Republican Party, all agreed, would be no more successful than any past political machinery in solving the city's problems of poverty and exploitation. Painters' leader Patrick Keady catalogued Republican impositions on the working classes: the coercive new conscription law with its class-biased clauses, the inequitable apportionment of wartime taxes, and the inflated prices caused by the Legal Tender Act. Keady did not care "by whom the federal government was administered," so long as it refrained from discriminating against wage earners. Finally, he invoked the fifties' idea of a national trades union, predicted that "politicians would no doubt interfere with them," and concluded that such a national organization independent of the parties would be the only way to protect workingmen from obnoxious laws. The war had strengthened the main thrust of artisan thinking in the mid- and late-fifties—workers had to protect themselves against unjust and intrusive political authority. Only now this impulse was directed against a power vast beyond the imagining of most antebellum workers, that of the wartime Republican regime.[46]

That spring the revived craft associations, led by Patrick Keady and Daniel Walford, began to challenge associations in the degraded artisan trades and allied middle-class reformers and politicians for control of the labor movement. The pre-war dispute over the participation of the middle classes reappeared, far more fierce now that it was intertwined with the wartime issue of loyalty. But the craft delegates remained a minority voice in the new Workingmen's Congress. Only Keady and Walford mentioned a need to keep party politics and politicians out of their debates. Only the durable Walford echoed the refrain of the fifties: no "ministers, lawyers and doctors." From the start the shadow of Fernando Wood and the Peace Democracy loomed over the assembly. The chairman of the March 24 meeting, Shoemakers' leader Charles McCarthy, had arranged for ex-mayor now Peace Democrat Congressman Wood to pay for the workers' use of the hall that night. M. D. Bucklin of the Hatters emphasized the "strength of the workingmen in the body politic" and associated zealous talk of national loyalty with capitalists "who wanted to prolong the war"—unlike Keady, he did not affirm his loyalty to the Washington government. Debate over the covert intrigues of Copperhead fraternal societies among the Workingmen intensified through April and May. Finally, in June, Keady and the Painters marched a small contingent out of the Congress in protest over McCarthy and Bucklin's Peace Democrat inclinations and formed the politically independent "United Organization of Workingmen." McCarthy and Bucklin's remnant renamed itself the "Workingmen's Central Association" and pledged itself to the election of workingmen to public positions (no doubt under Peace Democrat auspices)—it promptly received the endorsement of the Copperhead and pro-slavery *Weekly Caucasian.* With backing among the artisans as well as the striking longshoremen, the highly visible Peace Democracy was considered by many to be the voice of labor in New York City on the eve of the draft riots.[47]

Artisans' influence in the urban polity had increased with the war. In the mid-fifties they had not been able to make the theme of opposition to an interventionist political authority their own—it was advanced by many political movements—nor had that theme allowed them much influence in political debate. The war and the burden of Republican policies had the effect of bringing artisans and their reform organizations back to the center of public life. Reeling under the impact of wartime inflation and taxes and liable to a draft that singled out poor men, artisans now had special claim to the argument that a manipulative political minority (and so the Republican Party seemed from the perspective of Democratic New York City) was wielding the apparatus of government against the interests of the majority. It was a legacy of the mid-fifties that "independent" middle-class reformers, now ensconced in Fernando Wood's Peace Democracy, would voice that artisan grievance when the labor movement revived in early 1863. The main worry of the Copperheads, the encroaching power and centralization of the federal government, coincided with artisan fears. The more radical notion of the fifties—that workers could "turn their own lawyers" and pass legislation to moderate the economic injustices of urban expansion—was hardly evident in workers' thinking on the eve of the draft riots. The urgent task, as most artisans conceived it through early summer 1863, was to ward off the onslaught of that juggernaut the Republican government against the liberties and livelihood of the poor. This was the concern voiced repeatedly at the June 3 Peace Convention and the July 4 anti-conscription rally.

On the weekend of July 11–12, when it became clear that Copperhead politicians had failed to prevent or postpone the draft, it was the assertive craftsmen and not artisans in the degraded trades who were prepared to take matters into their own hands. The Republican government attempted to enforce conscription in New York City almost in the wake of the breakup of the Workingmen's Congress. Timing was everything. Walford and Keady were just beginning efforts to reorganize workers in building into trade unions. The question for most craftsmen in building and woodworking was not whether to oppose the draft, but how and where. When Keady issued a statement condemning the riots and praising painters who had refrained from violence, he spoke for the organized few. At this moment of political disarray and rudimentary organization in the building trades, many more workers in construction took to the streets. Some joined the demonstrations of the 13th with their shopmates and others with gangs and fire brigades. Most abandoned the violence after the first day and then joined the citizen anti-riot squads or patrolled the streets in their fire companies or *Vereine*. There is reason to believe that if the draft lottery in New York City had been delayed some months or even weeks, a more disciplined building-trades movement might have been won over to Keady's position of opposing Republican measures through trade organization. Regardless, these rioting craftsmen shared Keady's view that workers should act on their own, independently of party politicians, to right Republican injustices. Artisans in the degraded trades did not riot—were not willing to act independently—and deferred to the leadership of Democratic notables.[48]

The insurrection set labor reform on a path different from the one it had been following since the mid-fifties. The status of trade unionists in building was enhanced, and painters and bricklayers who had rioted were embarrassed. The violence also tainted the Peace Democrats and their labor adjunct, Bucklin and McCarthy's "Workingmen's Central Association," with treason. The Copperhead critique of the centralizing wartime government was discredited, and artisans searched for other political vehicles. The "Workingmen's Central Committee" drifted out of all labor columns except those of the Copperhead press. Keady and Walford's loyal United Organization of Workingmen, now renamed the "Workingmen's Union," moved to center stage. Loyal craft workers and not Copperhead politicians would now dominate the labor movement. But these craft workers now found it crucial to include sympathetic employers and middle-class reformers in their political ventures. The post-riot direction of the artisan movement could be seen in Daniel Walford's sudden loss of influence in the Workingmen's Union during late fall 1863.[49]

The machinists' strike that began in October 1863 triggered the events that brought on Walford's decline. Five thousand machinists walked out after their demand for a standard rate of wages and 25 percent increase in pay had been refused. The Machinists' Protective Association of New York represented less than a hundred of these men, with the unorganized remainder conducting their strike through shop committees and running up credit at the groceries. Finally, as winter approached, the hard-pressed machinists applied to the Workingmen's Union for funds. Would the Workingmen support the tiny trade union or the mass? Walford insisted that the trades assembly adhere to its Constitution and aid only the several dozen strikers who belonged to the MPA. He was jeered at machinists' rallies and reprimanded by his own Plasterers' Association, which revoked his seat in the Workingmen's Union and instead contributed two hundred dollars to the unorganized strikers. The victor in this controversy was E. P. McDermott, the trades assembly president and the Plasterers' second delegate. The conflict exploded at a December 15 Plasterers' meeting when Walford branded McDermott a "traitor" to the Workingmen's Union who "wanted the boss Plasterers to come in as honorary members among the journeymen." McDermott was a "friend of capitalists": he had not only funded "non-society men" but had tolerated "bogus delegates" in the Workingmen's Union (perhaps, we can imagine, sympathetic employers like those he admitted into the Plasterers' Society). Both Walford and McDermott sought to distance the labor movement from the major parties and especially from the machinations of Copperhead politicians. Both thought workingmen obligated to use "their organizations," as McDermott put it, "to elect men from their own ranks to represent them in the halls of legislation." But after the riots Walford's belief that only an institution composed exclusively of wage earners would allow for an authentic majority rule lost influence among craft workers. Now the view of McDermott and of Benjamin Price and James Bassett before him was ascendant—labor's political ventures should be guided by wage earners but could also include sympathetic employers and reformers. Though the issue of loyalty to the nation was never directly mentioned in these late 1863 altercations, it may explain the triumph of McDermott's thinking. After the riots loyalty to the nation became a prerequisite for political legitimacy. To ratify their loyalty, artisans in search of political authority found it imperative to form alliances with a loyal middle class. As we shall see, liaisons with the middle classes would be essential to artisans' post-riot efforts at reform—the Citizens' Association and the movement for an eight-hour work day.[50]

Like the artisan craftsmen, industrial workers and common laborers sought an authentic popular rule and were willing to take political action on their own, independently of party leaders. These workers did not preside over labor congresses and so shape the way New Yorkers defined reform; they did not command authority on the artisan scale. But they did have much influence in the narrower domains of workplace and neighborhood. Such prerogative, along with a deepening sense of vulnerability as the war progressed, brought industrial workers and laborers to the steps of the Ninth District Provost Marshal's Office in July 1863. In their own distinct ways, they too came to perceive the wartime Republican government and its draft as unjust and intrusive authority.

Industrial Workers

New York City metalworkers contributed little to artisan debates over the terms of exclusion of the middle classes between 1850 and 1863. By circumstance and by choice, metalworkers had already cordoned off much of their own social and geographic realm from that of middle- and upper-class New Yorkers. The very condition of life in the uptown factory districts was one of working-class separation and exclusion. Genteel sojourners to the East Side factory district felt the change in social climate when they crossed Second Avenue as acutely as a sudden shift in wind. Agents of the Association for Improving the Condition of the Poor believed the very air miasmic in the unlovely Eleventh Ward. In a special study of that "infected district," the poverty reformers were surprised "not that many sicken and die, but rather that the mortality is not greater where the laws of life and health are so flagrantly violated by large masses of the people." Other visitors commented on the immigrant aspect of the area. Wrote an 1853 observer, ". . . you would think yourself in an old German City." The Irish character of the dwellings along the East River just north of the Eleventh was no less marked. Tourists to the area's enormous Novelty Iron Works described the journey as one might a voyage to an exotic foreign province.[51]

Because of the labor requirements in the shipbuilding, metal, and machine manufacturing industries, a vast work force of craftsmen and laborers had by the 1840s congregated along the uptown waterfront. Some craftsmen were hired permanently by the shipyards and factories; many more craftsmen and nearly all the laborers formed a floating army of workers, drifting in and out of the shops as the volume of employment dictated.[52] In this period of relatively unspecialized production, many journeymen who had completed their apprenticeship passed from yard to yard in order to learn the techniques of the various master artisans. Laborers were engaged on an even more temporary basis. Once a thirty-five-ton bed plate (to support a ship's engine) had been cast at the Novelty Works, dozens of laborers might be hired to move the mammoth object through the yard and hoist it into a ship's hull. To be eligible for such temporary employment, uptown workers had to live close enough to the factory to appear at the early morning shape-up. Consequently Jacob Abbott observed that the work force at Novelty in spring 1851 fluctuated "from one thousand to twelve hundred," "all of whom reside in the streets surrounding the works." The Upper East Side factory district had by the 1840s become the social terrain of industrial workers and their families. They conducted most of life's activities within sight of the derricks and chimneys of the waterfront shops.[53]

Much as many artisans viewed their organizations as a privileged social domain, industrial workers came to view the neighborhoods where they lived and worked as a special working-class place. Even the highly skilled and aristocratic machinists shared this idea of the neighborhood and its institutions as their own social realm. When in November 1863 Daniel Walford admonished machinists for relying on groceries to sustain their families through a strike, he was forced off the podium with cries of "Let the machinists speak for themselves" and "Did the bosses employ you to speak for them?" During the draft riots, the city's new telegraph system flashed reports among city officials; the factory-district saloons similarly served as proletarian switchboards. Workers in the industrial wards collected at the saloon each morning of the riots to share information about the status of the draft and plan that day's activities. Ellen Leonard wrote from her midweek attic perch just east of Second Avenue, ". . . the liquor store on the corner was thronged with villainous-looking customers," and noted the following day, ". . . the crowd increased rapidly in the street and around the liquor store." The immigrant East Side industrial workers and their families viewed the local saloon as a resource for class survival. Any assault on neighborhood institutions smacked of middle-class moral intervention. During a strike, even Walford's talk sounded suspiciously like a boss's subterfuge.[54]

Once a Whig and nativist preserve, the East Side factory district was by the fifties becoming increasingly Democratic. Through the decade this area witnessed protracted struggles between industrial employers and temperance reformers on the one hand, and immigrant workers and tavernkeepers on the other.[55] In February 1854, Alderman William Boardman, proprietor of the Neptune Iron Works, staged an Eleventh Ward rally in support of strict enforcement of the state law prohibiting the sale of liquor on the Sabbath. Appearances by a local minister, a German temperance advocate, and a police captain were doubtless intended to elicit sympathy or at least respect for the Sabbatarian law among immigrants. Nonetheless many workers and tavernkeepers continued to flaunt the law. Three months later Boardman revoked the liquor license of a German named Lutz who hosted the "German Sunday Union" at his saloon. Lutz enraged Boardman with his defense that "he knew a certain Alderman of the Eleventh Ward who allowed work to be done in his manufactory the whole Sunday, from morning until night." In such fashion issues of control in the factory and in the neighborhood became intertwined. In July 1857 Boardman and other Protestant employers attempted to intensify their reform campaign through the Republican Party and the new Republican-sponsored Metropolitan Police. The Metropolitans would be zealous in enforcing temperance legislation that had often been ignored by Mayor Fernando Wood's old Municipal force. Many industrial workers accordingly saw Wood's Democracy as a vehicle for opposing Republican employers' moral incursions. John J. Blair, leader of the Machinists' Protective Association during the fall 1863 strike, began his public career as one of Fernando Wood's Municipals. Assistant Police Captain Blair was "popular with the men under him, and with the citizens generally. He held the place until the [Republican] Metropolitan Police law went into effect, and though offered a Captaincy in the new force, decided on principle, and held to the old organization, until it was disbanded. Mr. Blair then returned to his trade." Such commitment to the prerogatives of neighborhood self-government made Blair a popular figure along the industrial waterfront.[56]

Rejecting a moralizing and interventionist Republican party, the heavily immigrant industrial work force spent the Civil War decade defining its relationship to the Democracy. By fall 1862 it was clear that Tammany would fare little better in the factory district than the Republicans had. During the sixties the industrial waterfront helped elect an assortment of anti-Tammany Democrats, including Peace Democrats such as James Brooks and John Winthrop Chanler, independent German Democrats such as C. Godfrey Gunther, and machinist John J. Blair, whom it sent to the state assembly in 1867. Blair's entrance into politics was fortuitious. As union president during the 1863 machinists' strike, he pledged to his workmates that he would not return to the shop until their demands were met. Though circumstances forced many machinists to submit, Blair kept his word. He was nominated for the state assembly by an independent workingmen's organization and later endorsed by the Republicans and Union Democracy as an anti-Tammany candidate. "His election," wrote one biographer, "was regarded as a great triumph over Tammany Hall. . . . the workingmen were a unit for Mr. Blair and this settled the question." Blair's successful factory-district political career suggests that industrial workers, like some artisans, hoped to define some political alternative apart from the machinery of both major parties.[57]

How would industrial workers interpret the independence of their social domain of shop and neighborhood? Differences among the various metal-working trades may help explain how wage earners in the factory district came to define political independence.

Unlike the artisan crafts, the metal and machine manufacturing trades were new occupations whose traditions extended back no further than the second and third decades of the nineteenth century. The finesse of the blacksmith releasing the triphammer or the boilermaker flanging plates in a furnace had no analog in the centuries-old artisan crafts.[58] The work of the machinist, requiring mathematical reasoning alongside precise hand measurement and manipulation of materials, was closer to the artisanal activities of the carpenter or cabinetmaker; but still the machinist's trade required a mixture of abstract scientific and mechanical abilities without exact parallel in the building and woodworking trades.

The characteristic common to the metalworking and building trades (and not shared at mid-century by degraded artisan occupations like tailoring and shoemaking) was an empowering sense of craft and workplace control. Like many building-trades workers, journeymen machinists, boilermakers, iron molders, and blacksmiths hired and supervised their own helpers and paid them out of their own wages.[59] This sort of journeyman subcontracting was complemented in some shops by the "inside contract" system. Especially in interchangeable part manufacture (in New York, Singer's Sewing Machine Factory was a notable example), journeymen metalworkers bid among themselves to contract with the factory proprietor for individual jobs.[60] Both systems of subcontracting allowed these journeymen considerable managerial power. The hiring of assistants was so much a part of the journeyman's customary authority in metalworking that there was never any thought given to banning the subcontractor from the shop, as there was in bricklaying; it was assumed that journeymen who valued the prerogatives of craft over the acquisitive inducements of the marketplace could control the practice of subcontracting.

In spite of temptations to sweat helpers, work extra hours, and violate accepted stints at the furnace, journeymen metalworkers often used their managerial power to discourage acquisitive or "unmanly" behavior in the shop. Writing of his nineteenth-century apprenticeship at the Hoe machine shop, Charles Stelzle recalled the machinists' treatment of a "Yankee mechanic" who appeared at work a half-hour early "so as to get a good start before the strain of hundreds of machines was placed upon [the engine]": "He just wanted to be industrious and economical. But the men thought he was an ordinary "sucker," although that was the mildest term which they applied to him."[61] At least through the Civil War, New York metalworkers' sanctions were informally enforced by groups of journeymen on the shop floor and not through trade-wide strikes or union work rules.[62] Rarely were work rules a disputed issue in a Civil War-era metalworkers' strike.[63] But prerogatives developed in impromptu struggles over work rules doubtless helped define such workers' collective sense of self during strikes. Industrial workers were committed, then, to preserving the factory district as a working-class precinct and the shop floor as a zone free not of the subcontractor *per se* but of acquisitive individuals like Stelzle's "Yankee mechanic."

Nonetheless industrial workers disagreed as to how they might best defend and expand this realm of working-class autonomy. Machinists infrequently participated in strikes and almost never joined in the independent political experiments of the fifties and early sixties. There were few machinists and patternmakers among the draft rioters arrested in summer and early fall 1863. Contemporary accounts did not mention their participation in the riot. Iron molders, boilermakers, blacksmiths, and brass finishers, by contrast, turned out on strike more frequently and were usually among the workers active in citywide trades assemblies and independent reform movements. These four metal trades employed nearly all the industrial workers whose participation in the draft riots was recorded. The differences in political consciousness among these two groups of industrial workers seemed to stem from the distinct ways they perceived their social and cultural status.[64]

Each trade in the nineteenth-century machine shop had, as Charles Stelzle put it, a peculiar "snobbishness." At Hoe's, "the draughtsmen considered themselves much superior to the pattern-makers, the pattern-makers thought they were better than the machinists, the machinists looked down upon the tinsmiths, and so it went on. There were at least half a dozen different grades of 'society' among the men in the shop." Stelzle's recollection of the hierarchy of the nineteenth-century metalworking trades echoed a story told by

Terence Powderly, the Knights of Labor leader, of the Machinists' and Blacksmiths' Union's refusal to admit boilermakers to membership. A resolution introduced by Powderly in the early seventies to open the organization to boilermakers was almost unanimously voted down by the Union. The boilermakers were excluded in part because "their untidy habits and lack of neatness in dress would reflect no credit on our order through affiliation with it." Machinists changed out of their work clothes when the final bell rang; boilermakers went home in their dirty overalls. Powderly subsequently wrote a poem lampooning the machinist "aristocrats of labor" putting on "airs and graces."[65]

Machinists and boilermakers' differences in attire betrayed deeper divisions in these workers' perceptions of themselves, their work, and the community. Many machinists saw themselves as members of an exclusive club of practical scientists. They eschewed academic training and instead regarded their own shops as the laboratories of the American industrial revolution. Certain metropolitan area shops stood out as centers of innovation in the development of machine tools: Allaire's in the period before 1840, Richard Hoe's in the forties and fifties, and the Novelty Works in the fifties. Iron molders shared this sense of shop pedigree to a lesser extent: Seth Boyden's Newark shop, pioneer in malleable iron castings, was on a par with Allaire's and Hoe's machine rooms. John J. Blair referred to his apprenticeship at Allaire's as the formative event of his life; as a highly skilled machinist, he was asked by the Naval Department in 1862 to supervise the repair of machinery and gunboats at Hilton Head, South Carolina. Stephen Tucker, who rose from apprentice to part-owner of the Hoe Works, recalled the excitement of participating in the development of printing press manufacture at Hoe's in the late forties. No one better represented the scientific culture of the nineteenth-century machinist than Charles H. Haswell, who put his *Engineer's and Mechanic's Pocket-Book of Tables, Rules and Formulas Pertaining to Mechanics, Measuration and Practical Geometry, Mathematics and Physics* through seventy-five editions between 1853 and 1902. To a large extent, both boss and journeyman shared this appreciation of skill and innovation (though in the journeyman's hands it became a vehicle for imposing norms of cooperative behavior on the shop). No wonder, then, that the machinist changed his clothes upon leaving the shop. The machine rooms had an order and aura very different from the neighborhood outside the factory gates.[66]

Boilermakers, blacksmiths, and iron molders did not see themselves as practical scientists, though there was certainly an element of experimentation and innovation to their daily activities. Their work was heavy and physical. Nowhere was this more true than among the boilermaking squads of the extensive waterfront iron works. Jacob Abbott described the hammering and sweaty teamwork of the boilermakers at Novelty: "Some men stand inside [the boiler cylinder], holding heavy sledges against the heads of the rivets, while others on the outside, with other sledges, beat down the part of the iron which protrudes, so as to form another head to each rivet, on the outside. . . . One man holds up against the under side of the plate a support for the rivet, while two men with hammers form a head above—striking alternately upon the iron which protrudes." Skilled craftsmen though they were, boilermakers seem to have constantly feared their highly physical labor could be underbid and replaced by unapprenticed laborers. Differences in boilermaker and machinist attire that drew Terence Powderly's comment masked larger distinctions in the status of workmen in the two trades.[67]

While industrial workers of all stripes denounced the major political parties and their personnel, machinists avoided independent political assemblies, whereas boilermakers and other metalworkers frequently participated.[68] Machinists were not represented at the City Reform meetings of 1853 or among the forty-five delegations to the mass rally in support of striking painters that gave rise to the Amalgamated Trades Convention later that year. Machinists did conduct an unsuccessful wage strike that spring. At one meeting of the trade, one journeyman's denunciation of the city government and call for Mike Walsh and the land reformers drew cries of "We don't want any politics," "Keep to the point," and "This is a meeting of machinists—this ain't a political meeting." Machinists, it seems, were determined to keep their trade activities separate from workingmen's reform. By contrast, boilermakers elected "Charles Hoyle, from Morgan's Iron-Works [and] John Byron, from Neptune Iron-Works," molders, "Edward Wilson, from Hogg & Delameter's [*sic*] Foundry," and engineers, "Carman from Phenix Foundry" to represent them at the sympathy meeting for the striking painters. Iron molders' leader Robert Irving joined Daniel Walford at the podium of the meetings of the unemployed in the winter of 1854–55. Young boilermakers, tinsmiths, and brass finishers were arrested for anti-nativist rioting during the election of November 1857. These industrial workers, and not machinists, were sometime participants in the citywide meetings and street violence that comprised 1850s' labor politics.[69]

The Civil War may have heightened stratification between the labor aristocrats of the metal trades and other industrial workers. Skilled machinists like John J. Blair became especially valuable in the development and repair of naval machinery. When industrial employers told striking machinists in fall 1863 that the timing of government contracts would not permit them to make wage concessions, machinists—aware of their importance to the war effort—sent a committee directly to President Lincoln to explain their case. Meanwhile most other industrial workers, already feeling vulnerable, believed their economic status increasingly insecure. In February 1863, an attempt by the Novelty Iron Works to import cheap "contraband" laborers from Europe to undermine wages on the East Side waterfront provoked a mass protest meeting of "the iron workers of New York."

That meeting occasioned metalworkers' first recorded attacks against the

Republican Party in a trades gathering. Republican Horace Greeley, the city's acknowledged labor authority, was invited as the keynote speaker but nearly hooted off the podium when he neglected to concede Republican responsibility for the problem of "contraband" labor. One iron worker set the Novelty labor importation scheme in the context of the many Republican-sponsored wartime burdens on labor: "If they took proper action there would be no competition with capital, and no importation of foreign labor. He could not understand why white men should enslave themselves for any purpose. . . . The rights of the poor man were not recognized in any of the conventions of the day. . . . There are but two parties in the country—Aristocrats and Democrats." Young iron worker George Campbell attacked Greeley as a traitor and suggested that if the Novelty workers submitted to the importation of European mechanics, "they would be told at any time that if they dared to ask a cent advance in the price of labor, hordes of Europeans and negroes would be brought over to take their places. . . . There was a great effort being made to absorb the power of the masses in new ideas of government, and they should resist it." Finally boilermaker Robert McIlvaine introduced a resolution relating the importation scheme to the political views of their Republican industrial employers:

> that it be made known that when the unhappy war broke out, we were harangued by our employers, and told to organize and drill, and to volunteer, and after such had been done by numbers of our craft, they wished to have the remaining citizens drafted, and to bring foreign labor to supply our places; and also colored labor of persons in the South.

Daniel Walford made a final appeal to form "a general organization of all laborers in the country" but few of the iron workers wished to discuss the specifics of organization. At most the meeting accomplished the aim of McIlvaine's preamble—"to avoid strikes and other violent proceedings"—and directed the iron workers' economic discontent against the Republican Party.[70]

For these iron workers the Republican Party was liable for all the oppressive changes of the Civil War. The war focused the diverse complaints of the previous decade against party apparatus, Metropolitan Police, and temperance reformers. The wartime Republican Party was launching a multiple assault on these workers' status and economic security. The term "contraband" was used during the war to refer to a black slave who escaped to or was brought behind Union lines. After the Emancipation Proclamation the term expanded to fit two invidious groups: newly liberated black freedmen and European workers to be imported by the Novelty proprietors. In the words of one iron worker, "Contrabands would be brought here; and it was their imperative duty to organize as white men for their rights, to resist anything which might deprive them of their labor and a fair remuneration for it." Through these "contraband" laborers, iron workers argued, Greeley's

Republican Party and its local personnel of industrial employers were plotting an invasion of the journeymen's social terrain. The draft would be the stunning blow of the Republican assault, undermining the security of those workers who had not volunteered for the army. The attack on black labor in the February meeting was not based on fear of miscegenation; rather, the black freedmen represented an economic threat associated with the obnoxious Republican government. Finally, the iron workers' vocabulary—posing a national contest between "Democrats" and Republican "Aristocrats"—entailed more than political name-calling by Democratic partisans. The iron workers seemed to have understood "Democracy" not as the party organization per se ("the rights of the poor man were not recognized in any of the conventions of the day") but as an alternative, albeit vaguely defined, to the Republican industrialists' "great effort . . . to absorb the power of the masses in new ideas of government. . . ." Of all New York wage earners, these seemed to appreciate best the revolutionary implications of Republican wartime measures.[71]

With so many iron workers Irish-born and Catholic, it is curious that the issue of ethnic and religious conflict with Republican employers did not come up at the February meeting (of course, it may have been suppressed out of deference to iron workers who were not Irish Catholic). The problem did surface elsewhere. When Archbishop John Hughes returned from Europe in August 1862, he announced himself in favor of a general conscription to equalize the burden of defense and bring the war to a quicker end. What was Hughes's justification for a proposal that was sure to be controversial among the Catholic laity? Upon his return to New York, he had learned that

> the employers of . . . Irish Catholic citizens . . . in large establishments, immediately after the war broke out, suspended their factories and other departments in which human labor had been employed, to compel these Irish and Catholic operatives to enlist, in order that their families might not starve; and that all this was adroitly accomplished under the plea that war had rendered it necessary to suspend all manufacturing establishments; that this pretended necessity was only for the purpose of sending fighting men to the field, by which the neighborhood would be relieved from the presence of workmen of foreign birth; that, in point of fact, as soon as necessity drove that class away, their places were promptly supplied by other operatives. . . . Sooner than witness such mean and base tricks upon unfortunate laborers, I . . . am . . . prepared to approve of a thousand conscriptions openly appointed by the Government.

This tale of an anti-Catholic conspiracy may or may not have been true. To judge it fairly, we will have to acquaint ourselves better with the employers under indictment, as we later will. The story does call to mind the boilermaker McIlvaine's report of Republican industrialists who "harangued" iron workers "to organize and drill, and to volunteer" and who later tried to replace them. At the very least it suggests that during the war Irish Catholics who toiled under nativist (and probably Republican) bosses felt vulnerable. Irish Catholic iron workers employed by Republican industrialists must have thought their status fragile indeed.[72]

Being Irish or Irish Catholic did not necessarily predispose one against the war or the wartime government—to the contrary. Though the Irish were known for their commitment to the Democratic Party and enmity toward blacks, they were frequently deeply loyal to the Union. Carrying into battle both the green flag of Ireland and the Stars and Stripes, New York's Irish Sixty-ninth Regiment fought for the duration of the war. The Sixty-ninth appeared in the city during the draft riots and, alongside the heavily Irish Metropolitan Police, battled the rioters with sangfroid. Most Irish immigrants adopted America, with its political equality and legal guarantees for white men, as their own country.[73]

But for the Irish Catholic poor, such claims to American nationality were often found in amalgam with a religious culture contemptuous of the Republicans and their black "contrabands." The tone was set by New York's Catholic leaders who attacked Protestant abolitionists as both revolutionary sowers of chaos, "the American manifestation of the lawless liberalism" of Europe, and as infidels—"Deists, Atheists, Pantheists, anything but Christian." Slavery, Archbishop Hughes insisted, was best left alone. The *Irish American* hailed New York's Catholic clergy for declining "to preach the nigger from their pulpits." The shift in Northern war aims to include emancipation enraged local Catholic opinion. The Republican government, in the words of the vitriolic John Mullaly, had now made Irish army volunteers "subservient to the emancipation of the negro." The anti-Republicanism and racism of the Catholic press began to crescendo through winter and spring 1863 and reached fortissimo by early July. Industrial workers who suspected the Republican industrialists of conspiring to replace or devalue their labor found lively confirmation for their anti-Republicanism in the Catholic community. Late-week draft rioters accorded great respect and legitimacy to the Catholic clergy who sought to address them. It was the ideological stance of the Catholic leadership, and not merely the rioters' respect for clerical vestments, that explains such episodes.[74]

In sum, many industrial workers—boilermakers, iron molders, blacksmiths—saw their shop-floor conflicts spilling over into the neighborhood. These "iron workers" believed that in defending themselves against their employers they also had to guard against a host of Republican agencies in the community: black freedmen, the Metropolitan Police, and, in summer 1863, the personnel of the provost marshal. While the aristocratic machinists limited the jurisdiction of their craft to the shop floor, the iron workers had an expanded notion of the social and political terrain of their craft. These differences in each trade's political self-perception may have stemmed from differences in culture and status. Machinists' self-definition as scientific innovators was strictly a workplace role. They kept that capacity separate from community political involvements in the same way that they changed clothes before leaving the shop. The physically demanding labor of the iron workers created a different trade culture, one still based on the empowerment of craft but mixed with feelings of vulnerability and a greater need to rely upon community institutions and social networks. Boilermakers, blacksmiths, molders, and others viewed their workplace conflicts and their struggles against middle-class reformers in the neighborhood as part of the same project. Irish Catholic culture mattered deeply in the lives of many of the factory district poor. But in the case of iron workers that culture reinforced a sense of vulnerability and a suspicion that the Republican Party was designing against the interests of the poor on all social and political fronts. More likely than not, these iron workers and not machinists helped to build the barricades during the desperate last days of the draft riots.

Common Laborers

Karl Marx's advice to Siegfried Meyer and August Vogt—that uniting the Irish and Germans was the central task of a workers' movement in America—applied with special force to the common laborers of New York City. If Marx had paused to reflect, he might have added bridging racial conflicts to Meyer and Vogt's American agenda; still the case of New York City laborers would have been appropriate.[75]

The draft riots draw attention to a special group of white immigrant laborers: those who independently managed their own work. Among this group were longshoremen, cartmen, coal heavers, lumberyard workers, quarrymen, pipe layers, railroad stable men, street pavers or boulevarders, and the ditch diggers who built Central Park. It makes sense to focus on the longshoremen, who were mentioned prominently in many accounts of the riots and numbered especially high among the arrested rioters. These laborers did not share with building and metalworkers the empowerment of craft. But longshoremen had their own version of trade empowerment grounded in their historic position of importance in the functioning of the nation's largest seaport. Independence at the work site and this indispensable role in the commerce of the seaport combined to give these laborers a heightened sense of their own social and political power.[76]

Friendship and kinship networks were the mechanism at the core of both the bitter ethnic and racial antagonisms and the social and political empowerment of common laborers. These wage earners, like artisans and industrial workers, were preoccupied with questions of social and political inclusion and exclusion in the Civil War epoch. Beginning in the 1850s, the predominantly Irish organizations of longshoremen first created all-white organizations in order to establish standard wage rates in their trades. A close examination of the social world of these workers through the lens of the draft riots reveals that they used kinship and friendship networks to define the boundaries of their ideal polity. Such networks help to explain common laborers' reciprocal relations with merchant elites, their hostility toward "scab elements" and acquisitive *petit* proprietors, and their attacks on racial amalgamators and brothelkeepers during the draft riots.

Elizabeth Fox-Genovese and Eugene D. Genovese's dicta, derived from Marx, that merchant capital has "normally inclined to avoid any disturbance in productive relations" and that merchants have "generally prefer[red] familiar patterns of politics as well as trade," make up the theoretical starting place for a discussion of New York City's longshoremen.[77] The geographic point of departure is the New York City Customs House. Even as the local and federal governments changed hands between the major parties from the early 1850s through the 1870s, the political style at the Customs House—characterized by an ethic of friendship—varied little. In the Customs House merchants practiced "familiar patterns of politics" through the Civil War epoch.

The Collectorship of New York Port was by mid-century the most prized federal patronage post in the United States. The Collector and the Customs House officials he appointed had full charge of a steamer from the time it entered the port until it was cleared for its next voyage. Customs House officials granted the permits required before a ship could legally discharge its goods, and Customs House officers and watchmen supervised the unloading process. Needless to say, merchants demanded a sympathetic Collector and Customs House bureaucracy to insure that vexing delays in turnover time were avoided. Customs House appraisers were also legally entitled to take a percentage of each ship's merchandise to an Appraiser's Store to test the veracity of the ship's invoice. The routine exercise of these responsibilities gave Customs House officials an unequalled opportunity for graft and extortion. In late 1862, the Solicitor of the Treasury Department observed that, in New York, "nearly the entire body of subordinate officers in and about the custom-house, are, in one way or another, in the habitual receipt of emoluments from importers or their agents." One of the preferred sorts of corruption involved selling to political favorites the labor contract for hauling merchandise to the Appraiser's Store. Hiram Barney, Republican Collector of the Port during the early years of the Civil War, solved the problem by "hiring" his own firm to handle the Appraiser's haulage. This sort of extravagant corruption was an accepted part of the Collectorship and Customs House employment. The political struggles for control of the Collectorship and Customs House patronage were accordingly perpetual and fierce.[78]

Ideally, the Collector of New York Port would be a merchant whose high social standing, great wealth, and spotless character commanded the respect of capitalists of all political tendencies. Though few Collectors met this standard, the post was held in sufficiently high esteem for "Boss" John Kelly to give Augustus Schell's tenure as Collector prominent mention in an 1885 eulogy. President Buchanan had appointed Schell Collector in 1857 for several reasons: the lawyer was a personal friend of the President; he had for years been a faithful member of Tammany Hall; and his faction of the city Democracy had been loyal to Buchanan. But most important were Schell's vast wealth and his amicable relations with many of the city's merchants. Kelly remembered:

> The bond required of the collector for the faithful discharge of his duties was very large. He [Schell] did not have to seek bondsmen, but several of our most prominent citizens at once offered to go on his bond. This year [1857] is also to be noted from the fact that at this time he relinquished his very extensive law practice, never again to be actively resumed. The next five years we find him carefully discharging the duties of the collectorship, and winning favorable opinions from the merchants of the city, regardless of their political views.

Schell had the reputation, then, of being an amiable fellow who had clients and friends among merchants of all political persuasions. Merchant Simeon Draper, asked to replace Hiram Barney as Collector when the latter's corruptions became too embarrassing to the administration, resembled Schell in personality. The Republican Draper's ability to get along with Democratic and persistent Whig merchants invited the disdain of that unbending high-church Republican, George Templeton Strong. Strong called Draper a "scamp" in the fifties; by 1861 Draper's talent for compromise made his " 'Union Defense Committee' . . . stink in the nostrils of all good citizens." The affability that drew Strong's scorn made Draper an ideal candidate for Collector of the Port.[79]

The presence of men like Schell and Draper (and Barney) in the Collectorship meant that the political turmoil of the Civil War and Reconstruction would have little effect on business as usual at the Customs House. One historian has observed that the transfer of federal power from the Democracy to the Republican Party in 1861 little affected the kind or degree of corruption in the Customs House. The Chamber of Commerce, embracing merchants of opposing political views, also functioned smoothly through the sixties. Joseph Jennings, a longshoreman writing in the 1870s, noted how the four Walsh Brothers, New York City's preeminent stevedores, accommodated the fall of Democratic Tammany and the revival of the Republican Party in New York State in 1872:

> In the palmy days of the Tweed administration these four brothers were Tammany Hall Democrats, but when misfortune befell the great Tammany leader they forsook his banner and made peace overtures to Tom Murphy. By falsely misrepresenting their influence among the working classes, two of the brothers were allowed to join a Republican clique—thus dividing their influence for good or evil. With the disruption of the Tweed "ring," Apollo

> Hall sprung into existence, and the Brothers, always equal to any emergency, held a caucus, and it was resolved that one of them should belong to this new organization. . . . Two are Custom House Republicans, one belongs to Apollo Hall, and the other affiliates himself with the reconstructed Tammanyites.

Jennings nicely illustrated the malleability of political affiliation among those groups occupying the intermediate niches of New York City's commercial capitalist hierarchy. If merchant capitalism required "familiar patterns of politics as well as trade," the elastic politics of the Walsh family satisfied the dictates of both the Republicans of the federal Customs House and the different Democratic factions of city government. An ethic of friendship, at times facilitated by patronage and corruption, preserved stability in the Customs House, in the Chamber of Commerce, and along the waterfront amid the intense party contests of the Civil War era.[80]

One group in the local structure of merchant capitalism was not prone to political compromise: the immigrant laborers of the waterfront. By the 1850s the waterfront had become largely a white and Irish preserve. Black workers who had had at least a faint presence on the docks in the forties were increasingly denied employment by white Irish stevedores and work gangs in the mid-fifties. By 1855 few black men could find work on the docks outside of strikebreaking under police protection. German workers were only slightly less reprehensible than blacks to the Irish Longshoremen's United Benevolent Society formed in fall 1852. From then through the heyday of the Walsh Brothers in the early 1870s, the Irish Longshoremen's Society was strong enough to compel most dockworkers and the stevedores who employed them for the shipping companies to join the society and observe union wage rates. These Irish longshoremen were deeply committed to the Democratic Party and, often enough, to Tammany Hall. They furnished a solid political base for Fernando Wood in the fifties and for Tammany through the fall of the Ring and beyond. Typical was Joseph Jennings, a devoted Tammanyite who led the powerful Longshoremen's Union Number 3 in the early seventies.[81]

These Democratic longshoremen had their own working-class version of the merchants' ethic of friendship. The longshoremen's ethic, like the merchants', sought to preserve stable and familiar patterns of trade. Through the middle decades of the century, longshoremen believed that only if shipping merchants understood the nature of conditions on the docks—often enough, the exploitative behavior of their stevedore employers—could harmony be restored between labor and capital. When shipping merchants sought to reduce the wage scale during the business depression of winter 1855, the president of the Longshoremen's Society, Mr. John H. Williams, "argued long and strenuously in favor of a compromise with the merchants":

> He thought the merchants would not determine on the reduction unless they believed it necessary. . . . it would be highly impolitic to oppose the action

> of the merchants *in toto*. The Society should manifest a spirit of conciliation, that the merchants might see that the Longshoremen were susceptible of a calm discussion of the matter in which the two parties were interested.

The almost obsequious tone of President Williams's recommendations to the longshoremen in 1855 characterized longshoremen's organizations' posture toward the shipping merchants through this era. In fall 1873, Longshoremen's Union Number 3 raised one thousand dollars to send Joseph Jennings to Europe to complain to shipping merchants of the wage reductions and corrupt practices of the Walsh stevedores. Jennings told a "Mr. Hurst," the manager of the National Lines, "that if he discarded the Walsh Brothers, and worked his ships on behalf of the company, he could annually save a large amount to the stockholders"; Jennings conveyed the same message to a "Mr. Dole" of the Inman lines. In an attempt to discredit the Longshoremen's leader, the Walsh Brothers had meanwhile led the shipping merchants to believe that Jennings was an agent of the corrupt Tweed regime. No matter—Jennings continued to trust in the good faith of the European merchant capitalists: ". . . these European owners have had their eyes opened at last to the disreputable system of working their ships; the extortions and exactions by way of extra labor bills, and the wholesale plunder by means of falsified returns of the quantity of cotton screwed." At the end of his 1874 "expose" of the Walsh Brothers, the longshoreman reaffirmed his belief that "the interests of capital and labor are so closely interwoven as to be in a manner almost identical, and a little concession coming opportunely from either party would occasionally render the much dreaded 'strike,' and all the attendant evils which accompany it unnecessary." The longshoremen's confidence in the fairmindedness of the merchants eroded only when merchants refused to negotiate with the laborers' committees and representatives. Otherwise, longshoremen of the Civil War era believed that if merchants understood the true nature of working-class conditions, they would intervene, drive away exploitative outsiders, and reestablish equitable social relations.[82]

While longshoremen often adopted a deferential tone toward city merchants, they could be hostile toward middle-class offenders who disobeyed their waterfront regulations. Only a small social distance separated the longshoremen and most subcontracting stevedores. Most stevedores began their careers as longshoremen, and many longshoremen moved in and out of stevedoring as economic conditions permitted. One needed little capital to set up as a labor contractor. Indeed, the Walsh Brothers began their stevedoring firm during the Civil War years by handling the modest "extra labor" of the North River piers—landing mail, baggage, and other items "outside the regular work of discharging and loading." Because most stevedores were former or occasional longshoremen, longshoremen's organizations and their usages exerted a powerful influence on both laborer and small contractor. What made the Walsh Brothers' firm so controversial, an "aristocracy" in

Jennings's view, was its great size, its capitalistic pretensions, and its callous disregard for dockside work customs. Because the Walsh Brothers sweated their work gangs, cargo was damaged and the men (required to furnish bonds for the safe carriage of cargo) were "systematically robbed out of a part of . . . [their] hard-earned wages on pay night."[83]

The issue causing the most frequent conflict between longshoremen and their subcontractor bosses was some stevedores' practice of taking a few cents' personal stipend out of the longshoremen's weekly wages. "The principal complaint" voiced by the laborers during the fall 1852 longshoremen's strike was "their being compelled to pay one shilling a day to the head stevedores." The Walsh Brothers' practice of exacting a "tax" from their employees' wages similarly precipitated the massive strike of 1874. During the strikes of the Civil War era, longshoremen sent worker "committees" to each pier on the East and North riverfronts and physically forced offending stevedores to shut down. Stevedores who disobeyed the Society's regulations were roughly treated. An offending stevedore unloading the ship *Green Point* during the 1852 strike was "thrown on the dock and somewhat injured." While the Longshoremen's Society officially repudiated violence, intimidation and dockside beatings were the accepted means of enforcing Society rules.[84]

The social perceptions of white common laborers, like those of industrial workers, were shaped in the context of the neighborhood. Waterfront work was among the most erratic and seasonal in the city. To be available for work on short notice longshoremen had to live near the piers; like many industrial workers the dock laborer belonged to a floating army of workers that roamed the waterfront in hope of being chosen in the "shape-up" for a few days' employment. Longshoremen drifted into other trades when dockwork became scarce. Even into the early twentieth century they found jobs as teamsters, boatmen, and brickmakers in slack seasons. Inevitably weeks were spent idle until a ship arrived or the weather cleared. Then the laborers were thrown back on the neighborhood resources of saloon and family—few longshoremen could survive after mid-century without wage-earning wives and children. This sort of work environment explains why white longshoremen felt it necessary to monitor the workingmen who entered and left their home districts. Control over the neighborhood labor market was the only way to regulate the kinds of workers employed on the docks.[85]

The waterfront was also one of the few areas where black men could traditionally find a semblance of secure employment. Little is known about the employment patterns of black longshoremen in New York City, but they probably migrated among the same occupations as white laborers: teaming, boatwork, brickmaking, and carting. To find work on the piers black men, like their white longshoremen neighbors, had to live close to the docks.[86]

During the strike of 1855 and again in spring 1863 there was bitter racial conflict on the waterfront. In the earlier event black men were hired by merchants "foremost in the late movement to reduce the wages of the Longshoremen." The intimidation of the Irish Longshoremen's Society was sufficient to drive black laborers from the piers, while the presence of policemen discouraged the spread of violence. But in one instance a black longshoreman returning from work on an East Side ship was struck in the head by "an Irishman." The black man was armed and fired at his assailant; "as if by magic" several hundred white longshoremen appeared. This laborer managed to escape, but eight years later black dockworkers were less fortunate. In April 1863 longshoremen's attempts to enforce a standard wage rate and an "all-white" rule on the docks led to a protracted binge of racial violence that spread into surrounding tenements. For three days mobs of Irish longshoremen beat up black men found working along the docks and fought Metropolitan Police who attempted to save several blacks who defended themselves against lynching. May was quiet, but the strike resumed in early June. Again black longshoremen were among the scabs driven from the piers. Negotiations between strikers' committees and the shipowners failed, and the United States government intervened with federal troops to protect black and white non-union strikebreakers. Though the conflict was finally settled on the 18th with a compromise by the shipping firms, the strike was an omen. During the July draft riots, gangs of white longshoremen participated in some of the week's most grisly racial murders and announced that dockwork and stevedoring would be restricted to members of the Longshoremen's Association and "such white laborers as they see fit to permit upon the premises."[87]

The racism of Irish longshoremen requires close examination. Doubtless their efforts to exclude "cheap" black labor from the piers in 1855 and 1863 reflected an ingrained racial prejudice. Catholic pro-slavery sentiment may have figured in these racial assaults as well. But the most intriguing clue to motivation was the demand for an "all-white" waterfront. The Longshoremen's United Benevolent Society was an exclusively Irish organization—it marched annually in the Saint Patrick's Day parade. In the fifties and sixties its "all-white" provision seems to have meant all-Irish; conversely the "non-white" longshoremen barred from the docks included German as well as black workingmen. The use of the term "white" in this elastic fashion was continued into the early part of the next century. John Watson, a superintendent of a shipping company, told a Senate Committee fifty years later that the decrease in the number of "white men" on the docks had led to a decline in work quality. By "white men" Watson meant Irish- and German-Americans; excluded from the category were Polish and Italian workers. One Committee investigator asked Watson, "You may employ seven gangs of white men and one gang of Italians on a ship?" "Yes, sir." The ethnic character of these exclusionary work rules—though at no point does the Civil War-era LUBS proclaim an "all-Irish" waterfront—hints at the true boundaries of these laborers' "all-white" community.[88]

The draft riots afford an opportunity to investigate the "all-white" social world of the longshoremen. Among those arrested for "riot" was twenty-one-year-old Irish-born William Patten, a laborer on the Chelsea docks. Patten was taken into custody for leading "a large number" of rioters to the West Side molding mill of E. H. Sigler, "hooting and yelling" and demanding that Sigler close shop and release his workmen. Otherwise Patten threatened to mob and burn the factory. Also arrested was Dennis McDade, a seaman who resided at Patten's home address, West Twenty-seventh Street near the North River. When his case came before General Sessions in the second week of August 1863, Patten paraded a long line of defense witnesses into court. First was John Patten, his brother, a weaver who had roomed with him since he arrived in the United States in 1853. According to John, William had a reputation for "honesty, sobriety and integrity" and no arrest record. Next before the judge was Philip Farley, a Chelsea stevedore who lived next door to the Pattens, "frequently had said William in his employment," and knew him "for the past nine years intimately." Then Dennis Mannion, a liquor-store owner who kept shop a block from Patten's home and testified to his "temperate" character. Then Patrick Kernan, a workmate of Patten's for eight years who lived in Patten's tenement. John Blake, another grocery and milk store owner on the block (Patten a "quiet and peaceable citizen.") Then two other brothers, Neil and Michael Patten, who testified to his steady employment (and conceivably worked alongside him). Then John Horstmann, another saloonkeeper on the block, who had known William for ten years. John Wallace, a painter living down the block. Bernard Mitchell, a laborer living in Patten's house. Finally, the three who posted Patten's bail: William Scott, a "smith" who lived in the West Thirties, Alfred Budlong, a lumberyard proprietor who had intermittently employed the prisoner, and Cornelius Patten, undoubtedly another relative, who lived six blocks away.[89]

What emerges from the Patten court record is an elaborate male friendship and kinship network including small employers, tavernkeepers, workmates, and many relatives. All of these employers, friends, and relatives lived within a seven-block radius. Patten's stevedore employer lived next door. Dennis McDade lived at Patten's own address. Many of the defense witnesses seem to have been of Irish birth or parentage, but the presence of John Horstmann, a German grocer, suggests that the intimate social world of the young longshoreman could include non-Irish individuals as well. We unfortunately do not know where wives, daughters, and women friends fit into this family and friendship tree. Perhaps Alfred Budlong was married to a Patten cousin, or Philip Farley the friend of a Patten sister. It must be remembered that in the ideologically charged climate of summer 1863, testifying on behalf of an arrested draft rioter was an act of considerable courage and personal allegiance. William Patten's employers and workmates appear to have been not just neighbors but close and loyal friends and relatives.

A similar kinship and friendship network provided the foundation for the Walsh Brothers' stevedoring firm during the 1860s. Longshoreman Joseph Jennings saw the power of the four brothers rooted in their "endless retinue of cousins" serving as "workmen and spies" for the stevedoring family: ". . . the holds of the ships, the piers, the liquor stores, even the boarding houses are under the surveillance of these men." As Jennings mounted his campaign against the Walshes in 1874, John Griffin, financial secretary of the Longshoremen's Society and a Walsh cousin, obtained a Supreme Court injunction to prevent Jennings from spending Society money to travel to Europe to inform the merchants there of the stevedores' corruption. Affidavits supporting the injunction came from Patrick Burke, another Walsh cousin. For the Walsh Brothers, family and employment were one and the same enterprise. So too for Jennings and his longshoreman and stevedore supporters, who contested the Walshes for control of the labor market. Jennings hoped the Longshoremen's Society could wield the kinship and friendship network that structured the trade against exploitative shippers and contractors who debased the value of the longshoreman's labor.[90]

That network may help explain the consciousness of laborer draft rioters in July 1863. By symbolically extending the friendship and kinship circle into the community, common laborers included certain social groups and not others in their political domain. As amalgamators or potential amalgamators, black men were a social and sexual threat to a system of authority based on the family; hence the laborer rioters' highly sexual violence against black laborers *and* white women who married across racial lines. It is worth speculating that in districts where the longshoremen were well organized the definition of "white man" could become quite elastic. Perhaps a black docker linked to the white laborers' kinship or friendship network could be accepted as "white" ("he's one of us"). Conversely, along the unorganized piers of the Upper West Side, we find the most grisly and sexually charged racial murders. Here, one can imagine, the subtle discriminations of the Longshoremen's Society "white men only" rule degenerated into brutal racial killings by roaming Irish gangs assaulting all men of color.

Brothelkeepers who attracted large numbers of unmarried men into the laborers' districts were another threat. The single men who frequented brothels and waterfront boarding houses clearly could not be absorbed into the Irish family/employment network. Boarding housekeepers and brothel owneras catering to Germans and black men were particularly controversial. During the riots, young laborers invoked the charivari as a way of condemning such violations of "appropriate" family behavior, much as "youthabbeys" had played rough music outside the homes of morally suspect married couples in early modern Europe. It made sense that New York City laborers invoked a traditional mode of family regulation to enforce the "all-white" work rule in their neighborhoods.[91]

The war multiplied and politicized laborers' grievances against the Republican Party. Like other immigrant workers, Irish laborers had resented the moral intervention of Republican temperance reformers and Metropolitan Police in the fifties. But these workers were deeply offended by the willingness of the Republican wartime government to interfere in relations between employers and workers. A member of the Longshoremen's Society who addressed striking shipwrights and dockworkers in November 1863 captured the sentiment when he "denied the right of any government to take sides against the working classes, and bestow its favors exclusively on capital." Longshoremen who understood the ideal of class relations as a reciprocal fellowship between labor and capital were resentful and even hurt when the Republican federal government intervened on their employers' behalf to protect black strikebreakers in spring 1863. The Emancipation Proclamation and the draft strengthened laborers' fear that the Republican elite was unlikely to live up to or respect the ethic of friendship that characterized ideal relations between seaport laborers and paternalistic merchants: Republicans were adopting black workers as a clientele at the expense of the Irish laborer.[92]

The war experience also made Irish Catholic laborers more suspicious and demanding of their Democratic friends. During the draft riots, Democratic notables had to prove their loyalty to these laborers, to demonstrate that they too did not intend to upset the equilibrium of class relations. Laborers interrogated property-owners and sought to elicit gestures of friendship such as treating or reassuring statements of opposition to the draft and emancipation. Suspected individuals who made friendly overtures could often dissuade these rioters from violence. But even with such episodes of compromise, the riots demonstrated the ferocious pressure Irish Catholic laborers could exert upon the Democratic elite to protect them against those they perceived to be exploitative and meddling outsiders—in July 1863, Republican provost marshals, low-wage black and German workers, and Protestant reformers.

The draft rioters' rage against the Republican Party had its origins in workers' efforts to intervene in politics on their own terms during the 1850s. In 1850, artisan craft workers advanced proposals that envisioned city and state legislatures acting under the guidance of workingmen reformers rather than party politicians and regulating inequitable and oppressive economic relations. Between fall 1850 and summer 1863, this style of consolidation was redefined and ultimately redirected against the new Republican Party. In part this change in orientation was wrought by the altered political environment of the mid-fifties, which lent urgency to efforts to protect majority rule against the political machinations of powerful minorities. In part it was caused by the hard times and deepening sectional crisis of the mid- and late-fifties, which undermined the trade unions that had given substance to craft workers' hopes of political independence and brought southern-sympathizing Democratic politicians to the fore of the labor movement. But most important, craft workers' anti-Republicanism was a response to the wartime expansion of the Republican federal government. The Republican innovations in government, culminating in the discriminatory draft, placed heavy burdens on the laboring poor and gave the longstanding artisan complaint against exploitative and interventionist political power new significance and authority.

The draft riots were remarkable for the participation of so many different kinds of workers. The theme of opposition to unjust centralizing power was potent because it appealed to artisan craftsmen, industrial workers, and laborers alike. For Irish Catholic industrial workers and laborers the war brought on the political and social changes that led to draft rioting. The former saw Republican wartime policies such as importation of European labor, black emancipation, and the draft as a design by employers to degrade their economic status; the latter believed the Republican government hostile to an ethic of kinship and friendship that ideally governed relations between the wealthy and the poor in the seaport. The participation of these workers contributed to the extreme violence and spontaneity of the 1863 revolt and its success in involving unorganized workers, in drawing men and women out of their shops and kitchens. The war experience brought a large and extraordinarily diverse mass of working people to the center of political debate in New York City.

The war and the problems of political centralization and national loyalty it posed probably narrowed the possibilities for rigorous and ongoing challenge to the political order by the workers of New York City. With their plan to use government to supervise economic growth, the leaders of the 1850 Industrial Congress had conceived a radical alternative to the topsy-turvy course of development New York had taken at mid-century. Compared with that plan, the draft rioters' revolt against the Republican wartime government had less promise as a way of making New York's style of growth more just. But the insurrection did reveal workers in the riot trades taking political action on their own initiative and not at the instance of the party politicians. The ability of this immigrant poor to see connections between economy and polity and to challenge authority in both realms helped prompt the political crisis of 1863.

Rioters and militia exchange gunfire on First Avenue during the draft riots. Note the participation of women and the involvement of the entire neighborhood in the battle, from the roofs and windows of the surrounding tenements. The liquor store in the foreground served as the rioters' base of operations (*Courtesy of The New-York Historical Society, New York City*).

A mob lynching a black man on Clarkson Street. Black men, who some rioters feared would compete for their jobs, bore the brunt of much of the bloody violence (*Courtesy of The New-York Historical Society, New York City*).

The burning of the Colored Orphan Asylum. Rioters sacked and burned the splendid building on Fifth Avenue, but not before dozens of black children escaped through the back door (*Courtesy of The New York Historical Society, New York City*).

Crowds battle troops in front of the Union Steam Works, July 14, 1863. This drawing conveyed the grand scale of some of the confrontations of the draft riots (*St. Louis Mercantile Library Association*).

The Novelty Iron Works. The Novelty Works, which was closed by a crowd the first morning of the riots and may have employed many rioters, was one of the largest factories in the city, with over 1000 workers (*Metropolitan Museum of Art,* The Edward W. C. Arnold Collection of New York Prints, Maps and Pictures. Bequest of Edward W. C. Arnold, 1954 (54.90.588)).

The Twentieth U.S. Colored Infantry receiving their colors at the Union League Club House, Union Square, March 5, 1864. The Union League Club's presentation of flags to a black regiment it raised after the riots was a symbolic attempt to reclaim the public spaces of the city from racist draft rioters and their sympathizers (*Courtesy of The New-York Historical Society, New York City*).

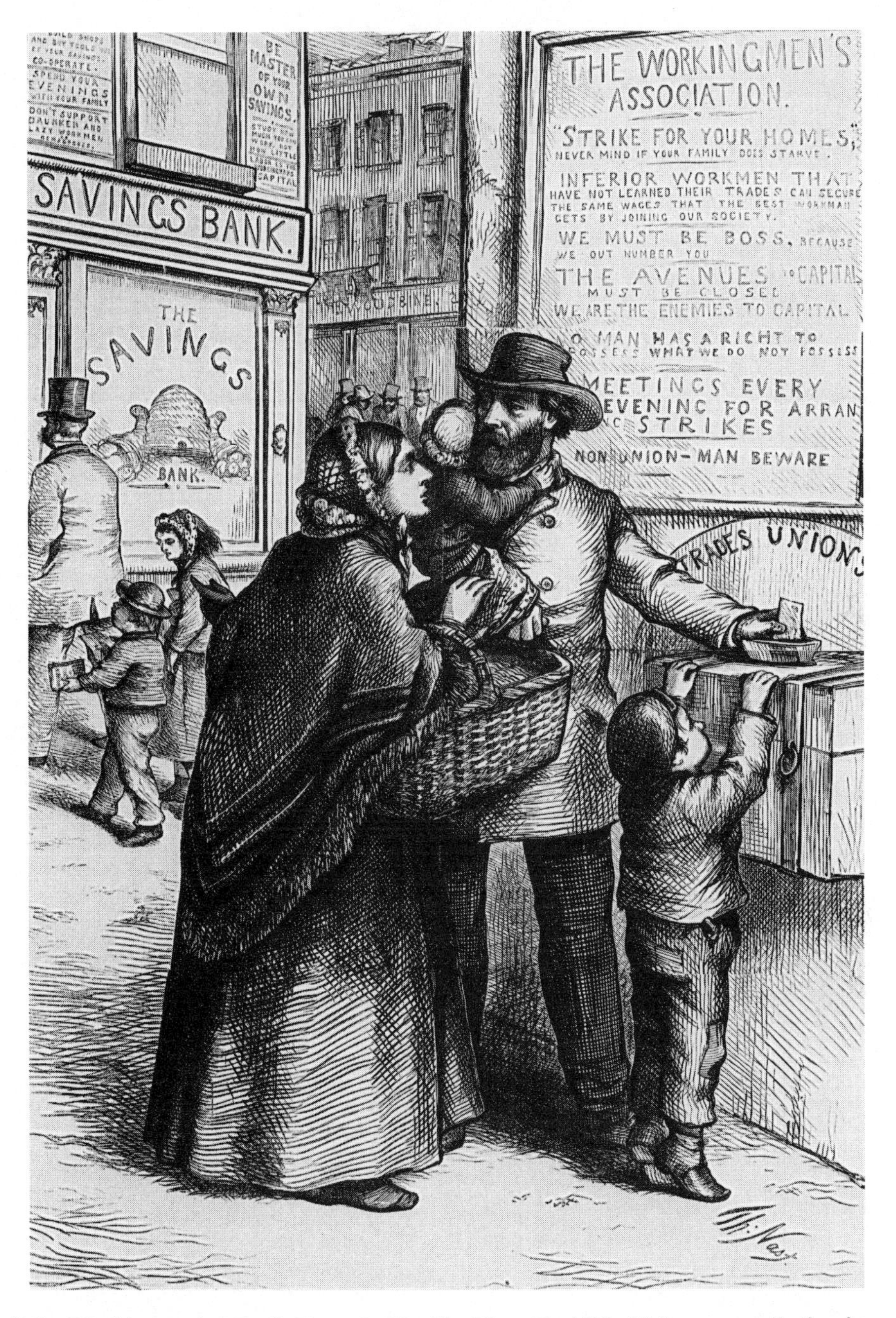

"The Workingman's Mite," ***Harper's Weekly,*** **May 20, 1871. This unsympathetic view of "The Workingmen's Association" suggested how controversial the power of organized labor was in New York after the Civil War (*St. Louis Mercantile Library Association*).**

"The Chinese Question," *Harper's Weekly,* February 18, 1871. Many New Yorkers saw the workingmen's anti-Chinese movement of the early 1870s as an echo, albeit a distorted one, of the anti-black racism of the draft riots (*St. Louis Mercantile Library Association*).

"The Usual Irish Way of Doing Things," *Harper's Weekly,* September 2, 1871. The harsh nativism of this cartoon reflected the hostility of many well-to-do New Yorkers toward Irish-American political rule after the Orange riot of July, 1871 (*St. Louis Mercantile Library Association*).

Cover illustration, *Civil Rights. The Hibernian Riot and the "Insurrection of the Capitalists." A History of Important Events in New York, in the Midsummer of 1871* (New York, 1871). The reform movement that emerged in summer 1871 to oust the Tweed Ring associated Tweed's version of popular rule with the excesses of both the Orange riot of that July and the draft riots eight years earlier (*The Library Company of Philadelphia*).

The Seventh Regiment Armory, Park Avenue and Sixty-sixth Street. The imposing medieval façade of the Seventh Regiment Armory, built in 1877–80, was intended to discourage would-be rioters. That architectural style well reflected the embattled mentality of the metropolitan elite during the seventies (*107th Infantry Regiment, New York National Guard*).

CHAPTER 4

Merchants Divided

Merchants were the most identifiable social group in New York City during and after the draft riots and, indeed, through the Civil War epoch. Roland Barthes's aphorism regarding the bourgeoisie—"the social class which does not want to be named"—did not apply to New York's lords of commerce. During the riot week they met on Wall Street to devise a "merchants' " response to the violence and form "merchants' " brigades. After the uprising some of these men created a "Committee of Merchants for the Relief of Colored People Suffering from the Late Riots." They sent "merchants' committees" to Washington to advise presidents and supervise legislation. They formed the "Society for the Diffusion of Political Knowledge" and "The Union League Club of New York" to publicize merchants' positions on the issues of the war. In all these instances, merchants made themselves and their programs plain to view.[1]

During the riots conservative and radical merchants debated whether to declare martial law. One wing of the commercial elite welcomed the early Monday protest against the draft (though not the insurrectionary violence that followed) and sought to preserve its own home rule by minimizing the federal role in restoring order. These businessmen were willing to negotiate with white draft rioters and ignored black riot victims. Another wing advocated what was by far the most drastic response to the riot. More draconian in outlook than even the industrialists, these merchants hoped to import federal military government, suppress the two uprisings, and institute summary tribunals for rioters and their well-to-do sympathizers. This group championed the cause of the devastated black community and broadcast this theme in a March 1864 ceremony that attempted symbolically to reappropriate the city's public spaces from the white rioters. During and after the riots, then, merchants disputed the role of the nation-state, proper relations between social classes, the status of immigrant and black minorities, and notions of urban space. The draft riots exposed a contest between two merchant camps, each seeking to demonstrate to the community its claim to social and political leadership. It was as if each group sought to identify itself as the true proprietor of the honorific "Merchants of New York."

The transformation of New York Port between 1780 and 1865 and the emergence of the Jacksonian mass party system in the 1830s and 1840s were the structural changes that provide the starting point for understanding the different positions merchants took toward the riots in 1863. The riot-week quarrel between merchants was a party affair between a mostly Democratic group, typified by August Belmont and his circle, and the predominantly Republican Union League Club. But the dispute also reflected differences in economic and political experience. The Democrats were often men of relatively recent arrival to the city who found themselves especially privileged by the economic and political changes of the Jacksonian period. The Republicans tended to be members of a more established patrician class with ties to an "Old New York" that predated the rise of mass markets and mass parties. Finally the conflict was also a cultural one between confident Democrats and anxious Republicans. Democratic merchants, promoters of the so-called "Young America" movement, believed that an expanding and agrarian-based market economy could absorb the shocks of social and sectional conflict. Republican merchants, by contrast, were chary of the economic and political expansion that so encouraged their associates. To be sure, the Republican Union Leaguers dreamed of American commercial empire. But that expanding empire, with its attendant social and sectional divisions, would have to be battened down by metropolitan culture and institutions to prevent anarchy. The riot-week conflict between merchants was so intense because it was only then, during summer 1863, that the battle between these two outlooks was fully engaged.

The Changing Marketplace and Polity

In colonial times, the New York City merchant was a versatile generalist. He loaned money to farmers, planters, and artisans, arranged for ships to handle their products (and was often part-owner of the ships), insured the ships and their freight, imported and exported, sold at retail and wholesale, and acted as agent for merchants in other ports. In all these transactions the merchant was personally acquainted with most of the individuals involved—hard as this may be for us to imagine from our modern perspective. In the manner of European traders for centuries past, the merchant of colonial New York preferred to have members of his own family serve as his agents in domestic and foreign ports; he knew all the other merchants of his home port, with whom he often collaborated in large ventures like insuring and owning ships; he was even acquainted with the hinterland farmers, planters, and country storekeepers with whom he dealt. Even with its far-flung markets, the business of the colonial merchant was a surprisingly intimate affair.[2]

By the early nineteenth century this personalized world of commerce was beginning to break apart. The most important impetus for change was the cotton trade. Though New York City did not lay on a direct line between Liverpool and the Southern ports, it attracted that trade with its well-established relations to the British mercantile houses, sheltered harbor, access to a large regional market for British manufactures, and capital for providing planters credits against the next year's crop. When after the War of 1812 British manufacturers unloaded their overflowing inventories in the New York market, city merchants began to specialize in response to the increased volume of trade. How else could merchants preserve the intimacy—the reliance upon a few trusted individuals—that was the stability and security of their existence? Expanding the firm and depersonalizing relations was out of the question. The alternative was to limit activity to one or two fields. The career of Anson Phelps, who began as an all-purpose merchant in 1812 but gradually restricted himself to the export of cotton and the import of metals, was typical. By the 1840s, specialized bankers, importers, jobbers, insurers, and brokers had increasingly rendered the generalist merchant of colonial times obsolete in New York and other Eastern cities.[3]

Even as specialization worked to preserve routine and personalized relations among merchants, the enormous rise in the volume of trade in New York City during the second quarter of the century undermined the orderly and familiar marketplace that merchants so valued. This was the era when New York emerged as the commercial capital of the nation and outpaced rival trading entrepôts and financial centers such as Boston, Philadelphia, and Albany. The completion of the Erie Canal in 1825 inaugurated this trend, but it was not until railroad trunk connections to the Midwest were completed in 1849–54 that New York's superiority was confirmed. A magnificent Merchants' Exchange was erected on Wall Street in 1841 to accommodate the expanded business community. By 1858 a new Chamber of Commerce was created to formalize the exchange of information. Gone was the tightly knit merchant aristocracy of the early century—the Gracies, Bayards, Leroys, Minturns, Astors, Stevenses, Lenoxes, Murrays, and Clarksons—evoked nostalgically by septuagenarian Peletiah Perit in his farewell speech to the Chamber of Commerce in 1864. Dry-goods baron William E. Dodge, Jr., included other names in the merchant patriciate of his "Old New York"—that of the 1820s: Griswolds, Howlands, Grinnells, Aspinwalls, Sturgeses. The early nineteenth-century merchants, Perit noted in the sixties, had "held a prominence which, at the present time, is not accorded to those of the same position. . . . I remember well the profound respect which the junior merchants of those days paid to these eminent men, and how scrupulously they bowed to them, whenever, in their offices or on business, they appeared before them." Such deference could hardly survive in the larger, more heterogeneous and anonymous marketplace of mid-century. The creation of the Wall Street money market in the fifties was the consummation of this process of specialization and fragmentation. The money market symbolized an economic reality in which, Alfred D. Chandler, Jr., has observed, "the merchant could no longer do everything himself" and "personal connections [were] no longer enough."[4]

If old Peletiah Perit had cared to, he could have reported a change in the structure of politics equally dramatic and roughly contemporaneous with that in commerce. In early nineteenth-century New York the wealthiest men—usually merchants—were the city's elected political leaders. Few poor men ran for office, their fellows preferred not to vote for them, and the poorest men of the city could by law neither vote nor hold office. Even into the early 1830s, after the suffrage and office-holding provisions in New York City and state had been broadened to include nearly all male citizens, merchants continued to dominate the urban political landscape. They ran the city council, served on ad hoc ward committees that distributed local charity during hard times, and even appeared at the head of political mobs, as in the case of Simeon Draper, the Whig nabob who led an 1834 election riot.[5] At the outset of Andrew Jackson's presidency in 1829, the assumption of the Founding Fathers that the community's social and economic leaders would also speak for the citizenry in political matters was still accepted by many New Yorkers. By 1850 this was no longer the case. A new class of professional politicians, many of them risen from the ranks of the humblest mechanics, controlled the city's elective offices and the apparatus of the Democratic Party, the mass party that organized political life for a majority of New York's voters. The electorate that elevated these new party professionals to power was also vastly expanded; by the early fifties it included many of the immigrant poor who flooded into the city in the previous decade. In the forties many wealthy merchants began to abandon or eschew elective office, or to distance themselves from public affairs altogether. William E. Dodge, Jr., pleaded pressing business and ill-health when the Sixteenth Ward Whigs tried in 1844 and again in 1845 to convince him to accept a nomination to the Common Council. When Fernando Wood was nominated for mayor in 1850, merchant Philip Hone, born during the last days of the American Revolution, despaired not just at the elevation of a scoundrel to public prominence but at a new mass politics—"our blessed universal suffrage"—which would "drive away Whigism, Conservatism, and good, honest Democracy as we formerly knew it." By mid-century, Jacksonian democracy had forever altered the city's political life, and merchants who sought political influence would have to confront or come to terms with the new mass parties.[6]

It is striking how many of the merchants and lawyers in the Belmont circle began their careers outside of New York City and moved to the metropolis in the late 1830s and 1840s, when the transformations of local commercial and political life were in full swing. August Belmont himself arrived from Germany as the appointed American agent of the House of Rothschild in 1837. Many others came from the small cities and especially the rural towns of upstate New York and New England: Samuel J. Tilden in 1834, George Law in 1837, Samuel L. M. Barlow in 1842, Edwards Pierrepont in 1846, Royal Phelps in 1847, Watts Sherman (and possibly William Butler Duncan too) in 1851, and John A. Dix probably not until the early 1850s. Southerners Richard Lathers, James T. Soutter, and Gazaway B. Lamar relocated to New York City in the forties and fifties. The leading Democratic businessmen of 1863 were frequently men of rural origin who arrived in the city when the more anonymous marketplace and mass politics of mid-century were largely accomplished realities.

Further, these businessmen felt comfortable working with the party professionals who dominated Democratic councils and closely cooperated with Tammany leaders in response to the riots during summer 1863. Augustus Schell and Edwards Pierrepont served as Grand Sachems of Tammany Hall, but more typical were men like Belmont and Barlow, who only occasionally appeared at party functions but revelled in the details of party organization, factionalism, and financing. The economic and political circumstances of the 1850s offered these Democratic men of means an extraordinary opportunity. They were not solely responsible for the innovations in the mass marketing of railroad securities that gave rise to the Wall Street money market in the fifties (some Republicans or Republicans-to-be were involved in these activities as well), although Barlow, Belmont, Dix, Sherman, Tilden, William Butler Duncan, and John J. Cisco were all prominent in this field. But these investment bankers and railroad attorneys were distinctive because they so often served as the captains and lieutenants of the Democracy, close advisors to the Buchanan administration, and strategists of the state and local party organizations. By the late fifties they wielded economic and political power in rare combination, with immediate access to the presidential ear in Washington, European capitalists, the New York merchant and banking community, and the Democratic politicos in the city government. At that moment these Democratic businessmen from New York probably enjoyed as much influence as any urban elite in American history.[7]

The Union League Club disparaged *arrivistes* such as Belmont and hoped to form itself around a core of "men of substance and established high position socially," in the words of founder Frederick Law Olmsted. It was ironic that Olmsted should designate himself spokesman for the city's native aristocracy—he and his co-founder Henry Whitney Bellows were New Englanders who arrived in New York in the late thirties and early forties, and Olmsted did not reside in the city for any length of time until 1857. E. L.

Godkin, editor of the *Nation* and a towering influence among the Union League set in the mid-sixties, was born in Ireland in 1831, arrived in the United States in 1856, and, though admitted to the New York Bar in 1858, supported himself through freelance journalism. But there was indeed a cluster of long-established "leading merchants," as George Templeton Strong called them, who helped bring the Union League Club into being. The shipping merchant Robert Bowne Minturn and tea merchant George Griswold, whom Strong repeatedly described as the socioeconomic nucleus of the Club, were active members; Minturn was the Club's first president. The Minturn and Griswold families had been prominent in New York City commerce since the eighteenth century, as had those of members William H. Aspinwall, John Austin Stevens, and future Club president John Jay. Many other Union Leaguers had come to New York City with the wave of young merchants who emigrated from New England in the decade after the War of 1812; Jonathan Sturges, Thomas Tileston, Moses Grinnell, and William E. Dodge, Jr., were themselves among the most established merchants of the sixties. Of course, even among merchants, there were exceptions to this distinction between the Belmont circle as newly arrived and the Union League Club as native or long established. Two of Belmont's allies, the conservative and Democratic bankers Moses Taylor and Henry G. Stebbins, were both born in the city to families that had belonged to the early nineteenth-century merchant elite. The Union League Club included men such as the millionaire stove merchant Isaac Sherman, an ex-Democrat born near Troy, New York. Still, all told, the Belmont circle would have been hard put to demonstrate that it spoke for a native aristocratic class of New York City, while the Union Leaguers could make that claim with relative ease. Many Union Leaguers had ties to the early nineteenth-century merchant elite that Peletiah Perit evoked in his farewell address or had lived in the city long enough to recall a local commercial world much more intimate than that of mid-century.[8]

By the fifties, then, the two groups of merchants at odds during the riot week were situated very differently with regard to the great changes of the second quarter of the century. Belmont and his circle, newcomers and outsiders by the measure of the "Old New York" patriciate, found themselves privileged by the altered economic and political circumstances of the forties and fifties. With their prominence in the ruling national and local Democratic councils and their importance on Wall Street, Belmont and his friends were now the dominant voice of the New York merchant community. Put simply, the Belmont circle of 1850 was a new class in social terms but an established group in cultural and political terms. The wartime Union League Club represented an attempt by an older merchant and professional class to challenge Belmont and his friends and reconstitute the urban gentry.[9] The Union League Club sought to create a new culture out of the established stock of "Old New York" in amalgam with new elements: the Olmsteds, Bellowses, and Godkins. These intellectuals, immersed in new scientific and literary developments, would enliven the authority of New York's men of "established high position" and give credence to this elite's new claim to recognition as a national intelligentsia. Of course it also worked the other way round: migrant intellectuals such as Godkin and Olmsted hoped patrician merchants would anchor them in society, sponsor their cultural ventures, and give added authenticity to their ideas about authority. In contrast to the Belmont circle, the Union League Club represented an established class in social terms but an emerging group in cultural and political terms.[10]

To understand how Belmont and his friends, so well positioned in the fifties, were challenged by the Union League Club and why tensions between the groups peaked in summer 1863, we have to trace the cultural aspects of this conflict in some detail. How was it that the Belmont circle came to oppose martial law, conciliate the white and Irish rioters, and neglect black victims? That the Union Leaguers came to champion military rule, recommend bloody punishment of the Irish rioters and their well-to-do sympathizers, and use the Club to raise a black regiment and march it through the city? The two groups of businessmen not only experienced the transformations of the thirties and forties differently—they also perceived them differently. Belmont and his associates viewed those changes with great confidence. As long as an agrarian-based market economy and all-white polity continued to expand without government interference, riots—and indeed all social and sectional conflicts—could be easily accommodated. The Union Leaguers regarded the same changes with worry. If that expanding economy and polity were not secured by an active state and a battery of new urban institutions such as the Union League Club, rioting and sectional controversy would quickly lead to anarchy. The Belmont circle's confident outlook, including a faith in free trade, commitment to a decentralized agrarian polity, international republicanism, government restraint in religious and cultural affairs, a paternalist attitude toward the white poor, and white supremacy, was dominant among merchants in the fifties. In that decade anxious patricians, usually of Whig antecedents, began to question those values and imagine a new culture. The intensifying social and sectional conflicts of the mid- and late fifties helped bring this new patrician culture to life. But it was not until the war that the patricians had their grand moment of opportunity and Belmont and his colleagues were thrown on the defensive. The draft riots occurred just as the new culture of the patricians, now articulated by the Republican Union League Club, was fully elaborated and mobilized against the Belmont Democrats and their outlook. It was this battle between two cultures that brought businessmen to the Merchants' Exchange on Tuesday of the riot week to debate a declaration of martial law—and made the stakes of that debate so high.

The Belmont Circle

On August 9, 1853, eighty-nine New York City businessmen invited August Belmont, the newly appointed Charge d'Affaires at The Hague, to a dinner in appreciation of his contribution to the "business and social relations" of the community. The subscribers to the dinner included a broad representation of Democratic merchants—Moses Taylor, Watts Sherman, Wilson G. Hunt, William B. Duncan, James T. Soutter, and Royal Phelps among them—as well as Whig shipping merchants Henry Grinnell and Marshall O. Roberts. Their letter of invitation commended Belmont's rise to "high position" as a merchant and banker and praised his "personal deportment in private life . . . that of an accomplished gentleman, who knew and practiced the social virtues which adorn the character of a citizen." The subscribers were acknowledging both Belmont's contribution to metropolitan commerce as a source of finance capital and foreign exchange and his aristocratic mien, which lent European pedigree and legitimacy to the upward strivings of the city elite.[11]

Belmont's reply to the invitation captured the ebullient mood of Democratic merchants in the fifties. Such were the imperial aspirations of New York and the nation:

> Our mighty empire, stretching from the rolling surf of the Atlantic to the placid waves of the Pacific [now] comprises a confederacy of thirty-one States. . . . Our steamers, after having successfully striven for supremacy with those of England, carry to every sea the proud flag of our country. . . . The far West is brought to our very doors by a net of railroads, carrying the products of its virgin soil to our seaports, from thence to provide the marts of the world. And the time is now at hand when . . . our possessions on the Pacific will be united to their sister States on the Atlantic by a public work which, in grandeur of conception and execution, will outstrip any human enterprise the world has yet seen. We shall then carry through the very heart of our teeming population not only the treasures of the California mines, but also the rich products of China, India, and I trust, Japan. The history of the world has sufficiently proved that any nation which can secure to itself the trade of the East is sure to command the trade of the world.

The Rothschild agent concluded with a glittering tribute to America's "beneficent Republican institutions." No group of nineteenth-century entrepreneurs was immune from anxiety over the moral and religious consequences of material prosperity and economic growth, but Belmont and his associates devoted little time to such concerns, to judge from both their public and private communications. Nor did the heterogeneity and anonymity of a large and rapidly growing mass society worry these businessmen. A fundamental confidence in the commercial and territorial expansion of an agrarian-based market economy inspired their cultural and political outlook. Perhaps because they were newcomers to the city and to the commercial elite and because they were Jacksonian Democrats, Belmont and his friends took the new mass markets and mass parties for granted and viewed them as unmitigated opportunity.[12]

August Belmont had reason to be confident when he sketched America's commercial future for his merchant colleagues in 1853. The prospects for a United States free-trade empire with New York City as its commercial entrepôt and exchange, the American South and West as its agrarian heartland, Great Britain at its industrial center, and mainland Europe, South America, and the Asian Pacific Rim as its periphery were bright by 1846 and luminous by early 1848. By the end of 1846 a Democratic government in Washington had enacted the low-duty Walker Tariff long demanded by New York merchants and seized military control of Northern Mexico and California, in one stroke expanding the productive area and nearly doubling the territory of the nation. Great Britain's repeal of the Corn Laws that year opened the world's first industrial nation to agricultural imports and unrestricted trade. The final condition for a free-trade empire was satisfied by the successful European revolutions of February 1848. The new European republics promised to sweep aside feudal commercial restrictions and unite the continent in a network of *Zollvereins*. The tide of political reaction in Europe during the early fifties only partially clouded this picture. Once the transcontinental railroad projected by Belmont was complete, the commercial vistas of New York would encompass two oceans and large regions of both hemispheres.[13]

It is difficult to tease out and examine the free-trade strand of the Democratic commercial elite, since most mid-century merchants, regardless of cultural outlook, endorsed international free trade and tariff reduction as a remedy for problems in the balance of trade or foreign indebtedness.[14] But unlike their merchant adversaries, Belmont and his associates gave free trade wide currency as a solution for urban questions. In October 1850, with the Industrial Congress still active and memories of the summer tailors' strike fresh, the *Democratic Review* addressed the questions of "Land and Labor Reform." The high-toned Democratic journal proposed free trade and tariff reform as a response to the Industrials' critique of the wages system. No attempt was made to reconcile the non-competitive solutions of the Congress—regulation or abolition of the subcontract system, cooperative production and exchange, and land reform—with the baldly competitive assumptions of the free-trade program. Instead, the *Review* looked to the promise of the nation's agricultural regions for answers to the decidedly urban problems posed by the labor movement. Stimulating the productivity of farm labor through low agricultural tariffs, the *Review* concluded, would allow rapid accumulation of "surplus money and capital applicable to manufacturing," rewarding wage earner and employer alike. As the rise of New York Port had proceeded in tandem with the expansion of commercial agriculture, these merchants believed, the countryside held the key to the future growth and prosperity of the city and the nation. The confirmation of the United States as "the granary of the world," in John A. Dix's phrase, promised deliverance from urban social tensions through untold national prosperity. In this laissez-faire view, any government interference with the economic circuit between American farmlands and British factories invited social chaos in the community.[15]

For Democratic merchants, the agrarian context of metropolitan development had moral significance. Some looked to the plantation South as a political and cultural point of reference. Richard Lathers invoked his early South Carolina experiences long after he committed himself to a Northern urban career. His admiration for the gentility of the Southern planter lost no luster as he rose in New York commercial circles during the 1850s and 1860s, becoming president of the Great Western Marine Insurance Company. Calling himself a "Union Southern Man," he regarded South Carolina's slave-holding rice planters as "gentlemen of culture," dispensing a "liberal and refined hospitality" and upholding "a remarkably high standard of public and private integrity." He saw, by way of contrast, "real danger that energy in the accumulation of wealth [would] bring with it the business frauds and corruption so prevalent in Northern cities." Lathers particularly admired the planters' relationship with the white yeomen who enjoyed "prosperity and refinement [under the antebellum] . . . patriarchal regime." Lathers and other Southern-born merchants such as Gazaway B. Lamar and James T. Soutter were not surprisingly New York City's most strident defenders of slavery, looking to the journalistic guidance of proslavery editors Gerard Hallock of the *Journal of Commerce* and Thomas P. Kettell, first of the *Democratic Review* and then of the *Merchants' Magazine and Commercial Review*. Lathers called Hallock "a fine exemplar of what was best in New England Puritanism" and applauded his contribution to the Southern Aid Society, formed when the American Home Missionary Society withdrew its support from slave-holding churches. It is safe to presume that such merchants had no quarrel with the opening premise of Kettell's 1860 *Southern Wealth and Northern Profits,* that "the history of the wealth and power of nations is but a record of slave products."[16]

Other Democratic businessmen, many of them "Barnburner" or Free Soil Democrats of 1848, looked to upstate New York rural experiences as their political and cultural lodestar. As Democratic State Committee Chairman, Samuel J. Tilden announced to a Columbia County gathering in fall 1868, "it was among the farmers in Columbia that I took my first lessons in politics." Some years later, Tilden declared that he was "for 300 years, descended of an unbroken line of farmers. . . . I became familiarly acquainted with their wants and their character, and early learned to share with them the class of opinions derived from Jefferson and strengthened by Jackson, which taught me to believe that the ideal state of human society was where the community was chiefly constituted of little proprietors of the soil." For this urbane Wall Street lawyer such statements seem to have been deeply felt—much more than passing campaign rhetoric reserved for rural audiences. Another New York City railroad attorney and financier, John A. Dix, echoed these agrarian sentiments in an 1857 talk before the New York State Agricultural Society at Albany on "the importance of the foreign grain and provision market to the farmers of the United States." Morgan Dix would later recall that his father "completed his preparation for the half century of responsibility and toil in the tranquility and comparative obscurity of rural life. To that life he was always devoted. . . ." Tilden and Dix's ideal society of "little proprietors of the soil" would, in the words of Tilden, end political and economic "centralism" and reestablish "the twin principles of local self-government and individual liberty. . . ."[17]

Similarities between Richard Lathers's agrarianism and that of Samuel Tilden and John A. Dix may have been more important than the obvious differences. No doubt the upstate yeomen whom Tilden and Dix idealized were a world apart from the genteel planters Lathers admired. Tilden and Dix did not share Lathers's unquestioning pro-slavery position, though they did not think slavery an unmitigated evil either (since the institution did acknowledge what Tilden and Dix regarded as the separateness of "an inferior race"). But these three—and other members of the Belmont circle—hoped to create or preserve a decentralized polity of small, independent communities that would set the moral and political tone for the cities and the nation. In the name of decentralization and local autonomy, these merchant Jeffersonians opposed any interference with the regional social and political control of Southern elites. This respect for Southerners' local prerogatives was confirmed at the December 1860 "Pine Street Meeting" convened by a committee including Dix, Lathers, Hallock, and Soutter to reassure Southern leadership of the good will of Northern businessmen. At the meeting Republican and Democratic merchants alike conceded that Southerners had a right to depart in peace should efforts at compromise fail. The size and multipartisan character of the gathering further suggest that, on the eve of the Civil War, this agrarian outlook was the dominant one among city merchants.[18]

Alongside free trade and an identification with agrarian interests and values, international republicanism was the Belmont circle's third article of faith. Agrarian-minded merchants rallied to the cause of "Young America" during the late 1840s and 1850s. Merle Curti correctly emphasized the frontier character of that movement. Stephen A. Douglas, the Illinois senator and attorney for midwestern railroads, was Young America's leading spirit. Douglas relied upon the support of congressional colleagues from Ohio, Mississippi, Alabama, Tennessee, California, and his home state Illinois.[19] Young America's immediate project was the construction of a transcontinental railroad, which Douglas and his supporters saw in its broadest context:

as an international avenue joining free-trade republics in both hemispheres. By breathing life into dying republican experiments in Europe and inciting rebellion in European dependencies in the Americas, the movement sought to create the proper international political climate for the commercial and territorial expansion of the United States. August Belmont was the leading merchant partisan of Young America in New York City, a close confederate of movement leaders such as Douglas and James Buchanan, an untiring advocate of aid to exiled European revolutionaries and the purchase or annexation of Cuba and Nicaragua, and a champion of a reorganized and expanded United States Navy. Belmont's enthusiasm was matched by merchant colleagues including Prosper M. Wetmore, Royal Phelps, and John A. Dix.[20]

In 1852 Young America failed to elect to the presidency either of its two favorite sons, Douglas or Buchanan—but it found a sympathizer in Franklin Pierce. President Pierce appointed the ultranationalist adventurer George N. Sanders Consul at London, former Secretary of State James Buchanan Minister to England, the Cuba filibusterer Pierre Soulé Minister to Spain, and the Southerner who coined the phrase "Young America," Edwin de Leon, diplomatic agent in Egypt. New York merchants were represented by Belmont at The Hague, by lawyer Daniel Sickles as Secretary to the London legation, and by publicist John Louis O'Sullivan as Minister to Portugal. Pierce asked John A. Dix to serve first as Secretary of State and then as Minister to France, but Dix declined both appointments lest his free-soil sympathies alienate the administration's Southern support. William Marcy, Pierce's ultimate choice for the State Department portfolio, was not inattentive to the claims of the Young America merchants. Pierce's successor James Buchanan encouraged a continuation of Young America's influence in the highest councils of government. Heady with the political possibilities of the moment, Belmont exulted to George N. Sanders that it would not be long before the United States found it necessary to throw "her moral and physical force into the scale of European republicanism."[21]

Belmont and his Young America associates were confident that the free-trade empire could buffer a variety of social and political conflicts. If freed from all constraints, America's market economy would act as a solvent against the hardened divisions that haunted mid-century European societies. This optimism underlay the Belmont circle's hostility toward state interference in religious and cultural affairs and paternalist attitude toward an assertive white poor.

The Belmont circle's opposition to a moralizing state had its intellectual origins in notions of religious tolerance and free thought that emerged during the Enlightenment of the eighteenth century. In his *Notes on the State of Virginia,* Jefferson had proclaimed that "the way to silence religious disputes is to take no notice of them," and that the duty of rational men was to oppose "tyrannical laws" interfering with religious belief and practice.[22] While Belmont was generally silent on religious and cultural matters, Tilden and Dix became leading expositors of the Jeffersonian position for their generation. As early as 1837, Tilden spoke against a bill presented to the New York State legislature requesting special privileges for the Society of Shakers in the entail of property to sect members. Tilden defended the right of the Shaker Society to maintain its communitarian theories of property under the laws of the state but refused to offer special legal protection and encouragement to the Society. As candidate for state attorney general eighteen years later, Tilden wrote a letter to representatives of the New York Prohibition Party opposing the recently enacted state temperance law. "Such legislation," Tilden insisted, "springs from a misconception of the proper sphere of government. . . . The whole progress of society consists in learning how to attain, by the independent action or voluntary association of individuals . . . [a] lessening [of] the sphere of legislation and [an] enlarging [of] that of the individual reason and conscience."[23] Dix painted the moralizing state in more threatening hues in an 1839 speech on "The Progress of Science." Christianity had positive influence on the political condition of mankind, he observed, but only through the silent and inscrutable workings of Providence. When religion had been made the "confederate of monarchs," the inevitable result was extirpation of "differences of opinion by fire and sword."[24]

The Republican Party, with its roots deep in evangelical Protestantism, was a Jeffersonian nightmare. The Pine Street merchants predicted failure for the new party: "It cannot possibly remain unbroken during the term of the incoming administration. The two chief elements—the political and the religious—can never harmonize in practice." Nor did the Democratic Party or the Catholic Church have license to mix politics and morality. The Belmont circle insisted that the Catholic Church, like the Shaker Society or the evangelical Protestant sects, receive no special state sanction. The Belmont circle even accommodated a few nativists within its confines—artist-inventor Samuel Finley Breese Morse, veteran of the Native American Democratic Association of the 1830s, most prominent among them. Morse held in equal contempt the Republican Party and the Catholic Church of Pius IX, which included in its assault on liberalism a censure of secular education and the separation of church and state. Making no exceptions, these mid-century Jeffersonians sought wholly to separate government from religious and ethnic affairs.[25]

The most eloquent testimony to the confidence of these businessmen—doubtless related to their tolerance of religious and ethnic division—was their paternalist attitude toward an increasingly assertive white poor. Free trade and the expansion of commercial agriculture were, as we have seen, the Democratic elite's solutions to the labor uprising of 1850. The financial panic and depression of fall and winter 1857 and the demonstrations and rioting that resulted would further exhibit the free-trade empire's tolerance for class conflict.

By the end of October 1857, 25,000 city workers had lost their jobs, and businesses were failing at the rate of 150 a week. Over the next three months, the Association for Improving the Condition of the Poor reported that over 40,000 homeless and jobless poor sought shelter in police stations. Food prices remained high, as the collapse of the credit system prevented the shipping of Western grain. Eastern cities confronted scenes of destitution without American precedent. "Our city," wrote New York reformer Robert M. Hartley, "presented a more appalling picture of social wretchedness than was probably ever witnessed on this side of the Atlantic."[26]

"What shall be done for the poor?" asked the *Journal of Commerce* and scores of other writers and reformers as the depression worsened. Mayor Fernando Wood gave sudden focus to the gathering debate over poor relief with his October 22 proposal to the Common Council. Wood's suggestion that the city government take responsibility for hiring the unemployed poor was not unprecedented, but the broad scope of his plan was new. The mayor proposed that the city hire unemployed laborers to assist in the building of streets, engine houses, and police stations, the repair of docks, and the construction of the new reservoir and Central Park; workers would receive their wage in both cash and truck—cornmeal, flour, and potatoes—to guard against the "imposition of middlemen." The city would fund the project by issuing construction stock redeemable in fifty years. More controversial than the scope of Wood's proposal was its tone and political vocabulary. The October 22 message ended with a dramatic overture to the laboring poor that would reverberate in the memory of the metropolitan elite for years to come:

> Truly it may be said that in New York those who produce everything get nothing and those who produce nothing get everything. They labor without income whilst surrounded by thousands living in affluence and splendor, who have income without labor. . . . Now, is it not our duty to provide some way to afford relief?

In large part Wood sought to awaken in the better classes a sense of paternalist responsibility for the welfare of the poor. Public employment would allow "no man excuse for violence or depredation upon property" and might prevent a reenactment of the 1837 bread riot (an evening-long raid on several flour stores that had to be put down by the militia) or, far worse, a homegrown version of the Parisian June days of 1848. But Wood's assault on the indolence of the wealthy, proposed measures against middlemen, and loose invocation of the labor theory of value were directed not at a sluggish paternalist elite but at an aroused working class. Plausibly, Wood hoped to confirm the allegiance of the workingmen and -women who in July 1857 had collaborated in his unsuccessful opposition to a Republican Metropolitan Police. On the eve of the December mayoral elections Wood was reaching out to the increasingly vocal and visible poor.[27]

Though Wood's political fortunes rested more and more with immigrant workers, his social outlook in many ways reflected the interests and values of Democratic merchants. This elite cooperated with Wood through the 1850s and accepted the genteel ex-saloonkeeper as a man of their own ideological stamp. The Philadelphia-born Wood had prospered as owner of a dockside grog shop in the late 1830s, invested in sailing vessels, and by the late 1840s had emerged as a shipping merchant of small repute. Though Wood never ventured too far politically from the immigrant longshoremen who patronized his Washington Street grocery in the thirties and sent him to Congress in the forties, he was elected mayor in 1854 and again in 1856 as the favorite son of some of the city's most powerful merchants. During his first term at City Hall, Wood cultivated a reputation as an earnest albeit light-handed reformer. His mayoral program included a modest campaign against prostitution, support of the planned Central Park, and proposals for a municipal university and a free academy for young women. The Mayor even spoke briefly for enforcement of existing liquor laws before recanting in 1855 and joining Tilden in opposition to the "Act for the Prevention of Intemperance, Paupers and Crime" passed by the state legislature that year. He was also a champion of free trade and hard currency and fought for both causes in the House of Representatives well into the 1870s. In the fall of 1856, merchants, bankers, and other property-holders including William B. Astor, Peter Cooper, William M. Evarts, James W. Gerard, William F. Havemeyer, Wilson G. Hunt, Commodore Vanderbilt, and, incredibly, Horace Greeley gave Wood their imprimatur at a public meeting after Wood successfully battled upstate forces for control of the municipal police. Through the mid-fifties, then, Wood was the candidate of the Democratic merchants and not a few other business and political leaders besides.[28]

Businessmen and journalists of all political persuasions applauded the substance, if not the inflammatory tone of Wood's October 22 public works proposal. Earlier that month Horatio Seymour had begun to consult Western grain merchants and New York bankers to devise a way to transport Western grain to the metropolis in order to ward off the riots anticipated during a fall and winter of prolonged unemployment and high food prices. Gerard Hallock's *Journal of Commerce* denied that it was "the business of [the] city to provide employment for all people who chose to come and stay [in New York]," but approved efforts "to give out such work as can be done at this time as well as any other." Uneasy with Wood's denunciation of the wealthy, *Times* editor Henry J. Raymond—the voice of many Republican merchants and financiers—thought the Mayor's public works plan "eminently proper," as "a popular Government should not be behind the monarchies and empires of Europe in care for the people, especially in times of distress like the present." Businessmen of both parties agreed with George Francis Train's terse warning in *Young America in Wall Street,* written late that fall: "The people must be taken care of." Without some plan for liberal aid to the poor, ran the prevailing sentiment on the Merchants' Exchange, violence would follow.[29]

The merchants did not have long to wait. The unemployed who gathered in Tompkins Square in early November became more impatient each day that Mayor Wood's promised relief failed to appear. A "Workingmen's Committee" they dispatched to meet with Wood produced no tangible results. On November 6 a large crowd marched down Wall Street to the steps of the Exchange and demanded that bankers lend funds to businessmen who would employ the poor. Wood left the demonstrators unmolested but posted guards in front of government buildings and flour warehouses. On the 10th, the federal government moved to protect its property, the Customs House, with one hundred troops under Mexican War hero General Winfield Scott. The crowds could contain their frustration no longer: the next day witnessed an old-fashioned bread riot stirring memories of the charge on flour stores twenty years earlier.[30]

The demonstration and riot forced some merchants to reassess their support for Wood and his handling of the crisis. On October 27, Raymond was already drawing dark parallels between Wood's "communism" and that of the Parisian 'forty-eighters; after the confrontation on Wall Street he accused the mayor of exciting "the worst passions of the destitute and the ignorant" and flattering them "by promises he knows can never be fulfilled." The *Times* editor now began a search for a respectable Democrat to lead a nonpartisan challenge to Wood in the December election. He and a new "People's Union" Party found their candidate in the solid burgher Daniel F. Tiemann. Son of a poor German immigrant, Tiemann had risen to social prominence through diligent work and integrity. A nativist in the forties and a lifelong teetotaler, Tiemann could appease the American Party; but the foreign-born middle class also embraced him as one of their own. Son-in-law of the venerable reformer Peter Cooper, the paint manufacturer was sufficiently high-toned to satisfy even the fastidious elitist George Templeton Strong. Tiemann's candidacy was a well-crafted maneuver by merchants and their allies to reestablish social and political control after the disturbances of early November. Quite a few of the Democratic elite—John A. Dix, William F. Havemeyer, Daniel E. Sickles, Samuel J. Tilden, and John Van Buren—endorsed the People's Union movement. It surprised virtually no one that the unexceptionable Tiemann outpolled Fernando Wood by over 2200 votes.[31]

Whether or not they remained loyal to Wood on Election Day, Democratic businessmen applauded Wood's circumspect response to the November 6 demonstration and continued to favor use of the city government to administer poor relief. While Raymond clamored for the "cannonades of Cavaignac" on the 7th, the Democratic merchant Royal Phelps approved Wood's patient negotiations with the unemployed—so much so that he published a letter he received from Wood detailing strategies for responding to public disorder. In the letter, the mayor assured Phelps that a tolerant approach to the Wall Street demonstration would preserve civil peace:

> Whilst I regret that the persons composing these processions have thought proper to take this course in making their wants known, I do not participate in the apprehension of some, that the public peace is thereby jeoparded, or private property endangered; nor indeed that riot will ensue. I do not doubt my ability to prevent the one, and to protect the other, under a far more serious state of things than has yet occurred.[32]

Like Phelps, John A. Dix thought the Tompkins Square demonstrations required close monitoring but not forceful intervention by the authorities. An acute observer of working-class movements, Dix informed President Buchanan on the 10th that the Workingmen's Committee would run aground of its own weight: "This city is going through a somewhat threatening, and perhaps a perilous process of agitation growing out of the necessities of the unemployed laboring poor, but the extravagance of the demands of some of their leaders, as developed by the movements of the last 24 hours, will, I think, have a powerful influence in preserving the public peace." Fernando Wood shared with merchants such as Phelps and Dix a deliberative and forbearing approach to popular social movements. These men preferred to divert uprisings—beginning perhaps with an acknowledgment of some part of the complaint, followed with a period of waiting and observation, relying throughout upon a formula of accommodation and finesse to maintain public order. Close students of crowd behavior, these merchants remained confident that they could preserve their social and political authority without resort to coercion, even in the midst of a "perilous process of agitation."[33]

Dix provided the most systematic explanation of the Democratic merchants' paternalist response to the crisis in a letter to the American Industrial Association on November 9, three days after the march on Wall Street. He laid out three principles that most of the merchants at the Belmont farewell dinner would have applauded. First, the better classes and, by extension, the city government, had a duty to care for the poor during a business depression. President of two Western railroad lines, Dix acknowledged that he and his fellow merchants were partly to blame for the frenetic speculation and ensuing economic collapse and were consequently obliged to aid innocent victims of rapid economic expansion: ". . . no public or private effort should be spared to sustain those who are suffering from the prevailing derangement without having in any manner been instrumental in producing it." Second, a custodial elite had to set the limits of popular expectation during a period of crisis. The militancy of the unemployed workers could be explained by Wood's failure to draw "the boundaries within which the action of the municipal authorities [might] be properly solicited." Debtors would have to "settle their balances," and those who "lived beyond their means" would have to "pay the penalty of their extravagance and folly"; public works should be limited to those projects commenced or previously planned. With these restrictions, Dix saw no conflict between public employment of the poor and the dictates of laissez-faire. But he found "mischievous" Wood's claim that "those who produce everything have nothing" and his implication that government should relieve such inequality. The American social system was founded "upon the assumption that the general welfare would be best promoted by leaving with the governed the largest liberty compatible with the public safety." The diversity resulting from such laissez-aller policy represented a positive social value. In a loosely disguised criticism of Wood, Dix demanded that "if there are any . . . who by doctrine or practice would substitute for this great catholic principle of [republican] equality others more narrow, who would excite the laborer against the capitalist, the rich against the poor, or the native against the citizen of foreign birth, they are no friends of the republic and should be held by all lovers of the public order as fomenters of civil discord and disorganization." Finally, the free-trade polity could accommodate an ample measure of social conflict so long as popular anticipations were limited. In publishing his letter to the American Industrial Association, Dix sought to clarify the boundaries of legitimate paternalism: the government would not provide employment to all who lacked it, nor would it redress the maldistribution of wealth in the community. Had Wood first established the merchants' proper duties to the poor, Dix concluded, the equilibrium of expectation between the classes would have remained intact and the public order preserved.[34]

Dix and other Young America merchants sympathetic to Mayor Wood's program thus came to suspect his paternalist "technique" in late October and early November. Wood's extravagant and open-ended promises to the unemployed and his class-charged rhetoric deeply unsettled these men. Daniel E. Sickles, a foe of Wood in the scramble among city Democrats for federal patronage, wrote President Buchanan on November 20 of the "very strong desire among the best men in [the] party to get rid of Wood as Mayor, provided a better democrat and a reliable man can be got in his place." The Democratic businessmen who abandoned Wood for Tiemann also questioned the mayor's value to the party after so many leading merchants had repudiated him. On December 8, August Belmont confided to his relative, Louisiana Democrat John Slidell, "Wood is a d———d man, & no action of the Executive can galvanize him into political life again." Wood's penchant for political jobbery further distanced him from Dix and his associates. The Belmont circle hoped to find a political leader with Wood's bold vision of merchant paternalism but tempered, as Sickles put it, with "reliability." Though Wood recovered from the political defeat of 1857, he had begun to lose the trust of some of his merchant supporters.[35]

The confident paternalism of the Democratic merchants was accompanied by a white-supremacy proviso. In his Pine Street address, Hiram Ketchum assured Southern leaders that the best friends of the cotton states were to be found in the nation's free-trade capital, New York City: "We are neither a Northern nor a Southern city. Here is a field where all may mingle." The most meaningful connection between New York and the South, Ketchum continued, was racial. "If ever a conflict arises between races, the people of the city of New York will stand by their brethren, the white race. We will never suffer you to be trampled upon by those of another blood." This message was likewise the one Young America broadcast to its commercial outposts around the world. The British trade empire had degenerated, observed the *Democratic Review,* into "a den of Musquite empires . . . a vast farago of dependencies." The minority white cultures in Central and South America, the *Review* predicted, would be subdued or exterminated "unless shielded by a more wise and energetic race." The American free-trade empire, unlike the British, would bring the active domination of white and Anglo-Saxon culture to its commercial dependencies. "The United States," the *Review* concluded, "are the champions of the white man in the New World." The prerequisite for the cultural and social cohesion of the free-trade empire was racial homogeneity.[36]

The common commitment of many New York merchants and Southern leaders to free trade and white supremacy made the political choices of the secession winter of 1860–61 especially difficult. The plan for New York City to secede from the Union as a free-trade republic, publicized by Fernando Wood in a January 1860 speech to the Common Council, was debated openly by some Democratic merchants and considered privately by many more. Morgan Dix remembered his father's reflections upon the secession proposal: "On one . . . occasion he referred to the possibility that New York might become a free city, entirely independent, in case of a general break-up; not that he advocated the idea, but he placed it in the category of possibilities. It was his opinion that a separation, if sought by the South through peaceful means alone, must be conceded by the North as an evil less than that of war." President Buchanan also saw the secession of New York City as a real possibility and sought to reassure New York friends concerned for their commercial future. Buchanan wrote Royal Phelps in late December 1860, "I cannot imagine that any adequate cause exists for the extent and violence of the existing panic in New York. Suppose, most unfortunately, that the Cotton States should withdraw from the Union; New York would still be the great city of this continent." The President advised *Herald* editor James Gordon Bennett that "if the merchants of New York would sit down calmly and ask themselves to what extent they would be injured by the withdrawal of three or four Cotton States . . . they would come to the conclusion that although the evils would be very great, yet they would not destroy the commercial prosperity for our great Western Emporium." Buchanan most feared that "panic has even gone to the extent of recommending that the great city of New York shall withdraw herself from the support of at least twenty-five millions of people and become a free city."

His fears were not wholly groundless: at that moment there were indeed prominent Democratic merchants who entertained the idea of New York as a free city in the context of a generalized and peaceable breakup of the Union. The high-duty Morrill Tariff rushed through the new Republican Congress and the free-trade provision of the Confederate Constitution further caused some merchants to wonder whether the Republican North would have the city's best interests at heart.[37]

Events in Charleston Harbor on April 12, 1861, changed everything. Eight days after the firing on Fort Sumter, Democratic merchants led the "greatest popular demonstration ever known in America" in defense of the Union. What accounted for this sudden reversal of opinion? One historian has suggested that the city's merchants "all agreed that it was to their interest as businessmen to see that the laws were maintained, the government upheld, and the Union preserved."[38] There is indeed some evidence that merchants' decision to defend the Union was based on economic interest. A commercial attachment to the Western states doubtless contributed to the pro-Union choice of some merchants (though it is worth noting that merchants seldom defined their business interest as a simple preference for trade with Western rather than Southern partners, as Buchanan imagined in his letter to Bennett). In fall 1861 one speaker before the Chamber of Commerce envisioned a political atomization that would "break up the free traffic and unhampered exchange in the territory of the Union," should "the right of secession . . . be established." The attack on Sumter may have raised the specter of an era of prolonged political division and war that would undermine New York's commercial preeminence and reduce the free-trade empire to the status of a New World Poland. A generalized fear of political instability and disruption of trade may have explained the new pro-Union sentiments of some merchants.[39]

Still it is hard to account for the sudden reversal of merchant opinion after the attack on Sumter without some reference to mid-century nationalism. In theory, at least, free-trade principles were quite compatible with free cities and political decentralization. In a December 1860 letter to a Southern businessman, August Belmont pondered at length the possibility of free-city status and concluded that the secession of New York might economically benefit the city's merchants:

> If we did only look to our own material interests and those of our city, we should not deplore the dissolution of the Union. New-York, in such a catastrophe, would cut loose from the Puritanical East and her protective tariff, and without linking her fortunes with our kind but somewhat exacting Southern friends, she would open her magnificent port to the commerce of the world. What Venice was once on the sluggish lagoons of the small Adriatic, New-York would ere long become to the two hemispheres, proudly resting on the bosom of the broad Atlantic, and, I am afraid, sadly interfering with the brilliant but fallacious hopes of the Palmetto and Crescent cities.

Nationalism and not business interest prevented Belmont from endorsing the free-city scheme and accounted for his "warm and undying attachment to the Union." With a final flourish, Belmont informed his Southern friend, "I prefer . . to leave to my children, instead of the gilded prospects of New-York merchant princes, the more enviable title of American citizens."[40]

Not all Young America Democrats relinquished the free-city idea or were swept up in post-Sumter nationalism. James T. Soutter and Gazaway B. Lamar moved south. Soutter fought to maintain a pro-Union position while living in the Confederacy but was allowed to return north only after John A. Dix and other merchants convinced President Johnson to sign a pardon. Lamar became a Confederate blockade runner and the state paymaster of Georgia. Publicist John Louis O'Sullivan moved to England and now lobbied for British recognition of the Confederacy and mediation of the sectional conflict—much as he had worked for American intervention in European political affairs through the fifties. With an optimism no longer grounded in reality, O'Sullivan campaigned for a restoration of the free-trade empire to its antebellum status.[41]

But with these few exceptions, Young America enthusiasts of the fifties became supporters of the Union war effort in 1861. As president of the Union Defense Committee, formed in April 1861, John A. Dix led a unified merchant community in a drive to raise funds, supplies, and army volunteers. Committee member Moses Taylor joined with August Belmont and other bankers in New York, Philadelphia, and Boston to loan the federal government the $150 million necessary for the prosecution of the war. In his capacity as American agent for the Rothschild interests, Belmont labored to persuade British political leaders to support the Union cause actively, and then as Northern military fortunes sagged in late 1862, not to aid or recognize the Confederacy. Democratic merchants remained unconditionally committed to the war effort so long as its aims were restricted to restoring the political integrity of the nation.[42]

Nonetheless, the Civil War interrupted the ambitious program of the Belmont circle and threw these merchants on the defensive for over a decade. Cast from national power, Young America was forced to defer its plans for territorial conquest and watch its laissez-aller political arrangement destroyed by the new Republican government. Belmont and his colleagues opposed the Republican centralization of government through black emancipation, legal tender, federal taxation, and conscription; and, of course, they loathed the protectionist Morrill Tariff. But the first of these policies, the liberation of four million black slaves, became the crucial issue, emblematic of the other changes and threatening the very foundations of the free-trade empire.

The preliminary emancipation proclamation of fall 1862 galvanized the Democratic elite. It signaled the end of the patriotic unity in commercial circles that had characterized the first sixteen months of the war and initiated the contest among businessmen for rightful claim to the title "Merchants of New York." The first hint of the conflicts to come came in early October 1862 when a group of merchant supporters of General George B. McClellan resigned from New York's new "National War Committee." The NWC had been organized by anti-slavery Republicans Simeon Draper and John Austin Stevens, Jr., to promote volunteer enlistment and avoid a state draft. The resigning businessmen suspected that NWC volunteers would be diverted to regiments led by radical generals John C. Frémont and Ormsby M. Mitchell (the previous year Frémont had attempted military emancipation in Missouri). General McClellan, the resigning merchants' favorite, was pledged to winning the war without a revolutionary assault on slave society. August Belmont reclaimed the thousand dollars he had contributed to the NWC recruiting fund; Republicans (and former Whigs) Moses Grinnell and William E. Dodge, Jr., also abandoned the Committee.[43]

Black racial inferiority had long been an explicit component of the Democratic merchants' outlook. But a loud racism became the dominant if not the exclusive message of the Young America candidates who swept the congressional elections of fall 1862. Belmont himself nearly entered the congressional race in the uptown Eighth District but finally deferred to former Whig, Know Nothing, and Constitutional Unionist editor James Brooks. Proclaiming that "this Civil War, the calamitous condition of our country, the negation of the white race and the elevation of the negro over the white man have hushed up or exiled all our past political differences," Brooks presided over a coalition of merchants, ex-nativists and German Democrats and trounced Republican importer Elliot Cowdin by over four thousand votes. Of what might be called New York's class of 1862—the group of newly elected Democratic congressmen that included Brooks, the Wood brothers, Henry G. Stebbins, and John Winthrop Chanler—it was Chanler who made the most brazen racist appeals. A South Carolinian who moved to New York City and married the daughter of William B. Astor, Chanler reportedly asked one uptown audience, " 'Shall the white laborer bow his free independent and honored brow to the level of the Negro just set free from slavery, and by yielding the entrance to the great citadel of our nation, surrender the mastery of his race?' " The gathering of ex-Whigs and Constitutional Unionists under Young America leadership, the veneration of McClellan (especially after Lincoln rescinded his command of the Army of the Potomac in the first week of November), and, most important, the shrill rhetoric of white supremacy—here were the makings of the conservative constituency and position of July 1863.[44]

The primary vehicle for Young America's critique of black emancipation was the Society for the Diffusion of Political Knowledge. The Society was formed in early February 1863 at a fashionable Delmonico's dinner attended by Belmont, Barlow, and Tilden; the week before Belmont had urged Tilden that it was "high time for the conservative men to go to work . . .

to compel the Administration to a change of measures and men." Resembling in name and organizational style an early nineteenth-century British political education society, the SDPK was an old-fashioned effort at influencing opinion by businessmen used to exercising power, uncomfortable with their illegitimate status as a wartime opposition, and distraught as they watched the war pull the merchant elite away from its historic moorings. It was conceived not as an alternative to the Democratic Party, which Belmont now led as national chairman, but as an adjunct, waging a propaganda war against Republican policies with an emphasis on emancipation. The Society's anonymous Pamphlet Five, echoing Kettell's *Southern Wealth and Northern Profits,* argued that emancipation would bring to the American South the "ruin" of New Granada, Mexico, and Jamaica, while slave Brazil, Cuba, and Puerto Rico flourished as commercial marts. With slavery left intact, "THE MUSCLES OF THE NEGRO AND THE INTELLECT OF THE WHITE MAN THUS BECOME THE GREAT AGENCIES OF MODERN CIVILIZATION. The exchange of the products of the one for the other constitutes OUR COMMERCE, gives employment to shipping, erects our banks, lines our streets with marble palaces and makes a rocky island like New York, the seat of untold wealth." As late as 1864 Young America businessmen used arguments like these to justify the preservation of slavery.[45]

While many of these ideas were the stock-in-trade of antebellum proslavery argument, the SDPK attempted to fuse them into a distinctive racial position. Long a component of the Democratic merchants' thinking, theories of white supremacy were invoked as a political solution in the 1860s by merchants rallying to the defense of the free-trade empire. These merchants believed black emancipation to be the most serious of all the Republican attacks on the Young America program—tearing at the very fiber of economy and society. Belmont and his colleagues thought only a white and racially homogeneous free community could weather the severe social and sectional strains of commercial and territorial expansion. A belief in black racial inferiority, wrote the aging nativist Samuel F. B. Morse in one SDPK pamphlet, was common to the thought of such politically estranged men as the President of the United States, Abraham Lincoln, and Vice President of the Confederacy, Alexander Stephens: "Both President Lincoln and Mr. Stevens [sic] are in perfect accord in accepting and acting on the same great truth. President Lincoln accepts the *physical inequality* of the two races as completely as Mr. Stevens. . . ." If only men like Lincoln and Stephens would assert that common "cultural" concern against sectional division and radical measures, the Belmont circle argued in 1863, the expansive free-trade empire of the fifties could be restored.[46]

During and after the draft riots Belmont and his friends remained confident. They retained their faith that an untrammeled market could accommodate social and sectional tensions. But they were beleaguered by the Republican wartime government. The Republicans' aggressive interventions into the economy, the political affairs of Democratic localities, and, above all, race relations were disturbing the foundations of Belmont's resilient society. A Republican declaration of martial law during the riots would not only disrupt the Young America merchants' home rule but further hamper the free and decentralized development they believed necessary for social and political order.

The Rise of the Union League Club

At the same time, Frederick Law Olmsted and his Union League Club associates clamored for martial law and the blood of "Barlow and Brooks and Belmont and Barnard and the Woods. . . ." What Olmsted and his patrician constituency wanted, what gave meaning to the military arrest and punishment of such Democratic leaders, was a vindication of a new merchant outlook that challenged the views of Belmont and Young America.

The first inkling of a patrician challenge to Young America came during the Astor Place riots of 1849, when Whig merchants used the full power of the state to put down the crowd. The Whig elite's response to Astor Place, taken together with statements made in other contexts, revealed a deep anxiety about the rapid expansion and mass conditions of Jacksonian society. During the fifties and early sixties, the Belmont circle's concerns each received scrutiny and revision. Patricians developed a new perspective stressing the moral defects of free trade, the need for an active state and new urban institutions as a remedy, the legitimacy of government intervention in religious and cultural affairs, the assertion of American nationality against "treasonous" Irish immigrants and Europeanized merchants, and the sponsorship of the loyal black poor. The polarization of merchants during the draft riots was possible only after the patricians had fully elaborated their critique of Young America and given it institutional form in the Union League Club.

It should first be said that the new patrician outlook emerged alongside and, in some regards, from within Young America. Both views began from the premise that free-trade tariff policies were economically desirable. Both merchant groups searched for stabilizing means of influence as the sectional crisis disrupted the social and political order. To the extent that merchants of either camp came to endorse the destruction of slavery, it was because they saw emancipation as an inevitable consequence of social change and part of a program for military victory and national cohesion. Few merchants saw inherent value in a campaign to turn Southern society upside down. Both Young America and the patrician critique reflected Marx's dictum that merchant capital has normally been loath to disturb existing arrangements in politics and trade.[47]

That Young America would be challenged by a new merchant perspective was intimated during the Astor Place riots of May 1849. The British actor

William Macready's May 7 performance of *Macbeth* at the Astor Place Opera House was the occasion for an urban theater riot, as longstanding tensions between British actor William Macready and his well-to-do admirers, and the American performer Edwin Forrest and his more plebeian following, finally erupted. In the tumult, Forrest's supporters drove Macready from the stage. This "first" riot of the week proceeded according to the pattern of New York's anti-abolitionist riot of July 1834—including attacks on abolitionists, evangelical reformers and British thespians and omitting only the earlier riot's assault on the black community. The departure from historical precedent occurred not on the 7th but on May 10, in what might be termed the "second" Astor Place riot. On the 9th, forty-nine men petitioned the new Whig Mayor Caleb S. Woodhull to protect with force Macready's subsequent performances in the city. Led by editor Henry J. Raymond, the petitioners were mostly merchants and lawyers and nearly all were members of the Whig Party; only four would join the Belmont farewell entourage four years later. With the blessing of these influential allies, Woodhull guarded Macready's May 10 performance with three hundred police and two hundred state militia. When the throng outside the theater began to stone the building, the military fired—the first time in American experience that soldiers shot point-blank into a crowd—leaving twenty-two people dead. The contest for possession of the Astor Place Opera House occasioned what was in 1849 a new and controversial exercise of the power of the state.[48]

Two aspects of the elite response to the Astor Place riots merit attention. Though some businessmen decried the absence of a riot act to be read to the crowds on May 10, most applauded the action of the city authorities and state militia. Even the *Journal of Commerce* conceded that "there was no other alternative for the authorities but to order the military to fire." In a post-riot sermon, Henry Whitney Bellows offered his merchant congregants a justification for such official violence: "The Christian Patriot feels a greater obligation to the Constitution and to the Law, than to the caprices of a fickle public, and to the Gospel than to either. . . . If ever there was a proper occasion to use the whole power of the State, and to fall back upon the last resorts of civil order, it was when the most willful, least justified and most atrocious mob since the abolition riots . . . attempted to overthrow the laws. . . . Brethren, let us stand by Liberty under Law—whether of the Constitution or the Gospel." This celebration of the use of force against the mob and hint at a divine authorization of state power introduced a main theme of the patrician critique of Young America over the next decade and a half.[49]

The petitioners who backed the harsh measures of May 10 also sought to give new credence and authority to their own brand of urban culture. In 1834, rioters who mounted the stage of the Bowery Theater to protest a British actor were quieted by an apology from the theater manager, who then appeased the crowd with a black minstrel performance that redeemed the Bowery stage as an American place. In May 1849, a far more consolidated and self-conscious merchant elite was in no mood for concessions to what Henry Whitney Bellows dubbed "the caprice, ignorance and brutality of a few hundred lawless boys and men." After the boisterous "Forresters" seized the Opera House on May 7, Philip Hone wrote in his diary, "This cannot end here; the respectable part of our citizens will never consent to be put down by a mob raised to serve the purpose of such a fellow as Forrest. Recriminations will be resorted to. . . ." The Opera House was the jewel of the new upper-class "society" that first migrated north of Bleecker Street in the 1840s. The issue of whether patrician or plebeian notions of culture would prevail in New York had become one of great moment to Hone and his colleagues. The forties were the heyday of what one historian has called "New York's lower class world of rough amusement," featuring Bowery minstrel shows, the pretensions of irreverent B'hoys, the internecine street battles of fire companies, and the rowdy retinues of charismatic popular politicians such as Isaiah Rynders and Mike Walsh. Gazing out from the verandah of his Broadway townhouse on the evening of May 10, 1849, Hone must have felt he was witnessing the climax of a plebeian invasion of his own cultural precincts. What made Astor Place different from preceding riots, then, was the merchants' new determination to reclaim the public spaces of the city from the "caprice" and "ignorance" of demotic usages and assert the dominance of their own standards and institutions.[50]

But the merchants' response to the second Astor Place riot of May 10 may have resulted from the "internal" experiences of the commercial elite as much as from lower-class challenges from without. Merchants' draconian suppression of the Astor Place riot and their campaign for martial law during the draft riots fourteen years later were closely related to their misgivings about a fast-expanding and impersonal marketplace.

The merchants who looked to Henry Whitney Bellows for leadership in 1849 and joined his Union League Club during the war saw the commercial rise of New York Port differently from the subscribers to the Belmont dinner. Where the Belmont circle regarded economic expansion as unmitigated moral and cultural improvement, Bellows and his merchant congregants had their doubts. New York City, Bellows told the mourners at the funeral of merchant Jonathan Goodhue, was a precarious place to live and work. Because of its rapid rise to commercial preeminence, the city, and particularly its merchant quarter, had become a place of high moral risk:

> Commerce is dangerous precisely because of the magnitude of interests involved in it. Money is "perilous stuff," just because it is the representative of all other physical and of much intellectual and moral value. This community of business interests and businessmen is a dangerous and difficult place to dwell in, because those exclusively occupied in dealing with that, which most

> nearly and universally touches the present welfare of millions, feel the passions and wants of the nation pressing back upon them, and shaking with convulsive energy the nerves which they themselves are. . . . Let it be understood that the merchant occupies a post of peril. . . .

Where Belmont's businessman was, a priori, a force of order and stability on the globe, Bellows's businessman teetered on moral catastrophe. If Belmont's marketplace exercised an orderly, centripetal pull on the community, Bellows's market resembled a spinning catherine wheel, its centrifugal energy generating moral chaos. It should not surprise us that Bellows's view of the market was accompanied by a rejection of the Belmont circle's passion for territorial aggrandizement. It was not merely that Bellows's Whig and later Republican comrades saw Democratic expansionism as a Southern ploy to expand the domain of slavery. These merchants also worried that the republic would not be able to withstand the moral pressures—the jingoistic popular and partisan feeling, the aggression and the lust for power—that would attend a war for territorial expansion. Whig merchant Philip Hone put it well in 1845 when he learned that Congress had approved the annexation of Texas: "The Constitution is a dead letter, the ark of safety is wrecked, the wall of separation which has hitherto restrained the violence of popular rage is broken down, the Goths are in possession of the Capitol, and if the Union can stand the shock it will only be another evidence that Divine Providence takes better care of us than we deserve." For Bellows, Hone, and their colleagues, the republic was a fragile vessel that expansion of any sort—commercial or political and military—might fracture.[51]

The patricians' response to Astor Place, taken together with Bellows's and Hone's views of expansionism, suggests that this group was something more than a displaced class nervous about declined status (though George Templeton Strong's view of the German-born Jew Belmont as "a mere successful cosmopolite adventurer and alien" does imply they were that). Bellows and his merchant congregants were not merely anxious about lost status—they seemed to find the very experience of commerce worrisome. It may make more sense to understand their anxiety as that of an established class that had experienced a full reordering of economy and politics in the second quarter of the century. Character—moral and personal—was the adhesive that bound together New York's early nineteenth-century commerce and politics. By mid-century, this patrician class, linked by individual memory and family lore to Dodge's "Old New York," confronted an expanded and impersonal society that seemed to ignore questions of character altogether. The defect of that mass society was an anonymity and heterogeneity that distanced the representatives of responsible authority from their proper clienteles and audiences and devalued moral and personal worth. The Astor Place crowd, that "fickle public" with its "ignorance" and "caprices," the mass markets with their "convulsive energy," and the jingoistic Democratic Party, so vulnerable to "the violence of popular rage," were all of a piece in the minds of the patrician elite. They all were symptoms of a rapidly expanding society verging on dissolution. The only hope for solving this problem of consolidation, this elite believed, was to reinvent their authority, that of a wealthy and cultivated class. This gentry would create new institutions, new arenas, for restoring moral contact among citizens and reasserting the importance of character in public life. Here, in the late forties, were the beginnings of the new patrician culture that would be wielded against the Belmont circle during the summer of 1863.[52]

In the late forties and early fifties Frederick Law Olmsted and other patrician intellectuals began to suspect the moral consequences—though not the economic utility—of free trade. From the rural outpost of his farm on Staten Island in the late forties, Olmsted associated with a decidedly urban network of reformers, among them Charles Loring Brace, Henry Raymond, William Lloyd Garrison, and George William Curtis. He also began to write about society and politics. After the 1848 Whig presidential convention, he expressed "general opposition to the besotted laissez aller (faire) principle (pretended) of Loco Focoism." An 1853 investigative journey to the slave South commissioned by Raymond's *Times* occasioned more specific analysis of the problems of Northern life. Perched between rural and urban careers, the future designer of city parks did his most fruitful thinking about the need for new cultural structures to bind a fragmented market society.[53]

In a December 1853 letter to Brace, Olmsted explored the differences between the world-view of the Southern planter aristocrat and the democratic outlook he sought to cultivate in the North. The Staten Island agriculturist saw much to emulate in the Southern elite. Olmsted was especially impressed with the Southern view, advanced by Nashville lawyer and slaveowner Samuel Perkins Allison, that the United States lacked a true genteel class outside of the South and a few large Northern towns. Allison, Olmsted conceded, was "a thorough aristocrat": ". . . the conversation [with Allison], making me acknowledge the rowdyism, ruffianism, want of high honorable sentiment & chivalry of the common farming & laboring people of the North, as I was obliged to, made me very melancholy." But if Olmsted privately admired the honor and chivalry of his Nashville host, he was deeply troubled by the unabashed "materialism" of the Southerners' views.

> It seemed to me that what had made these Southern gentlemen Democrats was the perception that mere Democracy as they understood it (no checks or laws upon the country more than can be helped) was the best system for their class. It gave capital every advantage in the pursuit of wealth—and money gave wisdom & power. They could do what they liked. It was only necessary for them, the gentlemen, to settle what they wanted. . . . All that these sort of free traders want is protection to capital. . . . But I don't think our state of society is sufficiently Democratic at the North or likely to be by

> mere *laissez aller*. The poor need an education to refinement and taste and the mental & moral capital of gentlemen.

In the Southerners' laissez-aller outlook, he observed, slavery was regarded as a benefit to commerce, and so "it was a corollary that the measure would be for the highest good . . . so completely had they swallowed the whole hog of Free Trade." Southern elites, Olmsted believed, carried the free-trade principle to immoral extremes by condoning an unrestrained economic individualism and, by extension, the most "materialist" institution of slavery.[54]

Though the materialism of such extreme free-trade views might be appropriate to Southern conditions, Olmsted concluded, it was highly dangerous in the Northern setting. In the more intensely commercial North, with its unruly immigrant poor, the State would have to become a more important social institution. Government, he told Brace, would promote "parks, gardens, music, dancing schools, reunions . . . so attractive as to force into contact the good & the bad, the gentlemanly and the rowdy." Olmsted and Brace believed that New York State's aid to schools and agricultural societies set the proper precedent for this expanded governmental role. Brace similarly argued in an 1857 pamphlet that industrial schools and other activities of the Children's Aid Society receive state funding in accordance with the state aid to public schools precedent. Olmsted, Brace, and Bellows agreed with British school teacher Matthew Arnold that the State alone, serving as an agent of culture, could prevent "mere anarchy and confusion."[55]

Only some New York City merchants would fully appreciate or endorse Olmsted's cultural critique of laissez-faire and Southern elitism during the 1850s. In 1853 Olmsted confessed to Brace that he was still "blundering" over his ideas and had yet to weave together all elements into a system. But after the Astor Place riot it was clear that merchants were beginning to experiment with an array of new institutions as vehicles for influence and reform. They invested new energy in plans to build a large public park in New York City and supported Fernando Wood's efforts to preserve the original broad boundaries of Central Park in the early fifties. Merchants also applauded Peter Cooper's founding of the Cooper Union for the Advancement of Science and Art in the mid-fifties. An imposing structure guarding the northern head of the Bowery, the free school sought to instruct working-class sons and daughters in self-help and the acquisition of "useful knowledge." The original plan for Cooper Union included an art gallery, a debating society, free readings in "polite literature," a School of Design for women, a night school of science and art, and a public reading room. Under the cultural stewardship of a board of trustees that included Cooper, his son Edward, Abram Hewitt, Wilson G. Hunt, John E. Parsons, and Daniel F. Tiemann, the school sought to provide New York City with a new kind of institution binding together the class-torn community. Meanwhile, in an 1857 essay on *The Relation of Public Amusements to Public Morality,* Bellows exhorted merchants to renew their interest in theater, both to expose a rough popular theater clientele to "higher, purer and less superficial tastes" and to fortify themselves against "the headlong sobriety and mad earnestness of business"—or, in Olmstedian language, against the atomizing effects of unbridled "materialism."[56]

Though this intellectual experimentation of the fifties did not yet amount to a full challenge to Young America, it did attempt to revise the Belmont circle's paternalist emphasis. Bellows, Olmsted, Jay, Strong, and others in this growing fraternity of reformers hoped to supplant the Democratic elite's management and accommodation of the white immigrant poor with an ensemble of cultural institutions—schools, pleasure grounds, and theaters. The new institutions would serve as arenas of class association and interaction, with the outcome of each social encounter as random as the operation of the free market itself. By his direct and partisan negotiations with the poor, Fernando Wood had become a witting mouthpiece for the "caprices" and "ignorance" of the mass. By contrast, the patrician fraternity sought both to regain contact with the poor and to veil their authority through the new communal settings. To the extent that the park or the theater disguised the cultural supervision of the merchant elite, these men believed, their authority was enhanced. As Bellows wrote of the theater, "The instruction to be got from the drama or the stage must always be incidental, and perfectly subordinate to the pleasure got from them." The commercial and political transformations of the North had fragmented the merchant patriciate as a coherent social class, elevated new mass politicians such as Wood, interested more in electoral power than subtler forms of education and influence, and placed the working classes beyond the reach and redemption of the "gentlemanly." Quietly and almost surreptitiously, the new institutions would bind together a white commercial society in danger of disintegration. As arenas for moral contact among the elite and between classes, these institutions would restore authority to the patriciate and character to social relations.[57]

In the fifties, the patricians also began to shift Young America's agrarian emphasis. As Olmsted and others grew critical of the "ruffianism" of the "common farming and laboring people of the North" and the "materialism" of Southern planter aristocrats, they looked to the city for cultural solutions and a new social type worthy of emulation. With his acceptance of the office of superintendent of Central Park in 1857, Olmsted the gentleman-farmer committed himself to the city as the preferred venue of a national cultural transformation. Bellows was coming to the same realization in the late fifties and early sixties. "Cities," he wrote in 1861, "are rapidly becoming the chosen residences of the enterprising, successful and intelligent." New York's Central Park, part school, part theater, part pleasure ground, provided "the first grand proof that the people do not mean to give up the advantages and victories of aristocratic governments in maintaining a popular one, but to engraft the energy, foresight and liberality of concentrated powers upon democratic ideas. . . ." The city, Bellows predicted, would nurture a new democratic elitism vindicating the American republican experiment at a time of European reaction against popular government.[58]

Religion would become the life blood of the new urban culture. For Olmsted, "religion" was what the "materialistic" Southern aristocrat lacked and what the democratic Northern elite would seek to cultivate among its own ranks and in the community.[59] As Brace explained some years later, urban religiosity would transcend questions of piety and center on social responsibility. The reformer, Brace insisted, "must ignore sects, and rest his enterprise on the broadest and simplest principles of morality and religion." The public would learn to associate urban Christianity with the "feeling of humanity and religion—the very spirit of Christ Himself." The new institutions—Brace and Cooper's free schools and charitable societies, the reformed theater, and Central Park—would become, in Bellows's phrase, nonsectarian "missions." The highly abstracted language of Christian influence preferred by the evangelical Brace, the Unitarian Bellows, and the High Church Episcopalian George Templeton Strong provided a common ground for these men of rather different religious backgrounds and faiths.[60]

Judging from the Wall Street revival of winter and spring 1858, ideas of urban religious renewal had wide purchase in the business community. Sectarian barriers fell away, and merchants, clerks, and their families flocked to lunch-hour prayer meetings and overflowing weekend masses. The timing of the revival made sense: in the wake of the financial panic and riots of fall 1857, city merchants were understandably nervous. One testimony to the success of the histrionic revival meetings was the indignant reaction of the High Church Strong. After observing one prayer meeting in the financial district's John Street Church, Strong noted in his diary, ". . . what I heard seemed to me no doubt well meant, but the profane and mischievous babblings of blind, foolish, shallow, vulgar Pharisaism, trying to hide its 'I am holier than thou' under certain formulas of self-contempt; proud of its own admission that it is humble. The great object of the meeting seems to be to drug men up to a certain point of nervous excitement and keep them there. . . . More people will be harmed by the revival than benefited, I think." Uncomfortable with the enthusiasm and social irreverence of the "business revival," Strong looked for a quieter but more permanent means of cultivating religiosity in the city. Strong, Olmsted, Brace, and Bellows all sought a religious influence more wholly controlled by men of their own ideological stamp and genteel social stock.[61]

During the fifties some business and professional men began to regard the government as an instrument of communal religious transformation. The New York State movement for temperance and Sabbatarian legislation in the mid-fifties was warmly supported by the merchants who sustained Macready in 1849. In part, the temperance movement drew upon a tradition of merchant leadership in evangelical reform that extended back to the 1820s. William E. Dodge, Jr., Anson Phelps, and Lewis and Arthur Tappan could be found on the membership lists of the American Bible Society, the American Tract Society, the New York City Mission and Tract Society, and the Young Men's Christian Association. But in the mid-fifties, merchant evangelicals were joined by other men seeking an enlarged role for the state in religion as part of their cultural reclamation of the city. No evangelical reformer, attorney Strong applauded New York's 1855 Maine Liquor Law and Ecclesiastical Property Act: "To be sure," he conceded, "it's a novel stretch of law-making power, but social reforms can only be effected by new instruments." With the blessing of men such as Tilden and Dix, Fernando Wood stymied the enforcement of temperance and Sabbatarian legislation in New York City. It was not until the outbreak of war in 1861 that Bellows and Strong were granted an opportunity to Christianize the state.[62]

In spring 1849, Dr. Bellows's claim that "there is no rightful authority which is not in a sense divine" was controversial; in spring 1861, it had become, for many merchants, self-evident.[63] After the firing on Fort Sumter, Strong noted with some surprise that John A. Dix, John Cisco, and other Trinity Church parishioners usually reluctant to mix religion and politics agreed to hoist the flag on the Trinity steeple. "This flag," Strong exulted, "was a symbol of the truth that the Church is no esoteric organization, no private soul-saving society; that it has a position to take in every great public national crisis, and that its position is important." On April 20, the day after the massive Union Square rally in support of the government, Bellows preached at All Souls on "The State and the Nation—Sacred to Christian Citizens." The Unitarian minister had only some months earlier written to his sister of the "possible insecurity of life and property, if secession and revolution should occur, driving our populace into panic for bread and violence toward capital and order." He was not alone in these fears. In a letter to Treasury Secretary Salmon Chase during the first week of April, financier John Jay had predicted that "the moment hostilities shall break out at the front, we will be in danger of *insurrection at New York*." By the weekend after the Union Square demonstration, Bellows and Jay had regained confidence that nationalism would hold the polity together. Proclaiming that "the State is indeed divine," Bellows sounded the patrician theme for the war decade. In his 1863 pamphlet *Unconditional Loyalty,* he carried the argument a step further. At a time of growing opposition to the Lincoln administration, Bellows declared the political leader of the nation a "sacred person": "File at the staple which God fastens to his own throne, in the oaths of office which make a man chief ruler of a people, and you loosen thoughtlessly every link in that chain of law and order, which binds society together." Though probably not all patrician merchants favored Bellows's extreme position, they did become the loudest advocates of suppressing opposition to Republican policies and a no-holds-barred war effort. The war, as George M. Fredrickson has demonstrated, was the great moment of influence for which Bellows and his circle had so long waited. After the firing on Fort Sumter, the Christian state had begun to supplant the Jeffersonian state as the dominant merchant view of religion.[64]

Not surprisingly, patrician merchants and professional men were intensely anti-Catholic and anti-Irish. If the Belmont circle was the most publicly racist upper-class group in the city, the patrician fraternity was the most openly nativistic. Strong acknowledged some shared values with the German Protestant community: he was quick to note that Germans had "behaved well and kept quiet" during the draft riots. The Catholic Church, by contrast, was a monolithic barrier to the progress of the Christian state. In the mid-fifties, Strong had predicted that the Catholic Church, animated by "its aggressive and exclusive features," would provoke a great religious war in the United States before the end of the century. The Irish Catholic working class, of course, received the large share of Strong's venom. More than any other group, the patrician elite viewed the draft riots as a pure Irish Catholic uprising. In his epitaph to the insurrection, Strong wrote, "For myself, personally, I would like to see war made on the Irish scum as in 1688." This vitriolic ethnic hatred did not ease with the passage of time, nor did the memory of the riot as an affair of an Irish criminal class. Leather merchant Jackson Schultz stressed to his Union League Club cronies twenty-three years later that "by all it was known as a riot instigated and carried on by the Irish Catholics." Strong and his friends would long see their differences with the Irish Catholic community as the most dangerous fault line in the Christian republic.[65]

The Union League Club of New York, founded in early 1863, demonstrated that Olmsted and Bellows's perspective had arrived as a coherent world-view with broad appeal in the merchant community. In Olmsted's "club of true American aristocracy," all of the developing strands of the new merchant faith—the cultural critique of free trade, the search for institutions to bind a disintegrating society and polity, urbanism, the glorification of the Christian state, and, finally, recognition of the Northern free black community—were at last united. While in the early fifties Olmsted had identified the Southern planter class as the most threatening agent of cultural anarchy, by the end of that decade he and others were beginning to view New York City's Democratic merchants as an equally dangerous foe.

With the war, Olmsted and his associates sought a national field for their activities. In December 1853, the landscape architect had proposed to Charles Loring Brace the founding of a national journal, a "Commentator, as an organ of a higher Democracy and a higher religion than the popular. . . . a cross between the *Westminster Review* & the *Tribune*. . . ." By early 1863 Olmsted, with the help of Godkin, Strong, and Bellows, was fully involved in the literary project, now dedicated to "upholding sound principles of loyalty and nationality" (and doubtless the blueprint for Godkin's *Nation*). Olmsted and Bellows also looked to the United States Sanitary Commission—led by Bellows and an executive board (Cornelius R. Agnew, William Van Buren, Wolcott Gibbs, and Strong) that would serve as the nucleus of the Union League Club—as an instrument of "higher religion" during the national crisis. The war gave Olmsted, Bellows, and their circle the opportunity to extend their project of cultural cohesion from the city to the nation.[66]

But as the patrician reformers rallied opinion behind their new institutions and program for a loyal national culture they met resistance from the proponents of a free-trade empire. Opposition to the patrician project from men such as August Belmont and Henry G. Stebbins (both of whom Olmsted had encountered on the Central Park Board of Commissioners) was the impetus behind the idea for an urban club of gentlemen dedicated to the cultivation of nationalism among the metropolitan elite. In the fall 1862 letter in which Olmsted first proposed the Union League Club, he expanded the notion of wartime "loyalty" to include not merely a certain view of the sectional crisis, but also a special upper-class outlook:

> To what are we loyal and they not? We agreed that Belmont and Stebbins must be of the other sort. To what are they not loyal? Both will swear allegiance to the Constitution. Stebbins within a year has declared to me that slavery must and should be abolished and the rebels exterminated. Supposing him sincere, I still could not suppose him to be sympathizing with what loyalty includes with me. I feel that liberty and Union is not all. Neither Belmont nor Stebbins could with any sincerity say, I believe, that they would not, if they could, have a privileged class in our society, a legal aristocracy. Both, I believe, hold in their hearts *European* views on this subject.

As it developed in the writings of Olmsted and others, the plan for a "loyal" Club had four articles: a rejection of Belmont and his allies' pandering to popular political sympathies and machine politicians, a repudiation of their suspect or conditional loyalty on war issues, a critique of their "materialist" or extreme "free trade" outlook, and a distaste for their orientation toward European "society." This was the "radical" program August Belmont would confront in his efforts to rally "the conservative men" of the city in January 1863.[67]

Not Olmsted but George Templeton Strong made the most thoughtful inquiry into the cultural conflict with Young America. Strong eliminated Charles Gould and Prosper Wetmore from the proposed Union League Club roster for their too-close connections to the machinery of the mass parties. The broker and the dry-goods merchant represented "a dirty set of false-hearted hackstump orators and wire-pullers." Strong also dismissed from consideration August Belmont, James Brooks, and Samuel L. M. Barlow. This trio of Democrats combined the sin of machine-tainted paternalism with suspect loyalty to the Union cause. Like Olmsted, Strong advocated summary tribunals and capital punishment for the Southern-sympathizing elite: "May they only bring their traitorous necks within the cincture of a legal halter!" The Club also excluded those who were, in Strong's telling phrase, "mere traders and capitalists"—who, like Olmsted's "materialist" Southern planters, believed that free trade was sufficiently cohesive to bind together the expanding commercial nation. By 1863, the patrician fraternity aimed a cultural critique once reserved for the Southern planter elite at the Young America merchants of their home city.[68]

Finally, in Strong's estimation, August Belmont and William Butler Duncan, the Buchananite bankers, could be stricken from the Union League Club list because of their sycophantic social relations with British capitalists. Of Duncan, Strong wrote, "We have repeatedly named as a representative and strongly-marked type of the men against whom we are organizing ourselves. I fear he is a snob and a 'squirt,' the dining partner of his banking house, and the toady of British aristocracy when they condescend to visit New York." It was not Duncan's British connection per se that offended Strong. The Whig petitioners of 1849 had championed British actor William Macready in 1849 exactly because they approved certain aspects of British culture. Strong privately admired the cultural authority of the British aristocracy (much as Olmsted had privately praised the aristocratic gentility of the Southern planter in the 1850s). But by the Civil War years, Strong and other Union Leaguers censured Belmont and Duncan's willingness to derive their social and economic status from the British upper classes at a critical moment in the process of American nation-making. In spring 1863 Strong bristled at a report that a young English aristocrat at a Belmont soirée "had been consorting with W. Duncan and Belmont and naturally thought sympathy with the rebellion *the thing* in New York." Unlike Duncan and Belmont, Strong and other Union Leaguers hoped to create a "true American aristocracy" which would adapt the cultural authority of the British elite to the distinctive social and political conditions of American democracy. The Union Leaguers' badge of commitment to America's democratic experiment became their unconditional nationalism and paternalist cultivation of a loyal black poor.[69]

In 1863, the Union League Club counted over 350 members. The Club reached well beyond the small group that had signed the Macready petition in May 1849 and included a number who had attended the Belmont farewell dinner in August 1853. In addition to a broad representation of Republican merchants, the elite society admitted many Democrats. Not all Club members embraced Olmsted's full program or heeded Olmsted and Strong's pleas for federal military rule during the summer of 1863. But most wished, in the words of Strong, to demonstrate that "the intelligent, cultivated, gentlemanly caste" would at all costs sustain the Washington government. Most also applauded the Club's protection and patronage of the free black community after the draft riots.[70]

As a defense of slavery and theories of black inequality were becoming the trademarks of Young America, the Union League Club began to take new interest in the rights and welfare of the free black community. As early as February 1854, Olmsted, like many of his merchant allies a gradual abolitionist, declared that the future of Northern capitalism hinged on Northerners' willingness "to deal justly and mercifully with the colored people in [their] midst." Olmsted believed that the question of the cultural attainment of free black people held within it the fate of his entire reform project. European critics of republicanism and American conservatives used the argument that free blacks could not care for themselves to denigrate all free labor. Olmsted saw slurs against Northern free blacks as a veiled disparagement of the manual laborers, white and black, whom he hoped to elevate "to the mental and moral capital of gentlemen." The landscape architect pointed to the Rochester Colored National Convention of July 1853 to prove "the possible development of the negro race to an intellectual equality with [his] own." Indeed, Olmsted and his patrician allies regarded a Protestant and pious black poor as a more receptive audience for their cultural ministrations than an inattentive or hostile white and immigrant Catholic working class. Olmsted's description of the qualities of the black delegates to the Rochester convention—"higher civilization . . . manliness . . . [possessing] the virtues and graces of the Christian and the gentleman"—was echoed in large part by Jonathan Sturges, chairman of the Committee of Merchants for the Relief of Colored People, in his praise for a church-going and respectable black poor in the wake of the draft riots. During the sixties many New York merchants came to view the cultural progress of the free black community as a test for whether an American working class would respond to their new democratic institutions. After the draft riots, Olmsted's perception of the free black community was shared by the Union League Club merchants who, we recall, aided black victims of the violence, raised the black Twentieth Regiment, and paraded down Broadway side-by-side with that regiment in March 1864.[71]

The discipline and loyalty of the Club's new black regiment was thus for many merchants a vindication of the patrician outlook. The Union Leaguers did not completely discard paternalism but merely changed the color of its clientele. Through a paternalist cultivation of black *"vrais ouvriers,"* the Union League Club sought to prove to Young America critics that an Amercan working class could become a full partner in the new Christian state. Nonetheless, as George M. Fredrickson has suggested, there were strict limits to merchants' tolerance of a democratic working-class participation. "If the 'inferior elements,' whether Negro or white, consented to be led by 'the best culture,' then their rights were assured; if, however, they struck out in directions of their own, democracy and equality might again be questioned."[72] The Union Leaguers' opposition to all phases of the draft riots and their campaign for martial law well demonstrated how these merchants responded to the political deviations of an independent working class. Even the smallest deviations had to be curbed lest they, in Bellows's words, "loosen . . . every link in that chain of law and order which binds society together" and plunge the Christian state into anarchy.[73]

The quarrel between merchants in July 1863 was thus the outgrowth of a decade and a half of debate over the meaning of the rapid expansion and resulting mass conditions of Jacksonian society. The draft riots exposed a conflict between two merchant groups situated differently with regard to these changes and coincided with the full emergence of a new patrician outlook challenging the once-ascendant views of August Belmont and his colleagues. Olmsted, Bellows, and their merchant associates clamored for federal military rule and the blood of "Barlow and Brooks and Belmont and Barnard and the Woods" during the riots, but only after a long process of cultural differentiation through which they distinguished themselves from the dominant elite persuasions of Jacksonian America—those of the Southern planter and the Young America trader. For Belmont and Young America, martial law would not only disrupt Democratic rule of the city but undermine the free and decentralized development they thought essential to social harmony and progress. Their conciliation of white draft rioters and contempt for black riot victims were rooted in a belief that an expanding, decentralized and all-white polity could weather much conflict; to enhance the status of black people, either through emancipation or philanthropy, would undermine the ability of that white empire to accommodate social tensions. For Olmsted and the Union League Club, federal intervention would not only extend Republican authority but also heighten the power of the state, necessary to prevent a fast-expanding mass society from drifting into anarchy. Treasonous Irish draft rioters and their elite sympathizers put themselves outside the framework of that sacrosanct state and its allied urban institutions, and had to be suppressed. The victimized black community, loyal to the state and potentially receptive to the Union Leaguers' reforms, deserved protection. With the clarification of such differences between Young America and the Union League Club, the merchant elite—the leading social class of the city—became deeply divided. The crisis of the riot week encompassed both a political challenge on the streets and bitter dissension in the corridors of power.

CHAPTER 5

Industrialists

More than any other group of New Yorkers, industrialists made the Republican Party controversial in July 1863. While the Union League Club merchants helped to give the Republican Party of New York City its aristocratic, nativist, and coercive style, it was the industrialists and their aggressive brand of Republicanism that the draft rioters knew best. Industrialists were often the employers of the midweek rioters. They were the Republicans the rioters encountered six days a week at the shop and perhaps on the Sabbath in the person of a visitor from the charity society. It stands to reason that the industrialists did much to shape the rioters' image of Republicanism as unjust and intrusive authority. How the industrialists came to be so aggressive—and how they made the Republican Party a magnet for political and social dispute—provide the final ingredients in the history of the origins of the July crisis.

Industrialists addressed the riot-week issues of relations with the poor, between the races, and with the federal government in the context of a sweeping reform program. In their shops, metal trades employers promoted company-based draft insurance funds and enlisted their employees in military brigades to defend factories against arson. In the community, the industrialist reform organ, the Association for Improving the Condition of the Poor, denounced Democratic "pseudo-philanthropists" such as Fernando Wood who made extravagant appeals to the working classes and favored indiscriminate public aid to the poor. After the riots the AICP also intensified a campaign for the education and discipline of immigrant working-class children and for health and housing legislation to counter the "demoralizing" influence of slum tenements on the poor.

Like the Union League Club merchants, industrialists encouraged a liberal aid to black victims of the mob, but, unlike the merchants, they sharply discriminated between two groups among the white working classes. The distinction between loyal and disloyal workers or a respectable and unworthy poor lay at the core of the industrialists' social vision. Company-based draft insurance, profit-sharing schemes, apprentice schools, factory military brigades, and health and tenement reform were all industrialist devices to cultivate a loyal and upward-striving poor. Federal martial law in the so-called "infected" riot districts, campaigns for a professional and state-run police department, truancy laws, and heightened surveillance of the poor were intended to supervise a disloyal and insurrectionary pauper class. This taxonomy of good and bad workers authorized employers, reformers, and the government to investigate the poor and intervene in the shop floor, neighborhood, and domestic lives of the unworthy; it provided the rationale behind the industrialists' efforts to make the poor over in their own image. The taxonomy encouraged the energetic style of reform that made the industrialists so controversial. These employers can best be understood as mid-nineteenth-century New York's most aggressive proponents of individualist values.

An opening clue to the industrialists' interventionist style can be found in the mid-century economy. Industrialists were a new elite, separate from, dependent upon, and not always in accord with the more established merchant class. In addition, machine-building and metalworking firms were disadvantaged as the local economy began to favor labor-intensive, small-scale manufacturing in the 1840s and 1850s. By the time of the war, some heavy industry had already relocated outside the city, and more would soon follow. Indeed, the industrialists who battled the draft rioters in 1863 were by their very presence in the city a stubborn lot. Because they could influence so few of the factors of production, their economic problems were especially daunting. One thing they could attempt to control was the culture of their employees, both inside and outside the shop, and this they did with verve and creativity. The industrialists' style, that is, their penchant for observing, classifying, cultivating, and coercing the poor, was doubtless shaped by these unpromising economic circumstances.

Horace Greeley's plan for "Association" allows us to trace the emergence of that style. Greeley was no industrialist. But the ideas of New York's most outspoken mid-century employer highlight what was new and controversial about the industrialist perspective. Around mid-century industrialists began to revise Greeley's assumptions and elaborate the taxonomy in the shop through the "individual method" of wage payment, technical schools, mutual relief societies, and profit-sharing schemes. At the same time the Association for Improving the Condition of the Poor pursued the industrialist project in the community: monitoring the poor in their homes, forging alliances with sympathetic and evangelically minded groups in other quarters of the elite, and censuring wealthy donors who indiscriminately aided the poor. By the mid-fifties, the AICP was looking to the state legislature as an instrument for both the reform and the discipline of the poor. The new Republican Party was the obvious vehicle for industrialists' efforts to use the government to transform urban poverty. Republicanism had a problem of legitimacy, however, even among the "loyal" and "industrious" poor, that was dramatized during the draft riots. The Citizens' Association formed after the riots proved a more successful means of cultivating loyal workers and effecting political reform.

A New Elite in Difficulty

Industrialists were a new elite in two senses. It was common to find them engaged in middle- and upper-class pursuits new to the nineteenth century. Metal manufacturing and machine building were foremost among these new occupations. Along with the inevitable iron founders and engine builders, other new groups—slaughterhouse proprietors, lumberyard and molding-mill owners—frequently turned up in the industrialist camp. Further, industrialists were often newly arrived as members of the elite, many having risen through the social ranks from a fairly humble start. They did not follow the pattern identified by Francis W. Gregory and Irene D. Neu, who found that railroad, textile, and steel executives of the nation's largest firms in the 1870s usually came from business and professional backgrounds. Instead, judging from the principal machine builders and metal manufacturers of mid-century New York City (not a sample but an influential group), it is evident that many worked their way up to managerial positions from beginnings as skilled workers and clerks. The preeminent manufacturer of steam engines in Jacksonian America, James P. Allaire, began as a mechanic and advanced because of his knack for assembling the new machine parts that came over from England. John Roach, who dominated American shipbuilding after the Civil War, arrived from Ireland at sixteen, learned the trade of iron molder from Allaire, and left the workmen's ranks when he purchased a small factory in partnership with other mechanics (he later bought out their shares). Cornelius Delamater, builder of the engines for the Northern ironclad *Monitor,* started as an errand boy in a hardware store. Robert Hoe stepped off a boat from England as a penniless carpenter in 1803 and became a well-known manufacturer of printing presses; after he died in 1833, his son Richard, reputed for his mechanical ingenuity in the shop, expanded the business into an international concern. Two of the leading carriage builders of the city, John Stephenson and John W. Britton, both began their careers as mechanics. Horatio Allen, president of the Novelty Iron Works, was unusual for his college education but like his fellows rose in his field because of technical creativity—Allen designed and built engines for the nation's first railroad companies. Many of these industrialists achieved substantial wealth, often from modest origins and almost always because of their ability as mechanical innovators.[1]

As a new elite, the industrialists were distinct from the patrician merchants with whom they shared an inclination toward evangelical religion and Whig (and later Republican) politics. In this regard New York machine-building and metal manufacture were different from New England textiles, where merchants such as Francis C. Lowell and his Boston Manufacturing Company ran industrial enterprises themselves. In mid-century New York City merchant and industrial capital were often in separate hands. Many machine and metal firms were owned and managed by the industrialists themselves, either singly or in partnership. John Roach, Cornelius Delamater, and George Quintard were the sole proprietors of their businesses. The partnerships sometimes included merchants but often, too, skilled master craftsmen such as the Hoe Company's Stephen D. Tucker, who began as an apprentice in the firm. Of course, the industrialists required access to the merchants, both for money and credit and for the supply of materials and the marketing of their finished product. But the economic interests of the two groups conflicted as much as they coincided: the merchants supported free trade; the industrialists yearned for tariffs that would scuttle the free-trade empire and substitute their own manufactured goods for European wares in the American economy. Merchants often had a distant relationship with the working classes (they were the Celtic throngs beyond Second Avenue), industrialists a far more intimate one (they were employees—indeed, every day Horatio Allen had to walk beyond Second Avenue through the grimy and proletarian Eleventh Ward to get to his office at the Novelty Works).[2]

The uptown industrialists also developed interests distinct from those of other employers. By mid-century industrialists were discovering that the course of city growth did not favor their style of enterprise. Heavy industry began to depart the downtown and center-island manufacturing center and relocate on the uptown periphery. Nearly all the principal machine-building and metal firms joined what one historian has called "the exodus": the Hoes left their lower Manhattan site on Gold Street for Broome Street as early as 1835 (though all operations were not consolidated uptown until 1844); the Delamater Iron Works moved from Vestry Street to a Hudson River location at Fourteenth Street in 1850; Singer Sewing Machines migrated north from Center Street to Mott Street in 1857 and to Mangin Street in 1869. John Roach held his ground on Goerck Street until 1867, then took over the large waterfront site and facilities of the Morgan Works. The high price of real estate (selling the land could be more profitable than operating the plant), the difficulty of transporting merchandise through jam-packed streets, and the lack of space for expansion drove many firms out of the downtown district to the periphery. As early as the fifties, some factories found it necessary to abandon the city altogether, relocating in New Jersey, Brooklyn, and Astoria. At the same time heavy industry was beginning to leave, labor- and land-intensive enterprises such as clothing manufacture were becoming characteristic of the Manhattan economy. Clustered in an emerging downtown, center-island warehouse district, these trades exploited the now vast reserve of poor immigrant workers, thrived on the easy exchange of information in the crowded center city, required comparatively little capital, and in some instances, relayed overhead costs to subcontractors. By the time of the draft riots it was clear that Manhattan was shedding any dim resemblance it bore to British Manchester or Birmingham, and industrialists who remained on the island would have to adjust to adverse conditions.[3]

One way for industrialists to anchor themselves in the city was to form alliances with merchants, especially those with whom they shared evangelical religion and reform. The relation between the engine builder Horatio Allen and the banker James Brown was revealing. For many years the Novelty Works was a partnership owned and managed by Allen and the master machinist Thomas B. Stillman. "Stillman, Allen & Company" flourished at first but, according to one account, "some heavy losses embarrassed the firm and they had to seek outside aid. Mr. James Brown furnished the firm with more capital, and when a stock company [Novelty] was organized he became a stockholder, and Mr. Allen was President." The association between the two men also extended to matters of reform. Through much of the fifties and sixties, Brown served as the president and Allen the vice president of the Association for Improving the Condition of the Poor. Industrialists like Allen benefited by having the nation's leading commercial capitalists so close at hand. They needed the city's well-established merchant class both to survive in the urban economy and to secure authority for their values. Yet some—perhaps many—industrialists did not share Allen's fellowship with the commercial elite. The remarks made by carriage builder John W. Britton before a Senate Committee in the 1880s conveyed an industrialist perception of merchants and financiers that may have been typical of the sixties: "I can remember that within twenty years a small money-broker with whom I had been in the habit of trading moved into a house across the street from mine, and he was very cool for fear that he would have to associate with me and my family. The small money-broker, with one-quarter of the capital that we employ, feels himself much better than a carriage-maker, and the public recognizes that fact. I can remember that when on one occasion a carriage maker was elected to a noted club founded by the "blue-bloods"—I don't know how he got in there—it was a matter of the greatest astonishment how he did get in." Elsewhere, Britton observed that talented apprentices had become hard to find as every bright boy "wants to be a merchant, a broker, a banker, or a railway man. The example of that class of men has been set before the boy. . . ." The comments of this industrialist reveal a sense of social distance from the merchants (and of remove from the centers of social prestige) that no amount of dealing between the two groups could fully erase. Resentments of the sort that simmered just below the surface of Britton's remarks may have hindered industrialists' ability to court and win the merchants' ideological support.[4]

Another way industrialists could counter their marginalization in the economy was to make their work force more profitable. Industrialists were fairly obsessed with the problem of labor, it being one of the few factors of production they could hope to control. When asked why the iron works of New York were no longer productive in 1869, John Roach pointed first to "the conflict between labor and capital, and the damaging effect of combinations among workmen." Here, as in their relations with the merchants, there were limits to what the industrialists could accomplish.[5]

Two stories about James Allaire's shop suggest what he and succeeding generations of nineteenth-century metal and machine manufacturers were up against. In 1829, a newspaper reporter visiting Allaire's plant noticed a sign reading, "Any person that brings, or drinks, spirituous liquors on my premises will be discharged, without any pay for the week." Allaire boasted to the reporter that he had discharged only two men in the nine years that the sign had been posted. John Roach told a story about the start of his apprenticeship under Allaire that also conjures up an image of an employer accustomed to intervening among his men. The overeager young Roach saved the molder's apprentice fee of fifty dollars and paraded into the foundry only to find himself ignored by the journeymen. Allaire learned of the treatment Roach had received (doubtless informed by Roach), "took the lad by the hand and leading him to the shop said to the men, 'You see this boy. I want him to learn this trade. Mind you I will have no interference with him.'" After Allaire left, the men hauled Roach over a barrel and looked on while "a big Englishman" paddled the young apprentice who had been so quick to complain to the boss. Roach was taken on by the journeymen molders only when he agreed to give them sufficient whiskey to pay his footing.[6]

Allaire's workplace injunction against drinking and Roach's rite of passage nicely illustrate the industrialists' difficulty. Allaire and his colleagues were wholly dependent upon the skills of their workers. While evangelical employers posted signs prohibiting drinking and encouraging discipline, they had to rely nonetheless on journeymen like the "big Englishman" to train and supervise new workers. Roach's story makes clear that the molders drank behind Allaire's back and more often than two dismissals in nine years would indicate. The only way industrialists could control production, short of mechanizing their craft workers out of existence (and Allaire probably sat awake nights dreaming of the machine that would end his woes), was to reform the behavior of their employees. Heightened surveillance during work hours was one solution. Another was suggested by a project Allaire completed in 1833. That year he and the builder Thompson Price constructed the first New York City "tenement" or building designed exclusively for tenant families. Located on Water Street, a few blocks from his factory, Allaire's tenement was a "single-decker" with each floor designed for one family. It is not surprising that Allaire took an interest in the housing surrounding his New York City factory; in the twenties he had erected an iron works and company town called "Allaire" in the New Jersey countryside, with dormitories, company farms, stores, and a church to accommodate his resident work force. His early promotion of a new kind of housing for the families of New York's East Side (and we can only guess at his intent) was at most the beginning of a search for new ways to influence the behavior of workers outside the shop. Allaire and his fellow industrialists always returned to the same problem—that of managing their independent skilled workers. As a patternmaker in one New York foundry put it in 1853, "The employers . . . regarded their men in the same aspect that husbands do their wives—they were necessary evils and could not be done without."[7]

The problem of controlling the foundry shop floor became noticeably more challenging by mid-century. The new machine shop proprietors of the 1840s, the Roaches and the Allens, confronted an array of new conditions: growing specialization of the trades with the introduction of new machine tools, increasing craft hierarchy in the shop, and the addition of German and Irish Catholic skilled workers to the older Protestant English, Scot, and native-born work force. Those shops engaged in the manufacture of steamship engines, boilers, and the enormous "bed-pieces" (on which the ship's machinery rested) witnessed an increase in the scale of production. The Novelty Works, employing an average of 1200 workers by the late forties, was at the time the largest metalworking and machine shop in the country. These new conditions resulted in an increasing social distance between industrial employers and their journeymen, new and clearer distinctions among skilled workers, and a widening chasm between the skilled work force and the growing mass of day laborers who now assisted the journeymen and conducted materials through the plant. In short, the status of heavy industry in the metropolitan economy was beginning to worsen at just the moment that the problem of control in the shop—the one domain where industrialists might hope to compensate for the high costs of location—was also aggravated.[8]

This mid-century predicament made it more urgent than ever for industrialists to find a productive and acquiescent labor force, that is, to remake the culture of their unruly craft workers. There is no evidence that industrial employers of the 1820s and 1830s tried to discriminate systematically between the kind of workers who ran afoul of James Allaire's drinking prohibition and the kind who strove to emulate the bosses' values. But by the 1850s such a taxonomy of the working classes had already begun to become part of these employers' thinking. The industrialist taxonomy was a way of interpreting the new environment of the mid-century machine shop amidst the new conditions of the mid-century urban economy. Such economic changes provided the stage setting for the industrialists' cultivation of "good" workers and disciplining of "bad" during and after the draft riots.

The industrialist taxonomy will be discussed, in turn, in the several contexts in which it evolved: the shop-floor associations in metallurgy, the community-based Association for Improving the Condition of the Poor, the

Republican Party, and the Citizens' Association. But it is useful to begin a discussion of industrialist views with Horace Greeley and his plan of "Association." Greeley was not a representative figure of the industrialists nor of any other wing of the metropolitan elite. But the city's best-known mid-century employer can serve as a point of reference in a discussion of the transformation of managerial ideology. Comparing the reform program Greeley developed in the late forties and fifties to attitudes emerging at that time among metal-trades employers and the membership of the AICP will illuminate what was new about the outlook of the draft rioters' industrialist foes.

Horace Greeley and "Association"

Horace Greeley was an eclectic and unsystematic thinker, a one-man switchboard for the international cause of "Reform." He committed himself, all at once, to utopian and artisan socialism, to land, sexual, and dietary reform, and, of course, to antislavery. Indeed Greeley's great significance in the culture and politics of Civil War–era America stemmed from his attempt to accommodate intellectually the contradictions inherent in the many diverse reform movements of the time. Nowhere was Greeley's accommodating style of thought more evident than in his "Association," a far-reaching program of social reorganization which borrowed from the utopian socialist writings of Frenchman Charles Fourier and Fourier's American disciple Albert Brisbane as well as from the cooperative reforms of the European radicals of 1848 and the New York artisan reformers of 1850. While Greeley did give utopian socialist and artisanal reforms a managerial emphasis, he also shared with the artisan reformers certain fundamental assumptions about work and poverty—assumptions the industrialists would soon attack.

From Fourier, Greeley derived the four basic propositions of Association: that the competing interests of industrial society produced urban poverty, squalor, and discontent, and a new, harmonious order could only be created through cooperative communities or *phalansteres;* that the new order would succeed where the present one failed by adjusting the social environment to agree with men's and women's natural instincts; that these natural desires could best be satisfied through a sophisticated division and alternation of labor, ending the drudgery of work and placing each individual in his or her own *métier;* and, finally, that in Association, each worker would receive a settled proportion of the price realized for his or her product.[9]

In the new communities, Greeley hoped to do away with the wages system and the accompanying evils of "unjust division of toils, unequal distribution of profits, isolation and opposition of interests." The "fundamental basis" of Association, he wrote in 1846, "is a proportional distribution of products to Labor, Capital, and Talent, according to the just claims of each." Recalling

Fourier, Greeley imagined the processes of labor divided "into as many minute varieties as they admit . . . to allow all laborers who are capable, men, women, or children, to engage freely in any branch they please." Workers would be allowed to choose their own superintendents, foremen and overseers and regulate their hours of toil. "Frequent and ready alternations from Manufacturing to Agricultural labor" would "preserve the health" of laborers and "all seasons would be productive" (no doubt an appealing prospect to employers and workers alike). Work was to be performed amidst the tempering and civilizing influence of the family, in the company of "men, women, children, parents, friends, colleagues, reciprocally bound by ties of affection," and not among hard-drinking and class-conscious male gangs, in "dingy and noisome workshops, amid low and indecent companions." "Attractive industry" would replace "isolated labor." Greeley was not as interested as Fourier was in the possibility of beautiful and pleasurable activity as an end in itself—if Greeley had visited one of Fourier's *phalansteres,* he probably would have ordered everyone out of the pool. Instead, Association would liberate cities from angry strikes and crowded almshouses and secure the voluntary participation of the worker in an industrial world increasingly dominated by the prerogatives of capital.[10]

Indeed, Greeley the employer was at times audible when Greeley the utopian socialist was speaking. A proposal in an 1846 lecture on the "Emancipation of Labor" for workers' proportional shares in the earnings of the Association best revealed Greeley as the master printer addressing men of his own class: "Let us suppose . . . that a manufacturing company, in addition to the payment of the usual wages, were to set apart a small proportion of its net earnings . . . to be divided among all its permanent workmen on some predetermined scale at the close and settlement of each year's business. . . . the effect of this system upon the industry, fidelity, and the skill of the workmen . . . would hardly be possible to exaggerate." Fourier's system was designed to enhance workers' autonomy by assigning them individually suited occupations and empowering them to make decisions of production and distribution. Greeley's reading of Fourier had a different emphasis: proportional ownership and discrimination among workers on the basis of a "predetermined scale" of reward might help cultivate loyal and industrious workmen. Fellow Whig editor Henry J. Raymond, seeking to demonstrate that individualism offered more hope than Association to both employers and workers, observed that Greeley's cooperative communities failed to challenge the primary structural features of the competitive order: capital, interest, and private property itself. "Proportional distribution of products to Labor, Capital and Talent," Raymond concluded, would inevitably lead to the domination of the largest holders of capital, who would "become the *owners* of all the real estate" and, by that circumstance, limit the prospects for the "Laboring Class" to rise "to the Class of

Proprietors." Raymond was perceptive: in Greeley's hands, utopian socialism could at times begin to resemble a kind of utopian capitalism.[11]

For all that, Association did have much in common with the program that the journeymen reformers advanced during the 1850 Industrial Congress. Like the journeymen reformers, Greeley looked to conditions in the trades rather than the behavior of a deceitful pauper class to account for the misery of the poor. Poverty was the product of environment as much as the idleness and immorality of the individual. Neither almsgiving nor the poorhouse would eradicate urban poverty—each, Greeley wrote, "is plainly calculated to render the evils it combats chronic and enduring." Where philanthropy was palliative of pauperism, Association was corrective. The origin of poverty in social conditions became one of Greeley's favorite editorial themes in the forties and early fifties. In an August 1845 debate over the nature of poverty with James and Erastus Brooks, editors of the Whig *Express,* he defended the *Tribune* against attempts "to ridicule [the] idea that the evils we deplore do not exist through the immediate fault of any man or class, but are the result of a bad system—a defective organization." Consequently Greeley supported the newly founded Association for Improving the Condition of the Poor in 1844, but from a distance. Such rearrangement of private charity would no doubt improve the efficiency of the city's philanthropies, but a charitable association alone could not serve as the basis for a new type of social relations.[12]

Trade unions, cooperatives, and labor congresses with their plans for a reorganized production, Greeley and the journeymen reformers agreed, offered a more promising institutional structure for the emancipation of labor. Granted, the *Tribune* editor's version of cooperative production at times resembled a scheme to cultivate devotion to the middle-class trinity of family, sobriety, and individual self-help. But Greeley was concerned less with making "good workmen" than with designing a harmonious work environment: his social categories distinguished between types of work settings—"attractive industry" versus "isolated labor"—more than they did types of workers. He believed that journeymen's associations and rules could succeed in improving the living and working conditions of the poor if endorsed by all employers in a given trade; he blamed the strikes of 1853 on grasping employers "who refuse to unite in any efforts for the systematic adjustment of Wages, but insist on fixing and adjusting such rate of wages as they choose, without reference to the established regulations or current usages of the vocation." Greeley imagined a community which placed limits on the accumulation of capital toward socially exploitative ends—and like the journeymen he defined capital broadly, including the overly acquisitive boss among his enemies. The trademark of the industrialist outlook—rigid moral distinctions between groups of wage earners—was only barely discernible in his thinking.[13]

The Taxonomy in the Workshop

About the time Horace Greeley was formulating his "Association," employers in metallurgy began to notice a new social type in the shop whom they called the "good workman." The earliest references appeared in spring 1853 when, in response to a threat of strike, employers refused to cede their "right" to discriminate among their workmen and reward only the highly skilled, industrious, and temperate with wage increases. One delegation of machinists reported to a meeting of their trade that Richard Hoe had announced "that if any of his men were worthy of the advance they would receive it." A patternmaker observed similarly that "their employers regulated the wages by the ability of their workmen; that they gave from fourteen shillings to two dollars per day, and that they would make any reasonable advance." A committee who interviewed Mr. Ayres of the Mott & Ayres foundry was told that "he would advance the wages of all his men who were worth it. He said that an advance of two or three shillings per day was nothing to a good workman." In spring 1853 employers in the metal trades successfully fought off a collective movement among their skilled workers for increased wages by insisting on the prerogative of distinguishing "good workmen" from bad. The employers' "good workman" valued individual preferment over collective allegiances to fellow workers.[14]

While Greeley aimed in the end to restrain an overzealous individualism, industrialists strived to cultivate workers who above all else sought personal advancement in the firm. At the same time, industrialists sharply distinguished such loyal and upward-striving workers from the disloyal who failed to respond to their overtures. Finally, these employers developed a battery of new workplace institutions to promote this distinction between good and bad workers and supplant the cooperative associations that Greeley and the journeymen reformers hoped would produce social order. From the fifties until well after the Civil War, Richard Hoe, James Brewster, and John Roach were the three most aggressive proponents of this new taxonomy.

Richard Hoe, called the "Colonel" by his employees, was known for his intimate involvement in the shop-floor affairs of his printing press factory. During the 1872 eight-hour strike in the metal trades, a molder named White who had been employed in the shop since boyhood bemoaned that Hoe was in London on business and said "that it was the universal belief of the workmen that if Colonel Hoe were home the demand for decreased hours of labor would be granted." But it was not merely the force of the Colonel's personality that won him the lifetime allegiance of some employees. In the mid-1850s, Hoe had established two institutions designed to cultivate the loyalty of "good workmen" like White.[15]

One of these was an evening school for Hoe factory apprentices. All of the firm's apprentices were required to attend three evening classes each week, two in mechanical drawing and one in mathematics. To allow workers time for the lectures, they were dismissed at five o'clock on the appointed afternoons, allowed fifteen minutes for "washing and other such personal arrangements" and then fed a meal before class, courtesy of the company. Classes were taught by the firm's older draughtsmen, who received extra compensation for their services. Though Hoe himself never said it, the apprentice school seems to have been regarded as a training program for future foremen and superintendents.[16]

The other institution was the "Mutual Relief Society of R. Hoe & Company's Works," Hoe's answer to the working-class benevolent association. Employees were allowed to choose between four different kinds of Society membership, each entitling the worker and his family to different benefits for death, sickness, or temporary financial distress. This complex organization was governed by a "Board of Representatives" consisting of the Hoe proprietors and ten employees elected annually, one from each of the nine departments of the factory and one at-large. The Board appointed "Standing Committees" responsible for investigating "House Accommodation," "Food and Fuel," and "The Prevention of Sickness." It was as if Hoe had created his own intramural "Association for Improving the Condition of the Poor." The House Accommodation Committee was instructed, for instance,

> to ascertain the prices and rents of houses in the most accessible and healthy neighborhoods; to learn the best modes of building, and the most healthful and convenient arrangement of rooms; and to aid "fellows" in securing pleasant and comfortable homes, as near as possible to their work.

The Prevention of Sickness Committee was enjoined to "put into the hands of 'fellows' sound information on sanitary subjects; to enforce on landlords, by legal measures, if necessary, obedience to the laws which regulate the drainage, ventilation and sanitary condition of houses. . . ." Hoe's Mutual Relief Society, according to one account, was explicitly designed to correct "the liabilities of organizations of working men, established and managed by themselves," organizations which readily became instruments of class conflict. In contrast to the trade union and working-class benevolent association, Hoe's Society would educate workers to their shared interests with the proprietors. According to the Colonel's plan, Hoe workers would look to the firm for the satisfaction of all their material needs and participate in a program for their own sanitary and moral reform.[17]

James Brewster, the carriage manufacturer, sought an answer to the same question that troubled Hoe—how to cultivate and engage the loyalty of the "good workman" to the firm? In 1869, Brewster's answer was a revised and elaborated version of the plan for profit-sharing which Horace Greeley had outlined two decades earlier. That fall Brewster unveiled to his employees the "Brewster & Co. Industrial Association." The Association, according to the "Constitution" Brewster drafted, would "divide a sum of money equal to ten percent of their net profits in their Broome street factory and Fifth avenue warerooms, during the year [ending on July 1, 1870] . . . among certain of their employees, in proportion to the wages earned by them respectively, and in addition to such wages (the persons so to share in said sum to be determined by the employees of said firm). . . ." The remainder of the constitution carefully defined how "certain" of the Brewster employees were to be chosen and what the relationship between the Industrial Association and Messrs. Brewster & Company would be.[18]

The constitution established nine new decision-making bodies in the shop: one called the "Board of Governors," empowered to direct the Association, seven others called "Boards of Control," to oversee the affairs of each of the factory's departments (i.e., smiths, wheelwrights and carriage makers, trimmers, body makers, body painters, and so on), and the Association members themselves, designated as all skilled mechanics over the age of twenty-one and employed in the shop for at least six months. Laborers were entitled to vote in Association matters, but were not eligible for office. "Clerks, salesmen, boys, porters, apprentices, [and] cartmen" had no voting rights. The Board of Governors consisted of the seven "Chairmen of the Boards of Control," an at-large worker representative of the factory, and a president, who was elected by the governors from among the proprietors of the factory. The president was given a veto authority over the actions of the governors, but could be overriden by a two-thirds' vote, with the proviso that "in all such cases . . . the names of the members voting for as well as those voting against [the veto], shall be recorded in the journal." With this limited authority, the Board of Governors was broadly directed "to make rules and regulations for the shop." The Boards of Control, each composed of three elected workers, were to enforce in their respective departments the rules and regulations made by the Board of Governors.[19]

The Brewster Association did little to constrain the managerial prerogatives of the owners and much to encourage heightened productivity and firm loyalty on the part of skilled employees. Article XI specified, "Neither this Association nor any member thereof other than its President shall have any voice or authority in the management of the business of Brewster & Co." Nor did the Association diminish the power of foremen in their respective departments of the factory. The constitution made special provisions for the many skilled workers who paid "helpers and finishers" out of their own wages. The skilled journeyman was to subtract the wages of his helpers from his company wage in determining the amount of money he would be entitled to in the profit-sharing plan—clearly adding to the journeyman's incentive to "sweat" his helpers. In its limited cooperative features and attempt to foster workers' allegiance to the company, the Brewster Association followed in the path of the 1863 company draft insurance fund programs of John C. Parker's Carriage Factory (which Brewster bought out and took over in the mid-sixties) and John Roach's Etna Iron Works. While Horace Greeley in his more visionary moments imagined the company profit-sharing scheme as a stepping-stone to the emancipation of labor from the wages system, and James Brewster may have fancied himself introducing parliamentary democracy to the shop, John Roach saw profit-sharing as a managerial strategy, pure and simple.[20]

Roach was more given to reflection than Hoe or Brewster about the sources of success in the metalworking business. By the eighties Roach was the lone survivor of New York's once-important marine engine and iron shipbuilding industry—perhaps such longevity invited contemplation. He considered his ability to manage workers critical to the making of his shipbuilding empire. His system was set forth on a sign posted during a June 1881 strike at one of his New York City iron works. The sign announced that

> the relations which had always existed between the proprietors of the works were such that men seeking employment should do so in their individual capacities, and that the wages were fixed by the foreman of each shop according to his estimate of the workman's skill, sobriety and industry. This principle was as much in the interest of the workman as of the employer and rather than surrender it the proprietors would close the works.

How did Roach classify the "skill, sobriety and industry" of his employees? He divided his men into three groups, based on criteria such as "how many of [the] men send their children to church or to school, how their families are clothed [and] how many of the men are drunkards," or whether "one man is more wasteful than another" or whether "one man is more careful and economical than another. . . ." The shipbuilder speculated that only one out of every five were "really very valuable men." These were the workers who spent their wages on the comfort of their families. Not included among the "very valuable men," but perhaps in the second group, Roach made clear, were those men who were "very good men when they are at work, but they are never to be relied upon." Workers in this suspect second group were more likely to spend income on "excursions, or gunning, or something of that kind" rather than on their families. The "inferior men" of the third group, the industrialist concluded, were those most irregular in their work habits, most fond of their liquor, and most inclined toward collective action and trade unionism.[21]

Unlike Greeley's scheme for profit-sharing, Roach's "cooperation," outlined in 1872, was unabashedly designed to discriminate between "good" and "bad" workmen and bind the loyal employee to the firm. Roach proposed to divide his work force into twenty-two departments, "each with a competent managing head." The superintendent and foreman would play a crucial role, informing the proprietor about the disposition and personal habits of each employee and his family. Nothing was so valuable to the industrialist as a loyal foreman: Roach took his foremen from his old shops in New York when he opened his shipbuilding plant on the Delaware River. Roach then "propose[d], by a system of cooperation, to make every man interested in the success of the works. Of course no idlers would be tolerated, because such drones only eat the bread of the industrious workmen. No profane or vulgar person will be countenanced in this establishment. . . . No drunkard will be endured, for he but drags to the lowest depths, and demoralizes the workingman's brain. . . ." The "individual method" of fixing wages (personal interviews with each employee) and profit-sharing were the best way of distinguishing drones and drunkards from industrious workmen. During the eight-hour movement of June 1872, a committee from Roach's "Morgan Iron Works" reported that "Mr. Roach said he only cleared ten percent profit, and he would divide that with the men." What Roach meant became clear in his 1883 testimony before the Senate Committee on Labor and Capital. "Men of the right stamp" would be taken on as "partners" in a limited profit-sharing arrangement. Roach's "cooperation" was not, like Greeley's, an attempt to reconcile artisan association with the presuppositions of competitive capitalism. It was, instead, an effort to uproot "combinations among workmen," supplant the cooperative shop-floor ethic of the journeymen with a more individualistic framework of rules and incentives, and guarantee the loyalty of at least the class of "valuable" workers to the company.[22]

Without a doubt, enough of Roach's "temperate" and "industrious" workmen were attracted to strikes and trade unionism to make the loyalty of these workers critical to the preservation of industrial order. Roach could never be entirely sure that his "first-class" workmen were not in fact members of the second group, "the best talkers," who enjoyed their whiskey behind the boss's back. The blacksmith who soberly signed a draft-insurance agreement today might be stoning the Provost Marshal's Office with a drunken crowd tomorrow. The only solution was to do away with skilled workers altogether, both good variety and bad. Roach, like Allaire, could not mechanize his workmen out of existence, but he could train his own labor force. In designing his Chester shipbuilding plant, Roach decided to do away with the class-conscious "English artisans" in the metal trades who plagued city industrialists through the century: "I made up my force from the shad-fishers of the Delaware and the farm hands in the country, and I have to-day as good a shipbuilding force as can be found on the Clyde. . . . They have had eighteen months' training by good foremen at labor-saving machinery. . . . Workers of iron worked under the minds of masters." Roach, Hoe, and other industrial employers believed that workmen trained by themselves or by their foremen would be far less likely to prove "impostors." A truly "valuable man" would not just behave like a loyal worker—he would *be* one.[23]

The Taxonomy in the Community

At the same time that the new taxonomy was articulated by employers in the shop, it was also elaborated by the Association for Improving the Condition of the Poor as a way of interpreting and ameliorating poverty in the community. As already noted, the Novelty Iron Works's major stockholder, James Brown, and its president, Horatio Allen, were for many years president and vice president of the AICP. Iron founder Noah Worrall, carriage builders John Stephenson and James Brewster, and molding-mill proprietor A. T. Serrell were prominent industrial employers among the Association management in the fifties and sixties. John W. Britton, E. A. Quintard, "Hogg and Delamater," "Scovill Manufacturing Co.," and "Hoe, R. & Co." were on the 1863 AICP membership list. Uptown lumberyard owners Isaac Ogden and William Menzies and droveyard proprietors Archibald, Charles, David, and George Allerton were also AICP members. Ogden's yard and the Allertons' two yards and hotel were burned to the ground by the draft rioters. Hoe's factory, Serrell's mill, and Menzies's yard were threatened by the mob. Allen's Novelty Works may have also been threatened; Allen's request to General Wool for scarce arms and ammunition was approved. These attacks and threats did not necessarily mean that the crowds were singling out AICP members: more likely, these were the Republicans the rioters knew best or these were the employers most hostile to the draft riot and/or most eager to get their shops back to work. Given how important AICP employers were in creating a Republican presence in the factory districts, it is not surprising to find quite a few of them among the rioters' targets.[24]

Patrician merchants such as Robert Minturn, James Boorman, James Lenox, and James Brown joined industrial employers Allen, Brewster, and Stephenson as managers of the AICP. Reform-minded physicians, small businessmen, and office workers filled out the AICP ranks—the upward-striving young clerk Richard Hunter, Esq., the end-product of Ragged Dick's metamorphosis in the Horatio Alger story, had the makings of an AICP visitor. What made the AICP reformers so influential and controversial a group was their ability to define the taxonomy in a way that drew the city's disparate evangelical elements together and transformed them into a formidable cultural and political movement. By the mid-fifties, these reformers had mobilized a diverse constituency, an extensive organization, and the power of government to recast the culture of the poor in an individualistic mode and to discipline forcibly an unresponsive pauper class.

Distinctions between worthy and illegitimate poverty and harsh discipline of an unworthy poor were, of course, not new to the mid-nineteenth century. While criticism of indiscriminate almsgiving (as a parade of false piety) predated Martin Luther, the confluence of the Protestant Reformation and the rise of a commercial civilization in early modern Europe led to a new emphasis on the punishment of paupers. Luther himself denounced the demands of beggars as blackmail. The English Civil War occasioned systematic thinking about what economic historian R. H. Tawney called "the new medicine for poverty." A 1649 Parliamentary Act for the relief of the poor and the punishment of beggars offered vagrants a choice between work and whipping and set other poor persons, including children without means of maintenance, to compulsory labor. But in the emerging transatlantic debate over solutions to poverty in a capitalist order, the sharp discrimination between a pious and sinful poor remained for many centuries a minority outlook. It was not until 1834 that British Poor Law reformers overthrew England's paternalist Speenhamland arrangement. Through the 1850s, Young America merchants articulated the dominant paternalist view of poverty in New York City.[25]

From this perspective, five new aspects of the industrialist reformers' outlook emerge. Their taxonomy was a response to the new working-class poverty of the Victorian city. They tended to express the taxonomy in secular terms: they sought not only to save the poor but to "know" the poor. To create a poor that wanted to be "known," they were determined as no group of New Yorkers had ever been to transform working-class culture. They resolved to unify the middle and upper classes behind their categorization of the poor. Finally, they looked to levels of government outside the community for their solutions to urban poverty and were willing to use the cultural apparatus of national loyalty and the institutional apparatus of the nation-state to draw and enforce distinctions between a "good" and "bad" white working class.

The industrialists' confrontation with Victorian poverty was an outgrowth of the revivals that swept the Protestant churches in the late 1820s and 1830s. The inspiration for a comprehensive organization to reform poverty first came from the city's Protestant merchants—Minturn, Boorman, Brown, Dodge, and others—who were stirred by the millennial enthusiasm of that period. They led church missions to the poor and organized the New York City Tract Society, which distributed evangelical literature and sought converts in the slum districts. A. R. Wetmore, a hardware merchant who later became an AICP vice president, recalled: ". . . in the revival of 1831, I became a member of Dr. Cox's church . . . and desiring to do what I could for the cause of Him whom I professed to love and serve, my attention was drawn to the City Tract Society. . . . I became a visitor in a district of about one hundred families, poor but respectable laboring people. Very few observed the Sabbath or attended church; only one here and there professed religion." For Wetmore and other Protestant gentry, the enthusiasm of the revival spurred a new interest in the moral and religious condition of the working classes. Pietism drew these evangelical businessmen into the precincts of the poor.[26]

Wetmore and other genteel missionaries became preoccupied with the domestic scenery of the slums. The lack of ventilation in cellar dwellings and the "congregation of different sexes in one room" were the barriers to religious conversion most often cited.[27] Wetmore observed how a new "neatness" of abode and "cleanliness" of person presaged spiritual awakening in the clients of the Tract Society. In the forties, Dr. John H. Griscom became the city's leading champion of sanitary reform as an instrument of religious transformation. Griscom saw no conflict between his evangelical project and his calling as physician and scientist: ". . . the coincidence, or parallelism, of moral degradation and physical disease," he wrote in 1842, "is plainly apparent to an experienced observer." The capacity of a poor family for virtue, Griscom declared in his *Sanitary Condition of the Laboring Population of New York,* was largely determined by the sinful or Christian aspect of its physical surroundings. Cleanliness indicated godliness.[28]

The Association for Improving the Condition of the Poor, founded in 1843 by the managers of the Tract Society and independently established the following year, drew upon the Tract Society's enthusiasm for converting souls and Griscom's pietistic interest in sanitary reform. But the AICP was formed not in the midst of a middle- and upper-class religious revival, as was its parent organization, but during a prolonged economic depression. Though much indebted to the project and personnel of the evangelical reform movement, the AICP was an attempt to broaden the scope of the Tract Society and other moral reform agencies to address the new social problem of the 1840s: the appearance of a new, widespread, and more permanent class of indigent working people that could not be reached by the evangelizing campaigns of individuals, church congregations, and scantily funded mission societies. The Protestant Episcopal City Mission Society, to cite one example, disintegrated in the years after the Panic of 1837 as a result of financial difficulties and its managers' increasing alienation from the poor. Such occurrences convinced many Protestant reformers of the need for a comprehensive, citywide institution to attend to the moral and social problems posed by the new working class and new poverty of the 1840s.[29]

Though the AICP's founding circle saw the new charity as an instrument for saving souls, they also justified it in more secular terms.[30] The AICP's *Seventh Annual Report* promised to provide New York's upper classes with a "thorough system of personal investigation" and furnish elite philanthropic donors with "security against imposition." The Association's "scientific" system divided the city into one district for each ward and subdivided every district into sections to which one or more "visitors" were assigned. The newest feature of the AICP, then, was its systematic effort to enter into the working-class community: ". . . it traverses every street, and lane, and alley, within the utmost bounds of this vast metropolis; it penetrates every cellar, and garret, and hovel, where the needy are found. . . ." Such a "penetration" of poor neighborhoods by a citywide charity institution would discourage the indiscriminate almsgiver, the AICP's arch villain. A dependent pauper class had emerged because an "impostor" poor had "deceived" private philanthropists. Deception was a feature of the industrializing city, where the well-to-do and the poor were not personally acquainted; it was less prevalent in the countryside, the AICP wrote in 1856, "where individuals are better known and imposition not so easily practised. . . ." The AICP thus represented a new way of thinking among charitable Protestants: in order to create a Christian community in the industrializing city, reformers had to enter into the domain of the poor and intimately acquaint themselves with the working classes. The AICP added the more secular concern of "knowing" the poor to the evangelical mission of saving the poor.[31]

The AICP's understanding of reform was consequently different from that of Horace Greeley or the Union League Club. Greeley believed that a reformed productive process was necessary for social harmony and progress, and wage earners' own associations could furnish the arena for a new kind of relationship between classes. In 1850 the AICP countered Greeley's plans for cooperative production by denouncing all attempts to regulate "artificially" the chronic underemployment of labor in large cities: ". . . no association of citizens is competent to change this state of things by creating permanent employment beyond the actual demand." The AICP adamantly rejected the working-class organizations Greeley thought held so much promise as instruments of consolidation. Union Leaguers such as Olmsted and Bellows believed that reform agencies—Central Park and Cooper Union were good examples—should be situated on the periphery of the city or on the boundaries of working-class districts. Such liminal spaces, belonging neither to the better classes nor the laboring poor, would provide an arena for the creation of a democratic culture. The AICP outlook was distinguishable from that of the Union Leaguers by its insistence on an aggressive infiltration of the working-class community. That is to say, the AICP reformers preferred to visit the poor on working-class territory rather than wait for the poor to seek out reform institutions on neutral ground.[32]

As the AICP set out to acquaint itself with an estranged poor, it had to find needy families that wanted to be "known" and would accept the Association's highly personalized "scientific charity." The historian of London's mid-nineteenth-century Charity Organization Society has observed that COS reformers feared that depersonalized charity would undermine the "voluntary sacrifice, prestige, subordination and obligation" inherent in a face-to-face interaction between the donor and recipient of aid. More than any other elite group in Civil War New York, the AICP reformers had to find or create a poor that acknowledged and mirrored their own outlook and lifestyle. If

Greeley's notion of *métier* sought the distillation of existing instincts and desires among workers, the AICP tried instead to create "new tastes, new desires, new activities and purposes" among a "respectable" poor. Of all the AICP's innovations, this zeal to remake working-class culture was the most strikingly new.[33]

Sitting in the sparse kitchens of applicant families, AICP visitors read aloud from self-help pamphlets such as "The Economist" (on household budgeting and food selection) or "The Way to Wealth" (a selection of Benjamin Franklin's homilies on the origins of poverty in idleness). While male visitors conducted tract readings and urged the children to attend Sabbath and public school, visitors' wives invited daughters to sewing groups. Visitors frequently reported German families to be the most responsive to their overtures. "They are industrious and saving," one Upper West Side visitor wrote, "their business is dirty but commendable, and infinitely above rum-selling. Quite a number of them having found work, they have come, and with much warmth of feeling bidden me good-bye." Reports thus often included some acknowledgment of the visitor's efforts by the recipients of aid. The Upper West Side visitor continued, "I think I have gained the good will and esteem of all who have come to me, even those I could not aid." The 1855 Annual Report proclaimed that "[the AICP visitor's] credentials bear a recognized seal, which gives him unquestioned access to the homes, and often to the hearts of the downcast and outcast. . . ." Finding a respectable poor willing to grant such "unquestioned access"—a poor willing to recognize the legitimacy of the AICP's brand of charity and the desirability of its middle-class values—was part of the Association reformers' own act of self-definition. It was almost as if the AICP reformers had to enter into the homes and lives of the respectable poor in order to discover or confirm who they, the reformers, were.[34]

Deciding who the worthy poor were included figuring out who they were not. The "debased poor," the reformers wrote in 1850, could be recognized by "their indolent and vicious habits . . . so firmly established, that they are not likely to be changed, except under a course of powerful and effective treatment. . . . They love to clan together in some out of the way place, are content to live in filth and disorder with a bare subsistence, provided they can drink, and smoke, and gossip, and enjoy their balls, and wakes, and frolics without molestation." Where Horace Greeley blamed social ills on an obnoxious work environment—"isolated labor"—the AICP understood "debased" poverty as the consequence of personal idleness and immorality. Though the AICP's environmental "sanitary" reforms were a precondition for the social elevation of the working classes, Association reformers ultimately tied all responsibility for social transformation to the moral agency of the individual. The debased poor were not likely to change, and it was all their own fault.[35]

Robert M. Hartley, the animating spirit of the AICP for three decades, was the city's most aggressive campaigner for, as he put it, "the establishment of a line of distinction between the pauper and the independent laborer. . . ." Hartley detailed his social vision in a mid-forties' proposal to set up a laboratory for poor reform on the East River Islands. A Common Council Committee had already suggested that indoor indigent relief be removed from Bellevue Hospital on Manhattan's Upper East Side to Blackwell's Island: in place of the simple plan to remove the poor to Blackwell's, Hartley proposed an elaborate system of discipline and relief that drew near-indelible boundaries between different strata of the poor. He imagined Randall's Island, with its many acres of arable land and freedom from criminal associations, as a work farm for the able-bodied poor of both sexes; here, too, the city could house both the infant poor, "among whom religious and moral instruction should be blended with the earliest rudiments of industry," and the aged and invalid poor, "who by sickness or infirmity are incapacitated for labor." Downstream on Blackwell's Island, Hartley envisioned "a house of reformation and employment for the confinement and discipline of vagrant and disorderly persons." What better place to rehabilitate the criminal poor than next door to the city prison? The Blackwell's almshouse, designed for needy New Yorkers unable to obtain a livelihood "in a respectable way," would be a system of walled yards intended to prevent contact among male and female inmates. While Hartley's Randall's Island farm did resemble a rehabilitation project of sorts, the Blackwell's establishment seemed better suited for the confinement and punishment of the poor than for their reform.[36]

Sharp moral distinctions between two species of the poor were thus enshrined in the literature of the AICP and the writings of its corresponding secretary and agent, Robert M. Hartley. It would be unfair to cast Hartley as a moralizing ogre who wished only to regulate the thinking, behavior, and spatial movements of poor families. Hartley dreamed of eradicating poverty and worked to reform relief institutions which by the forties failed to meet the needs of the vast new urban poor. Like Greeley, he sought new social structures to elevate public health and living conditions and, in so doing, bring community and coherence to the industrializing city. But the taxonomy of the poor Hartley helped articulate became increasingly a device for transmitting or imposing individualist values and controlling the working classes. Between the mid-forties and the war, the reformers attempted to unite the middle and upper classes against a new social enemy called the "pseudo-philanthropist" and became increasingly interested in using the power of the state government to nurture the "independent laborer" and discipline the "pauper."

While the AICP had from the outset condemned the "indiscriminate private almsgiver" who distributed aid without a moral investigation of the recipient, reformers in the mid-fifties began to define that social category more broadly and censure a wide array of "false friends and advisors" to the poor. The mass meetings of the unemployed during the depression winter of 1855 sent a tremor through the propertied classes. For the first time, the AICP associated the inappropriate relief measures of the "pseudo-philanthropist" with the overtures to the poor made by Democratic reformers and politicians: ". . . it is not easy to conceive to what fearful extent mischief would have ensued, if the laboring classes had been as reckless and unprincipled, as some of their misleaders. . . . That the views of these pseudo-philanthropists were fallacious and unworthy of confidence, was abundantly shown." Fernando Wood came to embody par excellence the combination of social and political traits the AICP reformers most feared. Wood's 1857 public works proposal was, of course, anathema. In contrast to Wood's blind distribution of jobs and aid to the poor, the AICP offered "in the completeness of its machinery, the number of its visitors, and in the extent of its ramifications, just the minute arterial system, which the exigencies of the city required." The AICP turned down three-quarters of its applicants for aid during the hard winter of 1857–58. Where Wood's liberal hand-outs would encourage the growth of a dependent debased poor, the reformers saw their private discrimination saving hundreds of applicants "from the deteriorating effects of unnecessary gratuitous relief." As the mayor and his Young America supporters began to use the city government to implement paternalist remedies to social conflict, the AICP saw new cause to enter politics and establish their taxonomy of the poor as the policy of the "better classes."[37]

If deposing "pseudo-philanthropists" like Wood were one reason to enter the political arena, the AICP's growing interest in the legislative reform of working-class living conditions was another. Beginning in 1846, the Association almost annually lobbied the state legislature for a law regulating the construction of tenements. The industrialist reformers believed that sanitary housing was a precondition for the moral development of the respectable poor and achievement of even a modest moral influence upon the unworthy poor. An 1853 report of an AICP "Committee on the Sanitary Condition of the Laboring Classes" resumed where Griscom left off in the forties and conducted a pioneering ward-by-ward investigation of working-class living conditions. The findings of the report spurred the chartering of the privately funded "Working Men's Home Association" in 1854 and the AICP's construction of the city's first model tenement on Mott and Elizabeth streets the next year. The eighty-seven-apartment tenement for black families included thick brick walls between rooms to encourage privacy, fireproof construction, and the absence of "rear suites" so as to allow each room proper ventilation. According to one account, the Mott Street tenement quickly degenerated into "one of the worst in the city." The model tenement also demonstrated to the AICP managers the inadequacy of private, joint-stock investment for housing reform and the need for public legislation to bring about a general change in the construction of working-class housing in the city. An 1856 committee appointed by the New York State Assembly proposed the creation of a "Board of Home Commissioners" that would have the power to enforce building codes against poor ventilation, cellar dwellings, and fire hazards, as well as prostitution, incest, and drunkenness—but no metropolitan housing law was passed before the Civil War. Industrialists had to manipulate and amend laissez-faire theory to justify such interference in private economic affairs.[38]

Industrialist reformers also looked to politics and the power of government in the fifties because they increasingly saw force as a legitimate part of the reform repertoire. Allusions to the need for coercion were never far beneath the surface of AICP discussions of the families of the unworthy poor: "To keep such families together, either by occasional relief or employment, is to encourage their depravity. . . . These nursuries of indolence, debauchery, and intemperance are moral pests of society, and should be broken up." By segregating the men, women, and children of the debased poor in the single-sex "departments" of Blackwell's Island, Hartley and his colleagues hoped to educate these families in the first precepts of bourgeois propriety. In April 1853 the AICP sponsored and the state legislature passed a law for the care and instruction of "idle truant children." The act's sweeping provisions authorized the arrest of children between the ages of five and fourteen years found vagrant in the streets and neglected by parents or guardians. Small wonder that the AICP was an aggressive supporter of the creation of a state-controlled Metropolitan Police and the deposing of Mayor Wood's Municipal force in July 1857. After the movement of the unemployed that fall, the AICP began to collect precedent for more ambitious interventions against the unworthy poor. In 1858, the Association cited approvingly Count Rumford's 1790 *coup de pauvre*—when Rumford swept the vagrant poor from the streets of Munich—as well as the Irish Relief Commission's similar action in the late 1840s. Republican Police Superintendent John A. Kennedy's arrest of five hundred street vagrants in one day in 1860 drew applause from the poor reformers. In such episodes, one might say, Kennedy was rehearsing for the harassment and arbitrary arrest of "disloyal" New Yorkers that would make him so controversial, indeed, notorious, during the war. The reformers' enthusiasm for Kennedy's *coup de pauvre* anticipated their radical alliance with the Union League Club in support of martial law in July 1863.[39]

The Republican Party and the Citizens' Association

The industrialists' problem of creating and courting good workers and disciplining bad ones was not adequately resolved by workplace training, benevolent and profit-sharing programs, or by the Association for Improving the Condition of the Poor. The "quiet and unobtrusive" influence of the

AICP, industrialists believed, would slowly win over working-class families to evangelical reform and middle-class values. But the ambitious agenda of legislative action contemplated by the AICP—truancy laws, tenement reform, professionalization of the police, health, and fire departments—required political power. If the AICP openly entered the electoral arena, it would sacrifice its "unquestioned access" into the homes and hearts of the poor (contested though that access may have been). AICP reformers wanted political power but they could not use the AICP to attain it. Company draft insurance funds, industrial associations, and profit-sharing doubtless helped solve the problem of cultivating good workers in the shop. Attempts to organize trade unions in New York City's foundries and machine shops during the Civil War were notoriously unsuccessful. But while the industrialists' workplace programs may have helped to retard the development of trade unionism, such programs were not sufficiently far-reaching to transform the moral life of the community.[40]

The industrialists' most important political project of the 1850s and 1860s was, of course, the Republican Party. The founders of the AICP were primarily Whigs, and by 1856 most had shifted allegiance to the new Republican Party. AICP reformers and industrial employers were usually not highly visible party officials or spokesmen. "Mr. Roach," the iron founder's obituary read, "had been identified with the Republican Party since its organization, and was a liberal contributor to its campaign funds." Like Roach, most industrialists were associated with the Republicans beginning in the mid-fifties and contributed money rather than leadership to the local organization.[41]

The Republican Party did fulfill some of the industrialists' expectations. The reformers successfully used the party to transfer control of the city's police department (in 1857) and fire and health inspection departments (in 1865–66) to metropolitan boards under the control of the state legislature and the state Republican organization. The reform of the police department in 1857 was an instance of especially effective political cooperation between Republican industrialists and merchants.[42] Though the movement to place New York under "commission rule" could be carried only so far in a city with so powerful a Democratic Party, Republicans did succeed in passing much of the legislation proposed by the AICP in the 1850s and 1860s. From the AICP perspective, the major achievement of the Radical Republican state legislature of 1866–67 was the passage of the Tenement House Law of 1867, the state's first law regulating the construction of working-class housing.[43]

While the industrialists successfully used the Republican Party to pass their legislative reforms, they discovered that the party of evangelical and nativistic tendencies was unable to engage the political allegiance of large numbers of "good workmen" in New York City. In 1857, city Republicans were beginning to notice that German workingmen, whom AICP visitors often praised as "industrious and saving," were not responding to Republican appeals; the problem already visible in the days of the *Arbeiterbund* was now full-blown. German voters tended to be Democrats, if they were loyal to either party. At an 1857 German Republican meeting, Charles Loring Brace observed that some Germans "were beguiled by the word Democrat," and one Mr. Tzchirner, anticipating the criticisms of his countrymen, "defended the measures of the Republican Party. The Metropolitan Police bill was rendered necessary by the circumstances of the case. The Temperance law was not a party measure. He claimed the name Democrat for the Republicans. . . . He did not ask any German to support any man whom he might suspect of Know Nothing tendencies." *Kleindeutschland* Germans preferred Democratic to Republican candidates and were especially likely to desert the Republicans when offered the alternative of an "independent" German candidate, such as C. Godfrey Gunther in 1863. Such Democratic or independent voting in the German community helped to undermine the industrialists' attempts to engage the political support of a "respectable" working class. Almost from the outset, it seemed, the legitimacy of the Republican Party was suspect in the very working-class districts where Republicans thought themselves most likely to win support.[44]

The Civil War greatly intensified the Republican industrialists' problem of legitimacy. As ardent nationalists, they were caught in a curious dilemma. At the beginning of the war they lectured workers about their patriotic duty "to organize and drill, and to volunteer." The following year the AICP congratulated "stalwart mechanics and patriotic working men" who refrained from "insurrections and riots" and defended "the honor and Constitution of their country." Such patriotic exhortation had the ring of sincerity, not of a plot to purge the factories and slums of Irish Catholic workers. The industrialists were certainly capable of deviousness, but that devious they were not. Before long industrial employers faced a shortage and turnover of labor created by army enlistments—they had exhorted, one might say, too well. Workers who remained behind took advantage of such conditions to demand higher wages; indeed, they were compelled to do so by wartime inflation. Novelty Works president Horatio Allen and his colleagues must have wondered how to cultivate "good workmen"—that is to say, a faithful, stable, and affordable labor force—under such conditions. "So great was the difficulty of getting men at that time, owing to the demands of the army and navy," according to one account, that Allen "went to Europe and employed a large number there who were brought over." The industrialists' patriotic appeals no doubt helped them to secure the loyalty of the employees who marched in anti-riot factory brigades in July 1863. But by early 1863 many iron workers had come to associate the Republican Party with both the hardships of the war and the oppressive behavior of employers who were trying to undermine the value of their labor with cheap immigrant and black "contrabands." The economy and politics of wartime distanced industrialists from the mass of workers who came to see their employers' shop-floor regime and the Republican national regime as related threats. The draft riots were the bloody culmination of the industrialists' long-term difficulties with the metropolitan working class.[45]

The Citizens' Association, founded in December 1863 almost immediately after the Republicans' defeat in the first post-riot mayoral election, was the industrialists' most successful attempt of the sixties to gain the political favor of city workers. The Citizens' Association was not wholly an industrialist venture nor was it exactly a political body. The Association's president was the popular War Democrat Peter Cooper, and its members included businessmen of diverse political allegiances, ranging from Democratic financier August Belmont to conservative Republicans Hamilton Fish and William E. Dodge, Jr. But what made the Citizens' Association unmistakably industrialist was its program, an elaboration of the reforms the AICP had been advocating for the previous two decades. The Citizens' Association blamed the draft riots on the political jobbery of Tammany Hall and contended that such uprisings of the "dangerous classes" could be prevented by running the city on "business principles." The Association sought to use a reformed city government to bring about what AICP founder John Griscom had advocated: a thorough study of the city's health conditions conducted by an army of professional sanitary inspectors. Cooper's reformers also attributed the draft riots to the squalor of tenement life—as the reformers put it, " '[The] closely-packed houses where the mobs originated seemed to be literally hives of sickness and vice. ' " In 1865 the Association sponsored a "Council of Hygiene and Public Health," whose *Report* of that year was the most thorough block-by-block sanitary investigation of the city to date. The 1865 *Report* and the reformers' lobbying efforts in Albany were responsible for the creation of a Metropolitan Health Department the following year. Finally, the Association argued in pamphlets such as *Work Is King! A Word with Working Men* that only in a city free of political corruption and unhealthful living conditions could wage earners of "honest, persevering industry" make their way up the social ladder. Above all, the Citizens' Association sought to create a political and social environment conducive to the moral progress and upward strivings of a loyal working class.[46]

While the Republican Party had failed to engage a class of sympathetic "respectable" workingmen, the Citizens' Association found a small but receptive audience in some quarters of the city's revived trade union movement. In September and October 1864, skilled artisans in woodworking, printing, and cigarmaking held rallies in support of the Citizens' Association, passing resolutions against Tammany and in favor of good government and health and housing reform. On September 3, five hundred German cigarmakers planned to participate in a Citizens' Association demonstration against official corruption; the *Arbeiter-Zeitung* detailed the Citizens' Association campaign against Tammany. The organized German working class, which had spurned the Republican reformers in the fifties, joined the anti-Tammany movement in 1864.[47]

Though the liaison between industrialists and artisan trade unionists was short-lived, the Citizens' Association embodied the desire of many employers and workers for a loyal reform agency capable of assuaging class tensions in the riot-torn city. The two groups made strange companions: after all, the artisans believed in cooperative trade organization to restrain the acquisitive employer, and the industrialists dreamed of a workplace free of all rules and associations that might impede the moral and material advancement of the individual. But there was one enemy the two reform camps shared: the grasping landlord. The new Tenement House Law of 1867, based on the findings of the Citizens' Association's *Report of the Council of Hygiene and Public Health,* was introduced by Patrick Keady, former president of the New York Painters' Union and now state assemblyman from Brooklyn. For the trade unions, the flirtation with the Citizens' Association represented a change in emphasis. Before the riots labor was preoccupied with the problem of Republican authority; the insurrection may have had the effect of shifting trade unionists' attention toward the material conditions of urban life. In 1864 trade unionists began to revive and rework that rich compound of economic and political concern that characterized consolidation as defined by labor reformers at mid-century. To prevent a recurrence of the draft riots, industrialists and trade unionists agreed, the homes and neighborhoods of the poor (or the "infected districts," in AICP parlance) had to be cleaned up. When industrialists represented their political project at a distance from the Republican Party, they were able to attract a segment of the organized working class also eager to patrol the slumlord and inclined toward reform experiments at remove from the major parties.[48]

The Republican Party had a troubled career in Civil War New York and failed to engage the loyalty of the metropolitan working class. In the factory towns of rural Pennsylvania, by way of contrast, the Republicans and their allied organizations—Wide Awake companies and the Union Leagues—were not only a legitimate part of working-class life but a resource for wage earners who opposed Republican industrial employers in the mid-sixties. In a June 1866 puddlers' strike against Daniel Morrell's vast Cambria Iron Works at Johnstown, workers marched in their Union League and Wide Awake brigades alongside the puddlers' and boilers' union. Not until the depression of the 1870s was Radical Republicanism discredited among Johnstown's industrial workers; only then did the Democratic Party become more visible in strikes at the Cambria Works. The Republican Party never so became a resource for New York City workers who challenged industrial employers. Already by 1863, many New York wage earners associated Republicanism with a repressive government apparatus and an unyielding defense of the interests and outlook of "aristocratic" reformers.[49]

One reason for this early failure was the intensity of religious and ethnic conflict between a predominantly immigrant industrial work force and Protestant industrialists who saw nativist legislation as a means of controlling the immigrant poor. Most of the puddlers in the 1866 strike at the Johnstown Cambria Works were Protestant Pennsylvania-born farm boys—not, as in the case of many New York industrial workers, Irish Catholic ex-peasants steeped in traditions of struggle against Protestant oppressors.[50] But the provocations of ethnic and religious conflict alone do not explain the New York industrialists' aggressive, interventionist, and embattled style, nor their political failures with the urban poor. Isolated and disadvantaged in the mid-century economy, these employers were spurred to new and creative efforts to reshape the culture of their workers and nurture a loyal contingent in the shop. Further, as early as the mid-forties, they had begun to concede that the opportunities for upward social mobility in the city were not limitless. While the ideology of the antebellum Republican Party presupposed that a competitive free market open to a nation of acquisitive and market-conscious citizens would naturally reward talented workingmen, metropolitan industrialists were, even before the Republican Party was formed, hedging their bets. The industrialists' rigid distinction between a respectable poor worthy of intense moral cultivation and a dependent pauper class requiring surveillance and coercion may have tempered in New York the expansive optimism characteristic of the party of emancipation elsewhere. Some "demoralized" workers (often but not always Irish Catholic immigrants) whom industrialists hoped to label and segregate were less likely than others to respond to the social and economic opportunities of citizenship in an open market society.

Republican industrialists did not give up on the demoralized poor. But they believed that the identification of such an invidious class legitimated drastic and controversial measures to control its behavior: *coups de pauvre* or police actions against vagrants, the breaking up of morally suspect working-class families, and, in 1863, the declaration of martial law in so-called "infected districts." Eric Foner has described the antebellum Republicans as suspicious of great wealth, corporations, and economic concentration, and favoring instead the values of the small producer and small community. But because New York's Republican industrialists had managed via the taxonomy to justify (to themselves) coercive police actions and sweeping sanitary reforms in certain working-class precincts, they were increasingly finding themselves arrayed against a significant portion of the community and allied with the wealthy patricians who had violently put down the Astor Place riots of 1849. New York's Civil War Republican Party was not yet the political organ of a "business elite," as it was to become in the Gilded

Age, but it was nearer to that than it was to the party of small producers Foner depicts. The industrialist taxonomy prepared the way for New York City Republicanism to become, early on, an instrument for coercive solutions to metropolitan social problems and dramatic intervention into working-class neighborhoods and households. By the standards of many mid-nineteenth-century New Yorkers, such aggressive and intrusive use of government was illegitimate and deserved to be repudiated, either through the Democratic Party or through dramatic community actions such as the 1863 draft riots. The draft rioters' goal of driving Republican reformers from working-class districts was directed at New York City's especially interventionist brand of Republicanism.[51]

The Debate over Consolidation Revisited

At the moment the Republican Party rose to national political power and confirmed its ascendancy in a bloody civil war, New Yorkers were engaged in a fierce contest of their own to determine the future social direction of the nation's business capital. This struggle originated in the late 1840s and 1850s, when many New Yorkers began to search for fresh justifications for their authority and new institutions of social and political power as a means of disciplining a conflict-torn urban society. Workers, who bore the brunt of industrialization, were especially assertive in this mid-century debate over justice. Beginning in the 1850s they conceived new institutions that would translate their economic power into political influence and so impose limits upon the rampant individualism of metropolitan capitalism.

At the same time some middle- and upper-class groups also began to experiment with projects of consolidation—efforts to bring coherence to the industrializing metropolis apart from the apparatus and personnel of the "old political parties." Patrician merchants devised a new reform repertoire of parks, schools, clubs, and commissions. Industrialists developed their own quasi-political institutions, the AICP and the Citizens' Association, that could both pass reform legislation and cultivate a worthy poor. By contrast, Young America merchants and their allies readily looked to the mass party organization of the Democracy for solutions to social problems. Such differences in political style reinforced conflicting attitudes. As patricians and industrialists began to view political centralization as necessary for social progress, Young America remained committed to a decentralized polity. While patricians and industrialists began to subject an assertive immigrant poor to moral scrutiny and coercion, Young America allowed that poor much latitude in moments of social crisis. Where patricians and industrialists respected the free black poor and even saw black freedom as the promise of American democracy, Young America understood black freedom as a threat to the stability of an expanding white American empire. The social contests of the 1850s revealed great and growing differences in interest and outlook among the "better classes."

The new Republican Party served as the elite reformers' avenue to political power in the fifties and became the vehicle for their aggressive efforts to remake the culture of the immigrant poor. But it was the Republicans' attempt to construct a nation-state during the Civil War that polarized city leaders into radical and conservative camps, charged local disputes with considerations of national loyalty—and created the possibility for July 1863. By that year, assertive workers faced an elite profoundly divided over the most basic questions of social and political rule. The centralizing policies of the wartime government heightened popular ill-feeling toward the already controversial local Republican gentry. Republican enforcement of the Conscription Act provoked a riot and a crisis in New York because it crystallized and gave sudden focus to these mid-century urban disputes.

Instead of outright winners and losers, the draft riots produced a complicated set of unresolved conflicts. The conservative alliance of Young America businessmen and Tammany politicos emerged from the riots in a position of advantage. Conciliation of the white and Irish poor (by a municipal appropriation to pay the commutation fees of poor conscripts), neglect of black victims, and preservation of local rule established the conservatives' outlook as the dominant one in New York after the insurrection. Radicals were stymied in their efforts to import military government and this was for them a major disappointment. Nor were they able to reverse the contraction and withdrawal of the black community after the riots, despite their efforts through the Committee of Merchants for the Relief of Colored People and the pageantry of the March 1864 presentation of flags to the black Twentieth Regiment. However, the radicals did get some of what they wanted—because they did not declare martial law, the riot did not spread, the draft continued in New York and elsewhere, the army got its men, the North went on to win the war, and the Republicans got the credit. Moreover, the Union League Club and the AICP would remain vigorous influences in the post-riot city. While they failed to bring Fernando Wood to a hangman's halter, they did see the Peace Democracy discredited and loyalty to the war effort established as the dominant political mood in the city. The radicals had suffered only a partial defeat.

Finally, and most important, the draft rioters themselves had a qualified claim to victory. The rioters did not drive Lincoln from office or undo the Republican changes in government. Yet for a week they turned the nation's largest city into a forum on the justice of conscription and prompted action on their behalf (or in accordance with their interests) by powerful local and regional political leaders. They virtually prevented an effective draft in New York City, and the draft they got defended the poor against the unfair provisions of the Conscription Act. The rioters succeeded in making the public life of the city a more noticeably white domain. And they managed to keep the Republican government and its martial law armies at bay. Though the crowds were scattered on July 15 and 16, 1863, it would be some time before the draft rioters met their final moment of defeat. Boss Tweed's post-riot Tammany regime was the creation—and ultimately the casualty—of this lingering disequilibrium in politics.

PART III

Resolutions of the Crisis, 1860s and 1870s

CHAPTER 6

The Rise and Decline of Tweed's Tammany Hall

William M. Tweed's Tammany Democracy was the most successful attempt to resolve the problem of political rule exposed by the volcanic social eruption of July 1863. Tweed and Tammany's domination of the Democratic Party dated from late 1864 and 1865. In the eighteen months after the riots, a loyal Tammany Hall gained ascendancy after the Peace Democracy was branded the provocateur of an insurrectionary working class and traitor to the Northern cause. Tweed's political domination began with the electoral victories of December 1865 and in the state with the victories of December 1869. His regime lasted until the Orange riot and revelations of fraud in July 1871. During this six-year tenure, Tweed's organization managed to negotiate tensions between contesting groups of wage earners and elites, an accomplishment which had eluded a host of Republican and Democratic predecessors during the middle decades.[1]

This account of the rise and decline of the Tweed regime side-steps both the demonology of countless narratives of the Tweed Ring as well as other historians' attempts to rehabilitate the Tweed Democracy as a successful albeit corrupt experiment in municipal government. The concern here is not with the extent of the Tweed Ring's peculations—Alexander Callow has argued persuasively that the theft of public moneys and fraud were vast. Nor is the aim here to demonstrate the success or failure of Tweed and his associates in administering an expanding metropolis and bringing public funds and services to the immigrant poor. Dispensing altogether with the Tweed chroniclers' cults of demonology and rehabilitation, this account seeks to situate the Tweed regime in its historical moment. The Tweed Democracy can be understood only in the context of the dramatic social and ideological conflicts of the Civil War and Reconstruction.[2]

At first glance, one might mistake the post-riot disappointment of Radical Republicanism in New York City and the advent of white supremacist and conservative Tammany rule for the dawning of the Gilded Age in the metropolis, as it were, a decade early. Tammany Democrats accepted the irrevocability of black emancipation but denounced any attempt to bolster black economic and political status at the expense of the white Southern elite. The Tweed circle opposed the use of federal power to enforce black suffrage and only reluctantly came to support black citizenship. As Morton Keller has observed, the non-ideological and bureaucratic "organizational politics" which appeared throughout the North beginning in the seventies had begun to make their debut in New York under the auspices of Tweed's post-war regime of the sixties.[3] But much as Tweed and his lieutenants would have liked to substitute a languid "organizational politics" for the bitter ideological contests of the Civil War and Reconstruction, they pinned their hopes to one of the main impulses of the Reconstruction era, that of popular rule. While in the nation at large the post-Civil War "springtime of peoples" occurred under the aegis of the radical wing of the Republican Party, in New York City that extraordinary political season unfolded under Democratic auspices.

The Tweed regime's idiosyncratic brand of white supremacist popular rule can be understood best as a product of the possibilities and limits of political control defined in the wake of the July 1863 riots. The political possibilities included loyal nationalism, Americanization of the Irish, debt financing, metropolitan expansion, local self-government, and the territorial expansion of a white polity. The limits were defined by the boundaries of the Young America outlook and constituency, the political independence of the Germans, the exigencies of the international bond market, struggles over religion and nationality among the Irish, and, lastly, the political independence and interracial cooperation evident among what was by the late sixties the country's largest and most powerful organized working class. In three cataclysmic episodes during summer 1871—the July revelations of fraud and subsequent "insurrection of the capitalists," the July Orange riot, and the September Eight-Hour movement in the building trades—limits overwhelmed possibilities and Tweed's Tammany collapsed. The problem of rule exposed by the draft riots remained.

Possibilities

During the 1867 New York Constitutional Convention, upstate Republican M. I. Townsend offered the following defense of a metropolitan police force run by non-partisan state commission: ". . . in 1863, when the streets of the city of New York ran with blood . . . but for the fact that . . . the control of its police was in loyal hands, in all human probability, what was a mere mob, continued for three days, employed in riot, arson and bloodshed, would have been a revolution, and we should have had to fight with the elements of rebellion here upon our own soil, as well as at the South."

Linking the containment of the draft riots to the loyal nationalism of New York's police force, Townsend also noted that "two-thirds" of the loyal Metropolitan Police were Democrats.[4]

The rise of Tweed's Tammany Democracy cannot be understood without first referring to the new possibilities offered by the sudden emergence of nationalism during the Civil War as a force in local and national politics. No one, not even the superpatriots of the Union League Club, could match the histrionic flag-waving of a Tammany Hall Fourth of July celebration in the mid-1860s. Grand Sachem Elijah Purdy's July 4, 1864, address to the Tammany membership was reported by one Tammany weekly as "so full of fire and patriotism as to rouse the meeting to the utmost enthusiasm and excite . . . almost continuous cheering." George Washington Plunkitt's famous description of the Tammany Fourth some decades later would have suited equally well the Tweed-era gatherings: "The very constitution of the Tammany Society requires that we must assemble at the wigwam on the Fourth, regardless of the weather, and listen to the readin' of the Declaration of Independence and patriotic speeches. You ought to attend one of these meetin's. They're a liberal education in patriotism. The great hall upstairs is filled with five thousand people, suffocatin' from heat and smoke. . . . Yet that crowd stick to their seats without turnin' a hair while, for four solid hours, the Declaration of Independence is read, long-winded orators speak, and the glee club sings itself hoarse."[5]

Tammany's ebullient Fourth of July celebrations originated well before the firing on Fort Sumter, but the Civil War and the draft riots transformed the meaning of Democratic Party patriotism and nationalism. On the eve of the war, the loyalty of the Democratic immigrant working class to the nation remained at the very least open to discussion. Further, the influential New York City Archbishop John Hughes, if pro-Union, was avowedly sympathetic to slavery. As the war progressed, Hughes's attacks on emancipation and the Vatican's willingness to entertain Confederate diplomats at the papal court made the issue of Catholic loyalty to the Union even more worrisome.[6] By the second year of the conflict, it nonetheless became clear that the Northern working classes, religious and ethnic tensions notwithstanding, were by and large enthusiastic supporters of the war effort. Tweed and Tammany Hall's unwavering commitment to that effort from the first days of the fighting, as Fernando Wood's Mozart Hall organization moved toward advocacy of a negotiated peace, served as evidence of the loyalty of a broad segment of Democratic voters in the nation's largest immigrant and Catholic city.

If the war itself joined Tammany flag-waving to loyal nationalism, the draft riots associated Tammany patriotism with a defense of the social order. The specter of treasonous revolution raised by the draft riots invalidated both the Republican Party and Fernando Wood's Peace Democrats as legitimate political leaders of the metropolis. Republican governance would surely invite new Southern-sympathizing uprisings; Peace Democrat rule might actively instigate such upheaval, undoing post-riot sutures and setting loose a working-class challenge and, many imagined, a bloodbath without modern precedent. Though upstate Republican M. I. Townsend was presenting a brief for non-partisan commission rule of New York City, he was forced to concede an argument accepted by many middle- and upper-class New Yorkers during the post-war years: unless the Democratic metropolis was controlled in large measure by loyal Democrats, it would soon witness revolts alongside which the draft riots would pale in comparison. Tammany Democrat District Attorney A. Oakey Hall and Recorder John T. Hoffman accordingly became the two most popular politicians of the era, launching future careers as city mayor and state governor on the basis of their patriotic indictment and prosecution of treasonous draft rioters in late 1863 and 1864. The debonaire Hoffman made Democratic rule tolerable to many upper-class Republicans who found Fernando Wood unacceptable. George Templeton Strong justified at some length his vote for Tammany candidate Hoffman in the mayoral race among Hoffman, Republican William A. Darling, and Mozart candidate Wood: "Darling stands no chance, and a vote for him would practically aid the King of the Dead Rabbits, so I voted for Hoffman, not willingly. Hoffman is in league with the Ring, and, according to his enemies, robs and steals considerably; but if so, he robs and steals with comparative decency, whereas F. Wood stinks in the nostrils of mankind—the congenial *canaille* of New York excepted. Before and during the war he was as traitorous as he dared be. . . ." In 1867, Strong saw a choice between a powerless and controversial local Republican Party, a loyal Tammany Hall, and a treasonous and insurrectionary Mozart Democracy. Like many other men of his social and political stamp, Strong became a reluctant supporter of Tammany's loyal nationalism and its promise of public order.[7]

The histrionics of Tammany flag-waving during and after the war may then be explained in part by Tammany's new roles as bastion of immigrant loyalty to the Union cause and as patriotic defender of private property and the social order. The rise of loyal nationalism in the sixties created new opportunities for middle- and upper-class groups searching for fresh justifications for their authority. Of these groups, the Tammany Democracy best exploited the new opportunities of nationalism. Wartime nationalism had served to heighten social antagonism in New York before July 1863. After the riots such nationalism, as defined by Tammany, became a vehicle for social compromise.

Tammany-style nationalism was appealing to the many Irish-American men and women who hoped to offer some confirmation of their loyalty after the draft riots. The Tweed years were the first time that one Democratic organization in the city could rely on the political support of virtually all of the Irish social and fraternal associations. For an Irish lower and middle class striving to lose the taint of proletarian treason, the stylish Tammany

Mayor A. Oakey Hall became the emblem of a domain of bourgeois fashion and sensibility open for the first time to the Irish community. Hall believed himself a great literateur; one sympathetic listener imagined that "his first message as Mayor, in point of perspicuity and attractiveness, might have been written by Thackeray." To polish this image of cultivation and elegance, Oakey even submitted his own custom jewelry designs at Tiffany's. He nonetheless still fancied himself a man of the people—in one of his early acts as mayor he abolished the traditional salutation, "Your Honor." The Harvard-educated Hall was not Irish (though he played up a story that a maternal ancestor was one of the regicides of Charles I—the Irish community looked with favor on a man descended from the killer of an English king). But the prosecutor of draft rioters did everything otherwise possible to clothe himself in sentiments of loyal Irish-Americanism. Hall was fond of joking with Irish audiences that his initials A.O.H. stood for the "Ancient Order of Hibernians." One Irish observer remembered the Saint Patrick's Day Parade of 1870 as a high moment of Oakey's career. As Tammany Irish dignitaries Sweeny, Connolly, Richard O'Gorman, Thomas Coman, and all of the city's Irish societies passed Hall's reviewing stand, the mayor saluted "in the supposed regalia of an Irish Prince. It was not enough for him to put a shamrock on the lapel of his coat. . . . to adequately typify his consuming love for the 'Exiles of Erin,' he wore a coat of green material and a flourishing cravat of the same inspiring color." When he attended the lavish 1870 Annual Ball of Tweed's Americus Club in green fly-tail coat, green kids, green shirt embroidered with shamrocks and emeralds, and eye-glasses "with rims of Irish bog-oak and attached to a green silk cord," Hall's unmistakable message was that "Irishness" was now acceptable in New York society. For those Irish men and women seeking inclusion in the world of elite nationalism after the Civil War, Hall's antic display provided a sense of legitimacy. If loyal nationalism was one side of the Tammany political formula, Americanization of the Irish was another.[8]

New kinds of economic activity and organization also promised new political possibilities. During the 1850s New York City became the capital of the new American railroad securities market and headquarters for the professional contractors who built the new railroads. These two events, the opening salvos of Alfred D. Chandler, Jr.'s "managerial revolution in American business," created a new locus of power in the metropolis. The capital requirements of railroad construction were far greater than those of any other antebellum commercial or industrial enterprise. New York City financiers became agents for both American railroad firms in search of purchasers for their stocks and bonds and European capitalists looking abroad for investment opportunities after the political turmoil of 1848. By 1855, railroad, bank, and municipal securities were being traded on the New York Stock Exchange, and by the eve of the Civil War, the New York financial district "by responding to the needs of railroad financing, had become one of the largest and most sophisticated capital markets in the world." One of the organizational innovations of the railroad boom, Chandler demonstrates, was the emergence of large contracting in railroad construction and soon afterward in municipal government. In the fifties, city mayors and councils had begun to imitate the railroads in the letting out of contracts for street paving, school construction, and the installation of water, gas, and sewage systems. Municipal bond offerings patterned after the new railroad securities now gave cities access to large capital for purposes of expansion.[9]

In New York City, 1852 was the year when the new techniques of railroad finance were introduced into municipal construction. Until then, the city had customarily required that payment of contractors for street improvements be delayed until local assessments were collected. Similarly, the city would not reimburse those whose land was taken in the opening of streets until the comptroller had received sufficient assessment moneys. The effect of this system was to limit contract work to the few large firms with enough capital to survive the wait for payment and to slow the pace of municipal expansion. In 1852, a law was passed instituting the practice of issuing assessment bonds in anticipation of the collection of assessments and paying contractors from the bonds' proceeds. By 1860, $1,898,200 had been raised through the new assessment bonds and a longstanding constraint on the growth of the built city removed. Equally important, the occupation of street and utilities contracting was opened to hundreds of men of small capital, many of them Irish grocers and laborers who could parlay their associations among the neighborhood gangs and fire companies and a little savings into a successful construction outfit. The innovations of railroad finance also can be said to have created new opportunities for fraud in municipal finance and construction, opportunities exploited a decade later by the Tweed Ring. An 1858 law seeking to prevent the overcharging of costs of improvements to property owners allowed aggrieved parties to appeal to the state supreme court to vacate assessments—but the loose wording of the provision made it an opening for uptown real estate owners to evade just obligations. All told, the 1850s saw the appearance of a whole matrix of new metropolitan economic institutions. The national railroad system, the Wall Street market for railroad and municipal securities, the expansion of municipal contracting, and the new opportunities for fraud all collaborated to expand the possibilities of political rule in the middle decades.[10]

The use of bonds to finance municipal expansion in the fifties enhanced the power of the city's political leaders in at least two ways. First, during the Civil War epoch many of the city's social and political institutions were being "mapped out" as contested terrain in the intensifying class debates between social groups. The draft riots, of course, reveal that by 1863 the process of identifying urban institutions with class interests was well advanced. But the new instruments of railroad and municipal finance had not yet been clearly associated with specific upper-class interests in the 1850s and 1860s.

Needless to say, with the strike wave of 1877 workers all over the country would see clear connections between railroad and municipal expansion, Wall Street, and upper-class interests—but not in 1857. Because in the fifties and sixties workers were only beginning to regard the railroad system and its related institutions as controversial, those institutions could still serve as the groundwork for popular rule.

Second, before the 1850s, railroads were financed exclusively by investors in the regions they serviced. If, with the mid-century rise of the New York money market, American railroads could now be capitalized by European investors, was it not also possible to draw upon distant sources of capital for financing municipal government? This question had to have been forming in the minds of New York's Democratic political leadership during the 1850s and early 1860s. Merchants such as August Belmont, John A. Dix, Augustus Schell, and Moses Taylor, New York's Democratic leaders, were the nation's premier railroad financiers. While any one of these men might have seen early on the possibilities of municipal expansion and political control inherent in the international securities market, August Belmont was doubtless the most brilliant political innovator of the Tweed era. As American agent of the House of Rothschild, Belmont became the conduit for millions of dollars of European investment in American railroad construction in the fifties and sixties; as Rothschild agent and national chairman of the Democratic Party, Belmont helped to sell millions of dollars of New York City municipal bonds to European investors in the late 1860s and early 1870s. Unlike Republican schemes to remove the locus of power from the city—schemes such as commission rule which offended New Yorkers' powerful aspirations for local self-government—the Democratic attempt to expand the city through foreign capital could operate in tandem with the preservation of local self-government and even the reduction of taxation. By no accident, the same men who discovered new ways to finance railroads in the fifties found analogous means to fund municipal expansion in the Tweed era.[11]

In New York City, the Civil War furnished lessons in the ways that loyal nationalism and debt financing through municipal securities could work together to preserve political rule. The accumulation and funding of New York City's war debt was a rehearsal for the debt financing of the Tweed years. Soon after the firing on Fort Sumter, New York's predominantly Democratic Common Council began paying bounties to the families of volunteers through the issuance of municipal war bonds. After the draft riots, the County Board of Supervisors, half-composed of Republicans by state law, first floated its own municipal securities to pay for damages to property during the riots. The Board of Supervisors, we recall, then formed a bipartisan committee composed of William Tweed, William T. Stewart, Orison Blunt, and Elijah Purdy to exempt poor men with dependent families and certain municipal service groups (police, firemen, militia) as well as administer the three-million-dollar bond revenue raised to buy substitutes for exempted conscripts. After the draft riots, Republicans and War Democrats on the Board of Supervisors hoped to transfer as much of the bounty system as possible to the jurisdiction of their own loyal officials and away from disloyal Peace Democrats on the Common Council. The Supervisors were not alone in regarding the three-million-dollar bond issue designated to pay poor conscripts' $300 exemption fee as a means of preventing future draft riots. Tweed's County Substitute and Relief Committee was commended by Secretary of War Stanton for its patriotism and determination to "aid and sustain the government in crushing the wicked rebellion and in vindicating the majesty of the law." At its peak in 1869, the municipal war debt accumulated from the payment of bounties to volunteers' families and substitutes for exempted men amounted to $14,597,300. This war debt dwarfed the other bonded debt of the sixties—at that time the next largest city debt, for the construction of Central Park, did not much exceed $10,000,000. The bonded war debt also exceeded the cost of the city's largest antebellum expenditure, the Croton aqueduct. It was unlikely that debt financing would have gained the legitimacy it did in the Tweed era without the precedent of the war debt. The Civil War educated Tweed and other loyal Democratic leaders in the ways that loyal nationalism and municipal securities could mutually reinforce political rule.[12]

While debt financing would furnish the Tammany Democracy with a means of financing its elections, rewarding Tweed Ring members, buying the allegiance of local and state political leaders and bringing funds and services to the immigrant poor, municipal expansion allowed Tammany to accommodate the political challenges of wage earners and elites evident during the draft riots. New York's building, utilities, and real estate interests became the fulcrum of metropolitan class relations in the middle decades: only by controlling these interests of municipal expansion could political leaders influence workers and elites on either side of the ideological division revealed in July 1863.

The Democratic Party had begun to form a close relationship with the city's building and real estate interests during the 1850s. Uptown builders and contractors were prominently involved on both sides of the November 1857 controversy over Wood's public works program: Democratic builder Gustavus A. Conover was a leader of the People's Union movement for Daniel F. Tiemann, and a number of other uptown contractors remained loyal to Wood. An 1863 memorandum from the "Finance Committee" of Tammany Hall listed under the heading "Miscellaneous Contractors" twenty-nine builders, master brick masons, stone masons, plasterers, and carpenters, at least half of whom were Irish-Americans and nearly all of whom did business north of Fourteenth Street. The document did not indicate whether these contractors were donors or recipients of Tammany funds but at the very least suggested that the Tammany leadership was closely tied to uptown building interests by the time of the Civil War.[13]

The events of the draft riots, we recall, reinforce the hypothesis that uptown building and utilities contractors and real estate developers occupied a pivotal point of influence in metropolitan society and politics. Contractor and Democratic Alderman Peter Masterson, his brothers William and John, and their Black Joke Engine Company Number 33 led the Monday rioters' charge on the Ninth District Provost Marshal's Office. Masterson soon after repudiated the spreading violence and worked with his fire company through the week to defend property against arson in his uptown district. Clearly a popular and important man in his neighborhood, Masterson received a public commendation from the "grateful residents of West 52nd, 53rd and 54th Streets" for his riot-week valor. Alderman Jacob M. Long's name appears in the draft riot indictment records both as the victim of a sidewalk hold-up on July 13 and as the employer of Martin Hart, a laborer who led a charivari procession to the stores of Harlem's Washington Hall district. Long, known as "Colonel" to the laborers in the uptown Twelfth Ward, was president of the Harlem Gas Company and a utilities contractor. In his affidavit Long told the court he was threatened by laborer Richard Lynch and a gang of men and managed to avoid personal harm only by appealing to Lynch's public responsibilities as a fire laddie and by paying the group two dollars for a round of drinks. Both the Masterson brothers and Jacob Long remained powerful in uptown Democratic politics well beyond the Tweed era: John Masterson and Colonel Long were listed in 1875 as "leading and prominent men" in the General Committee of Tammany Hall.[14]

The glimpse into the social world of Masterson and Long afforded in 1863 suggests that these men occupied a station of high political influence in the wartime city. Like the uptown workers they so often knew by name, Masterson and Long opposed Republican wartime policies and could support or at least abide the Monday uprising against the draft. But the two contractors were also quick to rally behind Tammany Hall's loyal midweek defense of property and the social order. It seems that by 1863 building and utilities contractors played a powerful mediating role between an unconditionally loyal Republican elite and an insurrectionary and anti-abolitionist uptown working class.

The increasingly intimate marriage of the Tammany Democracy to the interests of uptown development can be understood only in the context of the shifting social geography of the city during the 1860s. Changes in the value and use of uptown land in the 1850s and 1860s altered the context of the metropolitan real estate market. As the economic life of the city began to move away from the downtown seaport to a middle-island retail and manufacturing district straddling Broadway, the consumption of goods and services produced in New York City for local purchase and use became the new focus of the metropolitan economy. The rise of New York City as a national center of fashion and consumption by the Civil War was for nineteenth-century observers synonymous with the rise of Broadway. Henry James later recalled the great avenue as "the feature and the artery, the joy and the adventure of one's childhood," with its gleaming shops and department stores comprising a world of goods "heaped up for our fond consumption."[15] At the same time as a middle-island district surrounding Broadway became a center for the manufacture and retail of consumer goods, uptown property values and rents soared, encouraged by post-war inflation. One Isaac Kendall, the putative author of an 1865 pamphlet entitled *The Growth of New York,* led a chorus of boosters hailing the transformation of the uptown city into a high-price and high-rent commercial and residential area:

> Is there room for any doubt that Broadway will retain its character, preeminent character, as a business street from the Battery to the Central Park? . . . Trade is now erecting its palaces on Broadway. . . . Buildings of magnificent proportions, massive curved and sculptured fronts of marble, durable walls and foundations and elegant finish, modelled after the sumptuous palaces of Italy, are now the homes of trade. . . . These luxurious habitations and appointments were fittingly given to trade while New York was a great trading city and was making its money thereby. In the next stage of the city's growth, when it is becoming a metropolis, we shall build palaces, not only for business, but also for residences.[16]

Kendall and others hoped that Broadway would establish the tone for the post-war metropolis by extending its genteel precincts into the half-built city north of Fourteenth Street.

As the uptown city became an area for intensive real estate investment and development after the Civil War, the factories that lined the East and West Side waterfronts began to disappear. Many left the city, the final step in the migration from downtown begun in the forties and fifties; others went out of business. In the early seventies Singer Sewing Machines moved part of its operation to Elizabethport, New Jersey, and John Roach transferred much of his to twenty-three acres on the Delaware River outside Chester, Pennsylvania. The Novelty Works closed its doors in 1870, its East River real estate now "very valuable" and its tools and machinery "old and out of date." A *Times* investigation of uptown factories in October 1869 noted that "ten years ago the iron works of New York were the pride of her manufactures and splendid lines of steamers went forth from her docks, equipped with engines and machinery which could not be surpassed in the workshops of any nation. Now . . . [with one exception], not a solitary marine engine or iron steamship is in course of construction in the great naval metropolis." The Quintard plant limped on with vessel repair work and the great Allaire Works became a horse stable. It is not surprising, that some factories tried to convert to construction-related industry. Before failing altogether, the Novelty Works attempted the manufacture of architectural iron, as did Roach's old Etna Works in fall 1869. The Neptune Works became a sawmill. Isaac Kendall and John Roach would have painted sharply contrasting pictures of the transformation of the uptown city after the Civil War: Kendall saw a boom in residential and commercial real estate, while Roach, had he paused to consider, would no doubt have seen industrial consolidation and decline.[17]

One effect of this uptown transformation was to create a new political interest in the metropolis, that of the real property holder. Of course, since the eighteenth century investment in real estate had been an important way for city merchants to tap steady sources of income to weather the uncertainties of transoceanic commerce and consolidate their position as an urban gentry. Elizabeth S. Blackmar has shown that, by the 1840s, with the rise of a speculative building industry, the market in uptown lots was beginning to expand beyond an older rentier class to attract numbers of prosperous master artisans in the construction trades.[18] But only after the Civil War did metropolitan real estate owners become a self-conscious, organized, and coherent political interest. The signal events in this change were the creation of the West Side Association in 1866 and the founding of the East Side and East River improvement associations in 1868.

By the late sixties, many uptown property owners—among them industrialists, merchants, and prosperous master artisans—began to regard themselves as members of a distinct social and political movement. The West Side Association was formed in 1866 by William R. Martin and other owners of real estate north and west of Central Park to insure that state legislation passed to project the development of the West Side according to a rectangular grid would take into account inequalities of surface and the promise of the area as an upper-class residential district. Martin and his allies succeeded that year in placing West Side development under the control of the Central Park Commissioners and kept the Association in existence to insure that municipal and state officials would attend to the public works necessary for uptown expansion. The East Side Association was formed two years later, in late winter 1868, with the same rationale as its crosstown counterpart. Membership in these uptown organizations was open to all owners of property within the designated boundaries. The executive committees of both groups were dominated by merchants and lawyers who lived uptown or owned uptown lots. The West Side Committee spanned the political spectrum from Mozart Hall Democrat Fernando Wood to Tammany Democrat Daniel F. Tiemann (Wood's opponent in the mayoral race of 1857) to Republican merchant shipbuilder Marshall O. Roberts. Small manufacturers and retailers were more heavily represented in the association of the more commercially developed East Side.[19]

The East River Improvement Association resembled the West Side and East Side groups in organization and style but differed in membership and outlook. The by-laws and constitution of the East River group were "based on and correspond[ed] to" those of the West Side Association. Similarly, the East River Association sought to influence local, state, and federal legislators. An early meeting of the East River group planned to petition Congress for an appropriation to enlarge the Hell's Gate entrance to the East River and allow large-tonnage vessels northern access to the East Side waterfront. One difference, though, between the East River and the West Side and East Side Associations was in the sort of individuals who composed their respective leadership. The East River Association was primarily an industrialist outfit. John Roach was chairman of its Finance Committee and was joined on the Executive Committee by prominent retail-trades employers George R. Jackson, James R. Taylor, T. F. Secor, Thomas T. Rowland, and Erastus N. Smith. Finally, if the West Side and East Side Associations were the cheerleaders of the post-war movement for a genteel, residential uptown expansion, Roach and his East River colleagues sought mainly to enhance commercial access to an industrial waterfront already in decline.[20]

Despite these differences of emphasis between the three real estate associations, they shared common interests. The *Real Estate Record and Builders' Guide,* founded in 1868 to represent the city's newly professionalized real estate business, properly saw the various real estate associations as part of the same movement: "The happy results derived from the organization of the West Side Association, has led to the formation of an East Side Association, [and] also an East River Improvement Association. . . . We hope in time to chronicle a North-End Association, a Middle Section Association, and a Downtown Association. . . . Heretofore, our property owners have not acted together . . . [and] with the property holders organized as they should be, we will have good local government." With the inauguration of a five-year speculative building boom in New York City in spring 1868 and a Tammany program of improving docks, piers, wharves, and roads over the following years, members of all three associations stood to gain.[21]

William R. Martin was the most articulate and aggressive spokesman for the uptown real estate associations. A lawyer and real estate developer, Martin was a complicated character who supported the Democratic Party and abided Tammany rule but emerged as a defender of reformer Frederick Law Olmsted in the Park Commission political struggles of the seventies. Martin's discussion of the West Side Association's support for the Tweed Democracy, presented in a series of late 1870 and 1871 pamphlets, helps to explain further why the broad segment of the middle and upper classes represented in the post-war real estate groups was willing to endorse or condone Tweed Ring rule.[22]

As president of the West Side Association, Martin took credit and responsibility for the Tweed Democracy's rise to power. "They are, by our aid, firmly seated, and have now no need to give all their attention to holding themselves in; but are settled in their duties. . . . They are well paid, and paid in advance. . . . They hold certain prominent pieces of property which stand out, as one walks uptown, as so many 'Receipts of payment.' They were supported in the last election by many taxpayers, on the avowed ground . . . that the great public works on this Island would be vigorously pushed forward, even without our help." Much of Martin's enthusiasm for the Tammany Democracy was attributable to the Tweed-controlled state legislature's passage of a new city charter in 1870. The Charter of 1870 restored to the city much of the power which the commission rule legislation of 1857 had removed to Albany. In particular, the new arrangement invested sweeping powers of appointment in the office of the mayor, limited the powers of the Common Council and abolished the corrupt Board of Supervisors, eliminated overlapping jurisdictions between departments, and converted the metropolitan commissions of Police, Fire, and Health into local departments. Martin noted that the Charter of 1870 in many cases restored local rule to a "private corporation of New Yorkers" rather than "the municipal government" but, all told, approved the new charter because it invested centralized power over public improvements in a local body and eliminated the inconvenient older practice of petitioning Albany. Like Martin, many other upper-class New Yorkers saw the new charter as a simplification of rule which would limit the possibilities for corruption and enhance the opportunity for "public opinion" of people of their own sort to influence government. Peter Cooper's Citizens' Association gave the new municipal arrangement full support; leading citizens such as Moses Taylor, H. B. Claflin, C. L. Tiffany, and Andrew Gilsey petitioned the state senate for its passage. The Republican Union League Club opposed the new charter by a small majority, but the Club had already endorsed many of its provisions in the 1867 report of its "Committee on Municipal Reform."[23]

William Martin and the West Siders hoped that the Tammany-controlled state legislature would quickly pass a rapid transit bill funding construction of Moses Beach's Broadway Underground Railway. But March 1871 petitions by the West Side and East Side Associations in favor of the underground railway failed to counteract the opposition of downtown Broadway merchants such as A. T. Stewart, and Governor Hoffman vetoed the Beach Transit bill. Hoffman did sign into law a "Viaduct Bill," which in the spirit of the new city charter, invested powers of "route, plan and capital" for a metropolitan rapid transit system in a Tammany-controlled local body. The West Siders immediately shifted their focus from the Albany legislature to Peter Barr Sweeny, the Tweed Ring lawyer who promised to play the largest role in directing the planned transit system. In an April 1871 speech to the West Side Association, Martin demanded that Sweeny and his Tammany associates make the construction of the Broadway line a priority. Martin saw heady possibilities in Tammany's new sweeping and centralized powers over municipal improvements: "I think it better to look at this absolute power face to face. It enables us to have a vigorous government, and a local government. . . . If you look abroad throughout the country for the combinations which are the sources of the greatest wealth, the most pervading financial influence and of political power, you will find them to be the comprehensive railroad monopolies. As a field for such enterprise, the fifteen miles' length of this Island with its future millions of population and its great commerce is better than any fifteen hundred miles of outside territory. . . ." William Martin and the hundreds of uptown property owners he represented hoped to bring the large capital means, centralized organization, and galvanic commercial energy of the railroads to post-Civil War municipal government and expansion.[24] When Mayor A. Oakey Hall denounced the opponents of public improvements as "some rich old men who cannot realize that New York is no longer a series of straggling villages or that taxation is not so much a matter for the present as for the future," Martin and members of the uptown real estate associations approved the sentiment.[25]

William Martin and the West Side and East Side Associations stood at the center of a movement with a broad social and political periphery. Though the industrialists who formed the East River Improvement Association opposed Tammany's indiscriminate public hand-outs to the poor, as uptown property owners they applauded the Tweed Ring's program of dock, pier, and wharf improvement, street grading and paving, rapid transit, and tax reduction. The East River Improvement Association seemed to have regarded Democratic president of the Citizens' Association Peter Cooper and Governor John T. Hoffman as key intermediaries in its relationship with the Democratic city leadership: in a May 5, 1868, meeting, the East River group made provision to extend the western boundary of its membership and jurisdiction from Third to Fifth Avenue explicitly to include Cooper and Hoffman in its ranks. That the morally irreproachable Cooper sanctioned the new charter behind the Tammany public improvements program and the respectable Hoffman signed the Tammany municipal expansion bills into law no doubt brought Tammany rule much legitimacy. Republican James Francis Ruggles was another leading citizen whose approval of Tammany's new charter lent authority to the improvements program. As secretary of the West Side Association, James F. Ruggles brought the weight of his father's reputation to the property owners' group. The name of Samuel Ruggles, the distinguished real estate developer who created Gramercy Park and Union Square, meant one thing to mid-century New Yorkers: the transformation of the uptown city into an upper-class residential preserve. With this sort of support for its municipal expansion program, it was no surprise that the Tweed Ring found six merchants and financiers to review cursorily and endorse its Finance Department and Sinking Fund records on the eve of the November 1870 election. At least three of the six businessmen were deeply committed to Tammany municipal expansion: John Jacob Astor as the city's largest property owner, Moses Taylor as a substantial property owner and principal in the city's two leading gas companies, and Marshall O. Roberts as a member of the West Side Association executive committee.[26]

In sum, Boss Tweed's Tammany Democracy drew the allegiance of a new and self-conscious social and political interest in the metropolis through its relationship with the uptown real estate associations and the city's largest owners of property. Here it is worth noting that Tweed's alliance with real estate, construction, and financial interests and his use of city building to mediate urban social conflicts bore some resemblance to the policies of Baron Georges Haussmann in Paris after the upheaval of 1848 (though of course Tweed lacked both Haussmann's intention of breaking up working-class neighborhoods and his ability to orchestrate urban social processes—for that, New York would have to await Robert Moses and the twentieth century). For a time, the post-Civil War movement of "uptown property" promised to accommodate tensions between Tammany's leaders and its social and political opponents and provide ballast for Tweed's popular rule.

Tammany leaders especially hoped that metropolitan expansion would provide a solvent for class conflicts within the building industry. The massive bricklayers' strike of 1868 provided the first major test for the Tammany municipal improvements program. The backdrop for the strike was the unprecedented infusion of merchant capital into the city's building industry in early spring 1868, occasioned in part by post-war inflation and resulting high property values and in part by Erie Railroad directors Jay Gould and James Fisk's speculations in railroad stocks, which convinced many merchant investors that real estate and building provided a safer outlet for investment as long as the money market remained in a state of artificial fluctuation. The *Real Estate Record* reported in July that "owing to the unsettled condition of the money market, and the difficulty among capitalists for selecting investments, a very large portion of the pent-up capital of the city was being turned in the direction of building; and two or three months ago there were symptoms of such activity in this direction as was never before witnessed in this fast-growing metropolis." Meanwhile, on June 21 at least two thousand journeymen bricklayers began a strike for the eight-hour day with 10 percent reduction in wages. Construction on the new spring contracts halted, and the press made dire predictions for the future of the building boom. The New York *Sun* noted early on that "much, if not most of this money [the new capital invested in building] is borrowed, some of it at a heavy rate of interest and the question arises whether the lenders and the brokers will allow their investments to lie idle during the summer in case of a strike." The *Merchants' Magazine and Commercial Review* urged employers to "persist in refusing to comply with [the eight-hour] demand," and the *Real Estate Record* warned tersely that the eight-hour movement "kills the goose that lays the golden egg." The 1868 bricklayers' strike to reduce the work day from ten to eight hours arrayed the power of the organized working class against the city's building, real estate, and merchant capitalists as never before and exposed some of the fault lines in the Tammany political arrangement.[27]

The bricklayers' strike revealed a contradiction within that arrangement. The very same journeymen in building who responded to Democratic appeals for local self-government, preservation of an all-white polity, and metropolitan expansion were militantly defending workplace prerogatives that ultimately placed limits upon the breakneck pace of Tweed-style city building. Misgivings about the quality of Tammany municipal development had already surfaced in fall 1864, when a mass meeting of "workers in wood" endorsed the Citizens' Association and passed a series of resolutions critical of Tammany waste and corruption. The gathering complained of exorbitant city and county taxes, "a fraction of which, if well and honestly applied, would be amply sufficient to preserve order, cleanliness and good government." A Mr. Poer remarked that politicians talked "of new markets, sewerage and all sorts of improvements . . . every year just before election and what has ever come of it?" The sentiment of the meeting was that a municipal improvement program should be administered not by the two major parties but by frugal and responsible workingmen elected to the legislature by an independent political movement. After a bitter assault on "politicians," Poer urged his listeners to "walk up to the ballot box and cast your ballot for honest, unsophisticated workingmen." Though these initiatives had little political consequence in 1864, the building trades were issuing an early warning to the Tammany developers. Other signals came in 1865 and 1866, when the city Workingmen's Union, with the building-trades unions at the van, began to campaign for a state law mandating eight hours as a legal work day. Under combined Radical Republican and Democratic auspices, an eight-hour bill was passed in April 1867—the result of public pressure from Workingmen's Union rallies and picnics, threats of a general strike from a State Workingmen's Assembly in Albany, and a fierce statewide competition for workingmen's votes. In practice, the law was crippled by a provision that guaranteed freedom of contract: workers and bosses were free to negotiate terms of employment as they pleased. The bricklayers' strike of 1868 was an early attempt by one building-trades union to enforce the law on its own. The implications of this "political" use of the trade unions for the ongoing debate over consolidation and the outcome of the draft riots form a separate chapter of this story and will be discussed in turn. For present purposes, it is useful to consider the bricklayers' strike as an episode in a slowly mounting challenge to the city's building and real estate interest for control over the pace and quality of metropolitan expansion.[28]

Shortly after it began, the bricklayers' strike expanded into a referendum on an entire catalog of journeymen's work rules. The journeymen's proposal for an eight-hour work day immediately raised the question of productivity: could the bricklayers produce in eight hours what they had formerly produced in ten? The largest builders in the city immediately saw an opportunity to challenge the journeyman's regulation that in their view placed the most serious restriction on productivity—the 1000-brick-per-day stint. One master mason, Alexander Ross, was reported to have told a meeting of bosses that the journeymen's stint would halt the trend toward large investment of capital in building: "Men . . . have now plenty of paper money, and they are investing it, but they will not do so another year. When he was a young man anything less than 2000 brick laid per day was considered a mean day's work, but now . . . not more than 1000 brick are laid by the majority of bricklayers. He had measured work repeatedly and found that the average was 700 brick per day." Fellow boss Peter Tostevin echoed Ross's theme. "A few years ago men would lay 2000 brick per day for $2, now they lay not more than 1000 brick for $5, an increase of fivefold," Tostevin lamented. "The more employers are required to pay, the less labor they receive." To the thinking of the *Real Estate Record and Builders' Guide,* the answer to the problem of restricted output was piece work: bricklayers should be paid at a fixed rate per thousand bricks. Piece work, many employers believed, was the only sure way to undermine the journeymen's customary injunctions against sweating.[29]

The thousand-brick stint was one of an array of journeymen's regulations the bosses challenged during the strike. The master masons disputed the bricklayers' longstanding prohibition against bosses working on their own buildings unless they belonged to a trade union, their rule that no employer have more than two apprentices at any one time, and their customary regulation of the ages and indenturing process for apprentices. In an unprecedented move, the largest builders in the city formed a permanent "Master Masons' Association" under the leadership of John Conover, created a legal fund, and brought suit against the Bricklayers Unions' work rules under the state conspiracy law. In a trade where the interests of a few large employers were traditionally pitted against those of the mass of smaller employers, subcontractors (or "lumpers"), and journeymen, the formation of a Builders' Association and the fresh assault on journeymen's work rules were attempts to impose the prerogatives of the most highly capitalized builders on all parties.[30]

As the bricklayers' strike continued through the summer and fall, it attracted the attention and support of the local and national labor movements. Local building-trades organizations donated thousands of dollars to the bricklayers' strike fund; William Jessup, president of the New York State Workingmen's Assembly, feared failure for the bricklayers would mean that "the Eight Hour system is defeated for years." In August, the builders paid to import Canadian strikebreakers to the city, while the bricklayers' unions patrolled steamboat landings and railroad depots. Despite late summer defections by some German workers and a number of "front men"—the highly skilled labor aristocrats of the trade who worked only on façades—the ranks of the eight-hour men remained firm. By mid-fall, it had become clear that the strike was at least a qualified success: Local Number 2 reported 1,030 of its members working eight hours, 511 employed at ten hours, and 76 idle.[31] Equally important, the bricklayers proved their ability to enforce their sanctions against the most highly capitalized parties in the industry. In his testimony fifteen years later before the Senate Committee on Relations between Labor and Capital, bricklayer William C. Anderson recalled:

> . . . in 1868, Mr. Conover, who was then the leading boss mason of this city, spent . . . [one million dollars], with the backing of such men as Mr. Vanderbilt, to defeat the bricklayers alone. At that time the [building] trades were not amalgamated. The bricklayers struck on their own account and were defeated, but not until Mr. Conover was bankrupt. It was found impossible, though, at that time for the bricklayers to bankrupt Mr. Vanderbilt. It was a well-known fact that Mr. Conover was then the leading boss mason of this city, and from holding out in that strike in 1868 he died a poor man.[32]

From the perspective of 1883, when New York City bricklayers fought merely to preserve Saturday as an eight-hour day, the 1868 strike may have appeared a defeat. But Anderson's recollection of the financial straits of John Conover, the city's largest builder, prompts an opposite reading of the strike's outcome. In 1868, the bricklayers demonstrated that the city's largest builders and finance capitalists would have to contest workers for sway over the quality and pace of post-war municipal expansion.

The bricklayers' strike owed its qualified success to two factors: the willingness of many smaller employers and subcontractors to yield to the eight-hour rule and the reluctance of the largest builders to attack the eight-hour movement head-on. At a July 15 meeting of the master masons, Alexander Ross noted that "only the owners of small jobs, who never built a house before, can't bear the idea of waiting to get their houses ready to move into. [The bosses' committee] had some trouble with these. . . ." Typical of these small employers was one "Mr. McCormack, who has seven men in his employ, but over whom he has no control, having lumped out his work to two journeymen who had employed others, and they were working eight hours." Small bosses like McCormack and the journeymen subcontractors he employed were often forced to perform their contracts with eight-hour men. With their often tiny margins of profit, these employers were highly vulnerable to the vagaries of weather and could hardly afford to suspend operations for an entire building season. It stands to reason, too, that some small bosses and subcontractors not only tolerated but sympathized with the eight-hour movement.[33]

Before 1865, the balance of power in the city's building trades was indisputably on the side of journeymen and smaller contractors. But with the resurgence of speculative building after the Civil War, the ranks of large builders—men who like John Conover employed fifty to a hundred workers—began to swell. With the unprecedented large investment in construction during the spring of 1868, the men of large capital saw their first opportunity to challenge what they accurately perceived as the hegemony of the small producer at the point of production. Hoping the prevailing gospel of frenetic city expansion under Tammany would create a favorable climate of public opinion, Conover, Tostevin, and other large contractors sought to use the eight-hour conflict to gain broader control of production and confirm the building industry as a principal outlet for finance capital. The length and qualified success of the bricklayers' strike suggests that even in 1868 the independent interests of the *petit* contractor still posed a formidable problem for Conover and other large builders.

In truth, Conover and his "Master Masons' Association" were not yet ready to launch a frontal assault on the eight-hour movement. Master Masons' leader Alexander Ross told a meeting of builders on July 17 that he had recently had "a conversation with a prominent man of the country that very morning and . . . [told] him the nature of the controversy between the master masons and bricklayers . . . that it was not the eight hour rule, but the claim of the journeymen to limit the number of apprentices and forbid the bosses from working. . . ." A few days later, Ross conceded that "the eight hour men now at work are doing better work than heretofore at ten hours; but if the eight hour movement be successful they would stop and do much less." It was telling that Conover and Tostevin's conspiracy suits against the Bricklayers' Union during late August and early September focused on the union's apprentice restrictions and not the eight-hour rule. The journeymen's restrictions on the number of apprentices bosses could hire, Tostevin and Ross told their colleagues, "dwarf[ed] the ambition of young beginners at the trade." The journeymen's rule forbidding bosses from working on their own jobs also received a share of the master masons' wrath. The largest builders shied away from a direct critique of the eight-hour rule: they seemed to be afraid or unwilling to attack the eight-hour movement's claim that a shorter work day would produce better workmen.[34]

During the strike, journeymen grew more adept at wielding distinctions between good and bad workmen against their employers. The Bricklayers' Union agreed with the Master Masons' Association that the "good workman" was to be encouraged and the "bad workman" discountenanced; the two sides differed as to who should define the criteria for each of those fictional artisans. Early on, the journeymen wished to prove that no "good workmen" were employed at ten hours, since the master masons contended that the strike had failed to engage the support of any but the least-skilled riffraff of the trade. At a July 27 mass meeting, several journeymen bricklayers asserted "there was no truth in the statement made at the meeting of the boss masons, [and] that the only men to work on ten hour jobs were, with but very few exceptions, worthless fellows, regular 'scabs,' and the society was better without them, while their continuance at work was damaging to the reputation of the bosses." Three of the bricklayers' locals voted "to approach the good mechanics working ten hours per day and endeavor to induce them to quit work; also that an advertisement be inserted in the daily papers declaring that all bricklayers working on ten hour jobs after Thursday next would be regarded as 'scabs' and treated accordingly." Other bricklayers present then insisted that anyone who had worked a ten-hour job was by definition a scab, a bad workman (regardless of skill), and should be treated accordingly. Finally, the meeting voted to expel all ten-hour men from the union. Two weeks later, the Bricklayers' Union offered employers a compromise which nonetheless asserted the union's right to define criteria for good workmen in the trade: "The Unions would alter their rule in regard to apprentices so as to permit the bosses to have more than two, but they should be indentured as now, be approved by the Union and remain under their protection, in order that good mechanics may be made of them and the quality of labor be not deteriorated." The master masons rejected this offer, in part, perhaps, because it associated the good workman with the prerogatives of trade unions and their cooperative values.[35]

In these debates, both sides seemed to be exploring new intellectual terrain. Workers and employers attempted, sometimes haltingly, to identify shared, trade-wide notions of skill and "good workmanship" with their new agenda—organization, for the journeymen, and a workplace free of long-standing collective sanctions, for the masters. Most significant, at the end of the bricklayers' eight-hour strike, the Master Masons' Association had begun only a tentative effort to define good workmen as the upward-striving "young beginners" who rejected trade union sanctions. Only some years later, in an altered social and political context, would large builders mount a frontal assault on the eight-hour movement and its claim that the shorter work day bred better workmen. In 1868, John Roach and other industrialists' aggressive campaign to cast the "good workman" in an acquisitive mold found only uneven support among building employers.

Tammany leaders watched the eight-hour strike with intense interest but did little to intervene. The Tammany press maintained a prudent silence. On September 15, the union bricklayers employed on Tweed's new County Court House were notified that they would be dismissed unless they agreed to work on the ten-hour day. "The reason assigned," one journal reported, "was that public opinion is against the eight hour system." But if Tweed and other Democratic leaders were waiting for public opinion to swing decisively against the strike, they were to be disappointed. As a consequence of the bricklayers' strike, Tammany found itself in a tighter embrace of the organized working class and the eight-hour movement in the building trades. In a November 1869 interview, Tweed lieutenant Peter Barr Sweeny told a *Herald* reporter, "The democratic party is sound on all the questions affecting the laboring interests. The Eight Hour law is accepted now by both political parties. Eight hours to work, eight hours to rest, and eight hours for social, moral and intellectual improvement and enjoyment have become an established maxim. In regard to the conspiracy law, one of the first acts of the approaching Legislature will be to repeal this odious and absurd law. . . . Submission to strikes will, after a while, be a necessity, and the excesses, if any, in the claims made for the time being must be left to the after good sense and sober second thought of the unions." Sweeny's endorsement of the social program of the eight-hour movement and the legitimacy of trade unions and strikes suggested that in late 1869 the balance of class power in the building trades and many other occupations was well on the side of the journeymen.[36]

Local self-government was one of the final two political possibilities exploited by the Tweed Democracy. No group was more eager to preserve local rule in the late sixties than the city's German and Irish tavernkeepers, and no group was more loyal to Tammany Hall. In an 1868 *Address of the Liquor Dealers and Brewers of the Metropolitan Police District to the People of the State of New York,* immigrant tavernkeepers demanded that execution of the excise laws and collection of fees "be uniform in operation" and, most of all, "be made through local authorities in the several counties." In his November 1869 interview, Peter Barr Sweeny was accordingly careful to affirm Tammany's plans for uniform and moderate enforcement of the state Excise Law.[37]

By the mid-sixties, the ranks of an immigrant middle class committed to Tammany-style self-government also included handicraft manufacturers and proprietors in the prosperous Broadway manufacturing district. As a result of government war contracts and the rising consumption of luxury goods, the Civil War years were a time of optimism and expansion for many of the manufacturers clustered around Broadway. Peace Democrat candidates who dominated this downtown manufacturing area through 1862 were forced after the draft riots to run for reelection in more sympathetic uptown districts on the poorer periphery of the city. By the end of the war, the "Irish Fourteenth" Ward—the geographic center of the Broadway district and once a Fernando Wood stronghold—had become uncontested Tammany territory.[38]

By 1865, then, many immigrant manufacturers could be counted as political constants in the Tammany formula. A New York *Times* survey of election officials and polling places in October 1865 noted that "not a single poll is to be held in a liquor-store, and that very few dealers in liquors have been selected for election duty—a circumstance that was never before recorded of a New-York election." The absence of the politically controversial tavernkeepers on the inspector and polling place list did not mean that such individuals were no longer loyal to Tammany—their loyalty could almost be taken for granted. Rather, the appearance of many manufacturers on the list may have meant that Tammany now felt sufficiently sure of the manufacturers' loyalty to trust them with the crucial political task of managing affairs on election day. Such employers seem to have had a sure claim on the political allegiance of their employees—so long as they could keep a Republican martial law government out of New York City. Timothy L. Smith has written suggestively of the "mediating role of immigrant businessmen" in early twentieth-century Minnesota mining communities. The Slovene tavern-keepers and their middle-class allies, Smith writes, "all joined the chorus of popular complaint against Wall Street, while making for themselves a niche in the power structure of Main Street." In the same way that Minnesota immigrant businessmen attacked Wall Street to solidify their local prerogatives, so too, it seems, did some immigrant businessmen in New York fifty years earlier lead the chorus of complaint against a centralizing Republican state and federal government. These may have been the individuals most likely to applaud Tammany Hall's attacks on Republicans in faraway Washington and Albany and its defense of local self-government.[39]

Finally, the Tweed Democracy championed the *idée fixe* of Young America, the territorial expansion of a white polity. The relationship between the Young America movement and Tammany Hall dated back to October 1851, when a Tammany mass meeting hosting Ohio expansionist William Corry denounced American neutrality and proclaimed active alliance with republican struggles throughout the world. In January 1859, Tammany sent President Buchanan a series of resolutions urging the acquisition of Cuba; the *Leader,* a Tammany journal, advised all prospective presidential candidates for 1860 to declare themselves on the Cuba question. The war interrupted Tammany's expansionist agitation, but the organization sought to strike up the Young America theme in 1865 exactly where it had left off in April 1861. While A. Oakey Hall was the best-known post-war champion of Irish and Cuban republicanism in the Tweed circle, the jingoism of Tammany District Attorney Samuel Garvin was no less shrill. Garvin told one Democratic rally in August 1869, ". . . we will have Mexico, and we will have Cuba, and we will have all the islands of the sea." That November, Peter Barr Sweeny predicted, "We will have a revival of the old spirit of 'fifty four forty or fight' which grew out of the Northwestern boundary, and which elected James K. Polk."[40]

But Tammany modified the message of the antebellum Young America movement. Whereas most Young America supporters before 1861 could not abide the prospect of black emancipation, Tammany came to terms early with black freedom. In October 1863, the city's Tammany weekly announced "in regard to the negroes," that "fears are said to be entertained in some minds that if Democrats triumph, the negroes will be again reduced to slavery. Nothing can be more absurd. . . . Emancipation is a fixed fact, cordially accepted in both sections of the country." By the end of the Civil War, Tammany leaders were quickly trying to put the question of the status of the black freedmen to rest.[41]

Tammany's movement to nominate conservative Republican Salmon P. Chase as the Democratic presidential candidate in spring 1868 demonstrated the eagerness of the Tweed circle to bury the explosive issues of the Civil War. Democratic city financiers applauded Chase's opposition to Ohio Senator George H. Pendleton's plan to redeem war bonds and the national debt in greenbacks. Though Pendleton contemplated no abandonment of the gold standard, his proposal to fund the debt with paper currency and to tax war bonds as ordinary property alarmed August Belmont and other Democratic bankers. But while Chase, for his part, was sound on the currency question, his racial views were far too "abolitionist" for some New York Democrats. Chase was an old Jacksonian Democrat who had joined the Republican Party in the struggle against slavery and supported the Fourteenth Amendment. Samuel J. Tilden adamantly opposed Chase and any talk of black suffrage. Only the nomination of upstate Democrat Sanford E. Church, Tilden believed, would insure the restoration of absent Southern states "with their local government in the hands of the white population." It seems Chase made support for black suffrage a condition of accepting a Democratic nomination but was willing to regard the Fourteenth Amendment as the last word on the political status of the freedmen. Chase believed that "freedom and manhood suffrage" were "unquestioned rights," but they also were, as any good Democrat knew, questions of state jurisdiction. With his proposal for a state-enforced black suffrage, Chase was able to sidestep the proposed Fifteenth Amendment and attract the support of the influential Democrat Samuel L. M. Barlow and many Tammany leaders. Even the eventual Democratic nominee Horatio Seymour briefly endorsed Chase. The following year Peter Barr Sweeny recalled, "We were not strong enough to win with the democratic vote; but it was possible and practicable to break down the republican party by making an alliance. Judge Chase was the man for that time, and we of the city of New York were strongly for him. It appeared to us as plain as demonstration that it was the time for a compromise that would be accepted by the whole country, and that Chase was the man to compromise on." Though the Chase movement was defeated, it revealed Tammany's eagerness to find some means of reconciliation with the Republican Party on the issues of the war, if only better to lure Republican voters.[42]

Merchants were the group most likely to approve Tammany's efforts to put the racial questions of the Civil War to rest. Republican William E. Dodge, Jr., leader of the movement on the Merchants' Exchange to declare martial law in July 1863, began to rally support for sectional compromise on the issues of reconstruction as soon as the war ended. As congressman for his uptown district, Dodge spoke to the House of Representatives in January 1867 in opposition to black suffrage: ". . . if these Southern States are still to be kept year after year in this state of disquietude, we at the North, sympathizing with them in our social and business relations, must to a certain extent suffer with them. . . . I feel that we are now in great peril, and ought not to look simply to the immediate enfranchisement of the negro race, overlooking all the other great interests of the country which are dependent upon the legislation we may adopt." By the following year, most New York City merchants, Young America and Union League Club alike, were insisting on some sort of compromise on the black suffrage question. In an impassioned editorial of February 1868, *The Merchants' Magazine and Commercial Review* described merchants' "deep feelings of impotence" at the "suspended animation" of commercial relations with the South:

> Is it not time for us then to bring to bear the concentrated force of the quiet conservative public opinion of the country upon the imperative necessity of devising some plan by which there can be established throughout the Southern States such a well-guaranteed and efficient public order as shall restore confidence in the future of those States not only among the Southern people, but among the capitalists, and manufacturers and merchants of the whole country?

The Merchants' Magazine approved a plan of compromise on the proposed Fifteenth Amendment that would limit the franchise to black men who could pass a literacy test, prove $250 in personal assets, or show evidence of loyal service in the Union Army. Black soldiers in the Union Army had "not only made proof of their loyalty but gained certain advantages of culture so far denied to their brethren who toiled on the plantations as slaves during the war." Frederick Law Olmsted's novel position of the 1850s, that a cultivated black working class would prove the success or failure of the American republican experiment, had by 1868 become part of the mainstream of merchant thinking and furnished many merchants with a basis for sectional compromise. The *Merchants' Magazine* stance and the Chase position on the Fifteenth Amendment were similar enough to make the Tammany-endorsed Chase candidacy attractive to merchants of all political and cultural persuasions.[43]

Tammany Hall's modified Young America program was consequently well suited to accommodate ideological tensions exposed during the draft riots. Young America, Tammany-style, was a project more than a few Republicans could support. Many Republicans warmed to post-war talk of Caribbean annexation and conquest. Indeed, though President Grant had balked at American invasion of Mexico, he and his close advisor Orville Babcock were eager to annex Santo Domingo and help the Cubans in their struggle aginst Spain. Tammany, of course, would not tolerate the Fifteenth Amendment. But after swallowing the candidacy of Republican Salmon Chase and his state-administered black suffrage, the Tweed circle had much modified the "territorial expansion for white Americans only" proviso of antebellum Young America. Few New York City merchants could object to Tammany's resolution of the issues of the war and reconstruction. The

Tweed Democracy's program of loyal nationalism, debt financing, municipal expansion, local self-government, and modified Young America principles was in place by November 1869. With much justification, Peter Barr Sweeny could then say of the recent Tammany electoral victory, "It gives us a great opportunity for a long lease of power in the State, and to lay a substantial foundation for the democratic party of the country."[44]

Limits

No two individuals were more emblematic of the changes occurring in New York City's middle and upper class during the 1860s than Simon Sterne and Alexander Delmar. Sterne was born in Philadelphia in 1839, attended the public schools there and later traveled in Europe, where he studied briefly at the University of Heidelberg. Returning to the United States, Sterne received a law degree from the University of Pennsylvania in 1859 and the following year moved to New York City where he began practice.

In New York, Sterne found time between clients to develop an interest in the new field of "social science" and in 1863 delivered a series of lectures at Cooper Institute on economics. Two years later, Sterne joined forces with Alexander Delmar to found the *New York Social Science Review,* a journal devoted to free-trade principles and the popularization of Herbert Spencer's works on political economy. Two years Sterne's senior, Delmar had attended the Madrid School of Mines, but aspired to combine his career as a mining engineer with the vocations of economist and historian. During the war years, Delmar employed himself writing vitriolic pamphlet attacks against the Republican Legal Tender Act and black emancipation. Wage and slave labor, Delmar argued in one essay, were entirely compatible: "No great nation ever existed which did not possess the monopoly of trade with some tropical region cultivated by forced labor." Elsewhere, Delmar connected inflationary Republican financial policy to the immiseration of a white working class and the racial politics of the Union League Club. "The arrogant money aristocrats," whose paper money expansion created "an immense population of unemployed laborers," were also the class of men who formed the "loyal Leagues, such as those who send their wives to crown redolent Negroes with martial bays, who drum their patriotism with unceasing clamour in the nation's ear—a patriotism which, like the drum they echo it on, is both deafening and hollow." Though not so openly racist as Delmar, Sterne doubtless shared the young engineer's critique of Republican monetary policy and its social and political consequences.[45]

After the brief collaboration with Delmar on the *Review,* Sterne helped to form the Personal Representation Society, devoted to introducing the English idea of "cumulative voting" to the American context. "Cumulative voting" or "personal representation," in Sterne's view, would restore the influence of the political minority—those individuals whom John Stuart Mill, in Sterne's favorite phrase, referred to as "the highly cultivated members of the community." Sterne's plan, borrowed from the Englishman Thomas Hare, sought to strengthen a minority elite influence in an electorate numerically dominated by the masses. The reformers planned to revise state constitutions to discard small geopolitical units—and, by extension, localism and party. "The educated classes could then by no possibility be swamped," Sterne observed. "They would be represented in proportion to their numerical strength, and, by the weight, character and ability of their representatives, would wield an influence out of all proportion to this numerical strength." Sterne was joined in the political reform project by senior Republicans David Dudley Field, Sidney Howard Gay, and Robert B. Minturn as well as a group of younger Democrats including Edward Cooper and David G. Croly. Though the Personal Representation Society failed to rewrite the constitution in its own state, it had sufficient influence to induce the Illinois constitutional convention of 1870 to adopt cumulative voting. Sterne's movement was an attempt, during the post-war flood tide of popular democracy, to restore upper-class influence in politics without jettisoning the principle of universal manhood suffrage. The stakes were high. Tammany-style popular rule without the leavening influence of "the educated classes," Sterne believed, was "the one blemish which of all others gives to the advocate of aristocratic privilege the only valid argument which seriously hinders the growth and development of liberal political ideas and forms of government in countries other than our own."[46]

The years 1871–72 were an important political turning point in both Sterne's and Delmar's careers. With the revelations of fraud in the Tweed administration in summer 1871, Sterne became the secretary for the Committee of Seventy that ultimately overthrew the Ring. This experience paved the way for Sterne to join the inquiry into suffrage restriction conducted by Governor Tilden's 1875 Commission to Devise a Plan for the Government of Cities in the State of New York. After serving as director of the new Bureau in Statistics in Washington from 1866 to 1869, Delmar returned to New York and became active in municipal reform. Moving to Brooklyn in 1871, he joined the Liberal Republican movement and worked for Horace Greeley in the presidential election of the following year. Delmar soon became one of the nation's leading monetary theorists. In 1872, he was invited by the Russian government to attend the St. Petersburg International Statistical Congress, and four years later he was appointed mining engineer of the United States Monetary Commission.[47]

The careers and attitudes of Simon Sterne and Alexander Delmar revealed the boundaries of the Young America outlook and bespoke the emergence of a new upper-class constituency and outlook beyond Tammany's reach. There were initial similarities between Young America and Sterne and Delmar's circle. During the late sixties, both groups tended to be Democratic, ardent supporters of free trade (much of the Personal Representation Society membership also belonged to the American Free Trade League) and bitter opponents of black suffrage. But Sterne, Delmar, and the younger members of the Personal Representation Society shared few of the Young America businessmen's formative experiences. John A. Dix, Augustus Schell, Samuel J. Tilden, Richard Lathers, and other leaders of the 1860 Pine Street Meeting were men more often than not born and raised in rural areas of the country, in the first two decades of the century, without college education (Tilden the obvious exception on this last score). By contrast, Sterne, Delmar, David G. Croly, and Edward Cooper were born and bred in cities, usually in the third and fourth decades of the century, and inevitably attended college. Abram S. Hewitt straddled both worlds: born in rural Haverstraw, New York, he grew up working summers on the farm and spending winters in New York City public schools, and attended college and law school at Columbia, where he met Edward Cooper, his future partner in the iron industry. For Hewitt and Cooper, like Sterne and Delmar, the 1871 campaign against Tweed and the subsequent reorganization of Tammany Hall were crucial formative political experiences. It seems highly plausible that such differences in age, place of upbringing, and educational background helped to shape the diverging attitudes of the two groups.[48]

The Dix circle and the Sterne group differed most in their postures toward popular rule. Through the sixties, Belmont, Dix, Schell, Lathers, and others of their ilk remained committed to the popular impulse behind the local Democratic Party. These old Jacksonians would allow the Tweed regime to run its course. As Commander of the Department of the East after the draft riots, Dix had kept the federal government at bay and fashioned the political settlement that made Tammany dominance possible. As has been noted, Belmont provided access to the bond markets that financed Tweed rule. The indulgence of popular democracy and tolerant, paternalist approach to popular disorders that characterized the thinking of these Young America businessmen in 1857 and 1863 remained evident through the war decade. By contrast, Delmar and Sterne had a passion for "scientific" and "efficient" government which emphasized the rule of a cultivated few. In the post-war years, these younger men cut their political teeth on projects designed to restrict popular control of politics and restore the influence of an educated upper class. The valuable skill of a Samuel Tilden was the ability to move easily between the two generations of elite Democrats—this ability no doubt facilitated his leadership of the anti-Tweed movement in 1871 and his unification of the state and national Democratic Party in the mid-seventies. Tilden could trace his Jacksonian political antecedents back to the rural upstate caucuses of the Albany Regency but was enough the elitist social scientist to satisfy the likes of a Sterne or a Delmar. In sum, the post-war Tweed Democracy could count on the allegiance of the old Young America fraternity, and at least an arm's-length truce with Tilden. But the Tweed regime discovered early on that young, urban elitists like Sterne and Delmar were beyond its reach.[49]

Another group beyond Tammany's grasp was the German-American community. No one was more aware of the Democratic Party's "German problem" than Horatio Seymour, who from his rural outpost near Utica saw the ethnic mosaic of metropolitan politics most clearly. Two letters from Seymour to Tilden, one in winter 1863 and the other in summer 1871, frame this Democratic dilemma of the Tweed era. In the first letter Seymour advised Tilden to consult with *Staats Zeitung* editor Oswald Ottendorfer regarding the appointment of a new city Police Commissioner. "It appears to me that the great point at this time is to confirm the German People in our favor. They now seem the true conservatives of the Democratic Party and they will support a National policy unless they are driven off by unwise measures. They feel with justice that their strength has not been duly respected in the City of New York." Seymour's second letter to Tilden came after Ottendorfer and other German leaders began to join the anti-Tammany reform movement of late summer 1871. "It seems to me that there is reason to fear that the Democratic Party will lose the German vote. It is clear that the Germans are deeply moved by the state of things in your City. They have gone into the [anti-Tammany] City meeting and by doing so they will feel that they are outside at the pale of the party. . . . Unless they get some position which will make them strong in our ranks they will leave us. . . . Their organizations must be recognized. . . . I can see no other way of meeting the crisis in the affairs of the Democratic Party. . . ." The threat of German political independence was a constant in the deliberations of the Tweed-era Democratic Party.[50]

In the fall of 1863, the successful mayoral race of German fur importer C. Godfrey Gunther revealed the full extent of the "German problem." If Republican unconditional loyalty, Fernando Wood's demagogic Peace politics, and the Tammany program were the three major political avenues of the post-riot metropolis, then Gunther and his supporters represented a fourth alternative, a small but important byway. Gunther and the so-called "McKeon Democracy" that elected him had split from Fernando Wood's Mozart Hall after the draft riots. The German C. Godfrey Gunther and Irishmen John McKeon and John Kelly spoke for immigrant and, in many instances, Catholic small proprietors weary of the political changes and economic burdens of the war but adamantly opposed to Wood's treasonous appeals to discontented workingmen. The Reverend Adolphus Berckmann, president of the German Democratic Club, voiced the anxieties of such immigrant proprietors at a late December 1863 meeting: "What [German] can pause to reflect whether he shall join those who strive to restore the Constitution and the Union, that the country have peace . . . and that the greatness, glory, power and happiness of our country may again take the place of a bloody, pernicious and damnable war and its horrors and evil consequences, subjection and impoverishment? Assuredly not one, for every one is interested in an early reestablishment of right, especially the tradesman, the laborer, in a word, the middle class. . . ." Though not all immigrant small businessmen would have endorsed Berckmann's peace position, many would have agreed that the Republican war, inflation, taxes, and draft had had harsh economic consequences for the "middle class." Accordingly, Gunther's votes in the fall 1863 mayoral election came from the poorer immigrant manufacturing wards on the periphery of the city: the Seventh, Eleventh, and Thirteenth wards on the East River waterfront, *Kleindeutschland,* and the poorer wards above Fourteenth Street. The straitened immigrant proprietors who flocked to the Gunther movement lacked the confidence of the Tammany manufacturers. Many had fled from war, anti-Catholic religious oppression, and centralizing national regimes in the German states and the United Kingdom. As the Civil War dragged on through 1863 and 1864, these people wanted nothing more from the federal government than some assurance that they would be left alone.[51]

No one better articulated the fears and aspirations of the German small employer and proprietor in the sixties than Oswald Ottendorfer. On the eve of the 1863 mayoral election, the editor of the *Staats Zeitung* explained the urgency of elevating Gunther to City Hall. The greatest threat of Tammany rule, Ottendorfer suggested, was the effect of its corrupt administration and resulting high taxes on the city's commerce and manufacturing: ". . . the oppressive tax burden must finally push the commerce and factories to the outskirts of our city, in spite of the natural benefits that New York offers. Capitalists will seek places for their investments where they do not need to hand over the greatest part of their wealth to corrupt officials. . . ." An honest Gunther administration, Ottendorfer concluded, would lower taxes and discourage the process by which much of the city's manufacturing was being driven to poorer outlying districts. While some of the city's employing class approved Tammany-style expansion, the German *kleinbürger* who cheered Ottendorfer's pronouncements on local politics believed the Tammany tax burden was pauperizing the small producer. Declaring their independence from both the Republican Party and Tammany Hall, Ottendorfer and his confederates swept Gunther into City Hall.[52]

Ottendorfer was no radical, however, and would not allow political independence to carry him outside the sacred terrain of the Democratic Party. There was no problem as long as the German community remained united behind Ottendorfer's movement: in October 1864 the *Arbeiter Zeitung,* the city's German labor weekly, allied itself with the anti-Tammany Citizens' Association campaign and announced its preference for Gunther's McKeon Democracy over both Tweed and Wood. But after the war the German workers' movement began to pursue its own program. By fall 1869, the organized German trades—woodworkers, machinists, tailors, and cigarmakers—were no longer content merely to endorse anti-Tammany German

Democrats. In an October 13, 1869, mass meeting of workingmen in Cooper Institute, the German "Association of United Workingmen" committed itself to "sever all combinations with the existing political parties, and not support any candidate who does not wholly represent the principles contained in the Labor platform [of the National Labor Union], and . . . enforce the same and urge the immediate necessity of organizing in . . . respective districts for the purpose of sending representatives direct from the ranks of labor to represent workingmen in the councils of the nation and State." This statement, with its opening credo that "both of the existing parties are corrupt, serving capital instead of labor," was the battle cry of the city's new labor reform movement, created by a joint committee of the English-speaking Workingmen's Union and the German trades of the Arbeiter Union. In an editorial response to the workingmen's resolutions, an angry Ottendorfer condemned independent labor parties as the work of "crafty and greedy demagogues like Butler" (the ex-Union Army general, now Massachusetts Radical Republican politician) and warned German wage earners to "beware of the tools of Radical corruptionists who like to preach a new gospel in order to steal food from your mouth when you're not looking."[53]

The events of fall 1869 represented a new and dramatic conjuncture in the history of the New York City working class. Under the leadership of a "Joint Committee" of English-speaking and German trade unionists, the two great projects of the middle decades—the establishment of an independent working-class politics and regulation of the trades—now merged. The same mid-October gathering that gave birth to an independent workingmen's movement and pledged to hold Tammany Governor John T. Hoffman to the enforcement of the state eight-hour law also resolved to strike from the statute book the state conspiracy law prohibiting workingmen from convening to "make laws for the government of their own trades." It was significant that Richard Matthews, a leader of the bricklayers' 1868 struggle to defend work rules, emphasized that workingmen represent their own interests in Congress and the state legislature. Matthews denied that "William M. Tweed had a right to choose politicians to represent the cause of the workingmen." In the flurry of workingmen's meetings prior to the November election, John Ennis of the Plasterers' Union identified repeal of the state conspiracy law against the journeymen's associations and their work rules as the first item on labor's political agenda. In fall 1869, the workplace and political programs of the building trades were fully joined.[54]

Despite the workingmen's display of unity across ethnic lines and new coordination of political and workplace goals, Tweed's candidates swept the November 1869 election. The workingmen's brightest hope, Nelson W. Young—a printer backed by every political organization in the city except Tammany—lost his race for coroner by almost thirty thousand votes. But Tammany leaders, who had thus far supported the eight-hour movement and the entire working-class social program associated with it—were now forced to confront another limit to their political rule. Independent workingmen's parties, Peter Barr Sweeny told an interviewer three weeks after the election, could not be tolerated. Asserting that "the capitalists, the aristocracy of wealth and the bondholders who enjoy immunity from taxation" were all "in the Republican Party," Sweeny proclaimed the Democratic Party the party of labor. "There is danger of the labor movement being wrecked by being converted into a political machine. There are demagogues among them, as there are among all other associations of men, and there are selfish leaders, who would like to ride into power on the strength of the labor movement. There was an illustration, and a very ridiculous one, in the candidacy of Nelson W. Young for Coroner at the late election."[55]

The workingmen's movement of fall 1869 signaled two other new developments in the politics of metropolitan class relations. First, Peter Cooper's Citizens' Association was no longer at the fore of the city's anti-Tammany forces. Embarked on an ill-fated reform alliance with Tammany (leading to Cooper's endorsement of the 1870 Tweed Charter), the Citizens' Association was virtually absent from preelection councils. Now the German Association of United Workingmen, allied with the English-speaking Workingmen's Union, had seized the helm of the anti-Tammany coalition. By late 1869, it had become clear that the organized working class and not the Citizens' Association would eventually deliver the death-blow to Tweed's political arrangement. The workingmen had assumed control of Cooper's anti-Tammany movement of 1864 and given it their own class-conscious stamp.[56]

Second, Tweed and his associates began to search for new ways to court the intransigent German vote. The election of 1869 had revealed the full extent of the "German problem." Tweed and Sweeny's first choice for the Tammany mayoral ticket the previous year had been Ottendorfer. Only when the prudent German politician declined did Sweeny settle on "the strongest man in our organization," District Attorney A. Oakey Hall. By fall 1869 the situation seemed to worsen, as dozens of German political clubs sprang up around the city. Some of them, such as the "Arbeiter Clubber 17 Ward," were pledged to the workingmen's ticket, but others, such as the "Seventeenth Ward German-American Club," sought only that "the Germans, irrespective of party, should nominate their own candidates for the next campaign." German political independence was one of the first subjects Peter Barr Sweeny addressed in his post-election interview.[57]

Significantly, Sweeny understood the problem of German political independence in terms of "the negro question" raised by the proposed Fifteenth Amendment: "We ought to get rid of the negro question. It hurts us more than the negro vote could injure us. It introduces a moral issue—a sentiment of justice—and presents the captivating cry of universal suffrage, which carries away many votes, especially among the Germans, and prevents the legitimate political questions of the country from having their just weight before the people." Sweeny's observation that New York City's German immigrants intimately associated American citizenship with the right of suffrage seems to have been accurate. Tammany's opposition to the Fifteenth Amendment as an affront to the racial superiority of white workingmen may have lost more support than it gained in *Kleindeutschland*. Indeed, by spring 1870, *Die Arbeiter Union* endorsed the Fifteenth Amendment as an advance in the status of both black and white workingmen. The German labor paper regarded the demise of black slavery as an auspicious omen for the emancipation of the white working class from the wages system: if black emancipation seemed impossible in 1850, and in 1870 was accomplished fact, perhaps, too, the overthrow of the wages system had now entered the realm of historical possibility. Now "even ten years was too long to wait." *Die Arbeiter Union*'s views on the Fifteenth Amendment of course reflected the longstanding abolitionist sympathies of its editor Adolph Douai.[58] But the German labor paper also spoke for the many German artisans who supported independent labor candidates in 1869–70, and it should be noted that its position represented a complete repudiation of the racial outlook of the draft rioters seven years earlier. The *Arbeiter Union*'s editorials in tandem with Sweeny's statements suggest that by 1870, a growing segment of the metropolitan white working class had begun to expand its definition of "citizenship" to include black wage earners.[59]

The summer 1870 campaign against the importation of Chinese "coolie" labor contained Tammany's response to the changing racial outlook of the labor movement. That June, a North Adams, Massachusetts, shoe manufacturer's decision to employ one hundred Chinese laborers as strikebreakers provoked a roar of indignation in labor circles up and down the Eastern seaboard. Although available evidence suggests that Tammany sheriff James O'Brien actually paid money to Workingmen's Union President Nelson W. Young to help stage the Workingmen's Anti-Chinese rally of June 30, 1870, the labor movement hardly needed Tammany's help to arouse working-class sentiment against imported Chinese laborers. The list of speakers at New York's June 30 mass meeting was a directory of the city's most militant and class-conscious trade unionists. Richard Matthews of the Bricklayers' Association, Conrad Kuhn, president of the German Workingmen's Union, Robert Blissert of the Tailors' Union, James A. Burke of the Painters' Association, John Ennis of the Plasterers, and Alexander Troup of the Printers—the leaders of the October 1869 anti-Tammany coalition—now took the stand one by one to denounce the importation of "cheap" Chinese labor into the United States. While Tammany Mayor A. Oakey Hall made the opening address, German labor leaders were prominent on the platform. The *Arbeiter Union* editor Adolph Douai denounced the capitalists who brought Chinese laborers to America but maintained he had no ill-will toward the oppressed Chinese themselves. Less restrained was the young John Swinton, the eloquent and class-conscious labor leader of the 1880s for whom the Workingmen's Anti-Chinese movement was a grotesque political baptism. Swinton extended Douai's criticism of *"die Kuli-Einfuhr"* to the Chinese male laborers themselves. In a letter to the *Tribune,* cited approvingly by Oakey Hall at the rally, Swinton described the Chinese as an "inferior type" of humanity, bringing paganism, incest, sodomy, and the threat of miscegenation to American shores.[60]

The Workingmen's Anti-Chinese agitation of 1870 provided common ground for Tammany and the labor movement at a moment when the Tweed regime's anti-black appeals were losing effect. The June 30 mass meeting suggests that anti-coolie sentiment induced some German trade unionists into at least a temporary collaboration with Tammany. But unfortunately for Tweed and his associates, this distorted echo of the draft riots lacked sufficient impact to influence long-term political loyalties. By fall 1870 the threat of an invasion of Chinese laborers seemed to subside, and along with it the intense emotions of the June 30 rally.

The problem of German political independence continued to haunt the Democratic Party. In early 1871 Siegfried Meyer, a German refugee mining engineer informally serving as a New York City correspondent to the First International, wrote the London-based Karl Marx that the twenty-one *Vereine* represented by the German Arbeiter-Union had voted to ally themselves with the International Workingmen's Association, the organ of the First International in America. "On the whole," Meyer informed Marx, "the German workers are much more interested in the IWA than the American; when recently an American worker pronounced the IWA as a humbug, the indignation was general and sincere." In fact, Meyer may have slightly overstated the case: although the local German sections 1 and 6 of the IWA were those most committed to Marxist principles, the organization in the New York City area had by the early seventies expanded to twenty sections and had received the endorsement of the English-speaking New York Workingmen's Assembly led by the sympathetic William J. Jessup. To most of its working-class supporters, the IWA stood above all for the proposition that workers should break ties with existing parties and without regard to race or nationality seize political power in their own right. The Arbeiter-Union's affiliation with the IWA and the support of at least some English-speaking brethren could only have heightened Tammany anxieties about workers' independent political ventures.[61]

Perhaps the most telling sign of Tammany's discomfiture in spring 1871 was a meeting held by the "colored men of the West Side" and described by the *Sun* as "an indignant rejection of Tammany's bid for colored votes." In November 1869, Sweeny had called the black suffrage question "greatly exaggerated in importance" because of the small number of disfranchised black men in New York City. Regarding eligible black voters, Sweeny added, "Our boys understand how to get them." Did Tammany leaders amplify appeals to the black community after the ratification of the Fifteenth Amendment, as they saw white working-class sentiment shift toward an acceptance of black rights? In the May 19, 1871, meeting of the "colored men of the West Side," a group of black men who had joined the Tammany Excelsior Colored Regiment were condemned by the uptown black community. The Reverend Dr. Dennison of the Washington *New Era* praised black loyalty to the Union during the Civil War and urged black men "to stand firm against the seductive whispers of Tammany Hall." One N. F. Gaylord "deplored the ingratitude of the colored man who could sell himself, bones, blood, marrow, and all for a miserable uniform, and vividly recalled the scenes of the July riots to prove the sincerity of their new friends." With the events of the draft riots deeply etched in the memory of the uptown black community, Tammany's attempts to associate itself with the black soldier's loyalty to the Union amounted to little. But the Excelsior Colored Regiment episode may have revealed an eleventh-hour Tammany ploy to assuage growing tensions between the Democratic Party and its white working-class constituency.[62]

The Fall of Tweed

The Tweed Democracy collapsed in the wake of three dramatic events of summer 1871: the repudiation of the Tweed Ring by municipal bondholders after the July revelations of fraud, the Orange riot of July 12, and the rejection of Tammany by the building trades in mid-September. The exigencies of the bond market, Irish struggles over religion and nationality, and what has been called "the power of organization" among the city's working classes defined the last and most fatal limitations upon Tammany rule.[63]

The withdrawal of municipal bondholders from the Tammany alliance has been reviewed by another historian and requires little comment here. By spring 1871, many investors and political observers were beginning to express concern over the size and nature of the city's bonded debt. In April, a German financial journal noted that New York securities on the Berlin exchange were protected only by the reputability of underwriters Rothschild and Discounts Gesellschaft. After a June 12 publication of municipal "statistics" by Mayor Hall, the *Times* observed that "neither from the Controller's report, which professes to give the city debt down to January 1, 1871, nor from the Mayor's message, which purports to give the additions since that time, will the reader be able to get anything approaching to a correct statement of the actual debt of the City. As for details showing how the money has been spent, we have absolutely none." On the 20th, the *Times* suggested that the issue of bonds by the comptroller was "one of the means by which our Tammany politicians are to get their customary 'rake' this season." When discontented ex-Tammany sheriff James O'Brien revealed secret city and county accounts from the comptroller's books to the

New York *Times* in the second week of July, already-suspicious bankers and bondholders on Wall Street and abroad were quick to cut off the city's credit. The international bond market was the Tweed Democracy's Achilles' heel.[64]

Less well known is the way middle- and upper-class New Yorkers came to associate the July "insurrection of the capitalists" with another event, the Orange riot of July 12. The Orange riot was the legacy of the draft riots to a world little changed since 1863, that of the Irish Catholic common laborer. Together with the Tweed exposures, the riot on the anniversary of the Battle of the Boyne—Protestant King William's 1690 victory over the Catholic James II—convinced many New Yorkers that Tammany-style popular rule had failed.

The quarrymen's strike of May 1871 provided early signs that in certain uptown quarters, little had changed since 1863. On May 1, the quarrymen employed near the uptown village of Yorkville went on strike for an increase in daily wages. At noon, quarrymen from the various Yorkville work gangs, armed with clubs and axes, paraded down the East Side avenues, halted work at various locations, and drew laborers into the procession. At First Avenue and Forty-seventh Street, a crowd estimated at two thousand, "largely composed of women and children," had gathered to cheer the strikers' parade. Police Captain John Gunner, a veteran of the uptown draft-riot struggles, arrived with a company of officers and managed to disperse the workingmen and their families without violence. On May 2, the parallels to July 1863 grew stronger. Early in the day, the Yorkville laborers gathered in Central Park to decide on a course of action, while the Upper West Side police "lined Sixth Avenue, and prevented any knots of men congregating on the corners or about the bar-rooms." The laborers then visited a quarry on Avenue A near Seventy-seventh Street, where they persuaded most of the workers to quit for the day but beat one man for refusing. This led to a confrontation on the West Side with Captain John C. Helme, another veteran of '63, who drove the strikers down Ninth Avenue. Inspector George Washington Walling, the "hero" of the West Side forces eight years earlier, called out the police telegraph force and detective squad to protect communications lines and infiltrate the crowds. With the strikers using Central Park as a sort of demilitarized zone and the police seizing control of uptown street corners and saloons and offering protection to quarry contractors, the strike ended after two weeks in a draw. "Shovelmen" had resumed work early on at the old rates, but "no inducement" could influence the militant "rockmen" to accept the pay the contractors offered. Some of the laborers who remained on strike joined the "Quarrymen's Union No. 3," which met during the first week of the strike in an attempt to coordinate the impromptu actions of the uptown work gangs. The 1871 quarrymen's strike reenacted many of the scenes of the laborers' uprising of 1863: workers parading from site to site halting work, armed confrontations between workers and police, the threats to telegraph lines, the employment of detectives to impersonate workers, and even the use of Central Park as a rendezvous. Only the racial assaults of the earlier event were missing—even laborers, it seems, were beginning to abandon some of the draft riots' most extreme forms of racism.[65]

The quarrymen's strike of May was a dress rehearsal for the Orange riot of July. The July 12 commemoration of the Battle of the Boyne was a day of pride for Irish Protestant New Yorkers, whose Orange societies traditionally marched through the uptown wards in full regalia on their way to a day-long picnic celebration. Parading Protestants and uptown Catholic laborers and their families had come to blows on July 12, 1870, with two days of violence resulting in five deaths and hundreds wounded. All parties had twelve months to nurse injuries and plot revenge for the following summer's inevitable conflict. But while in 1870 Mayor Hall appeased public opinion by acquitting the arrested rioters with a warning, in 1871 matters proved less tractable. Images of the Paris Commune filled the daily press through May and June; in the words of Friedrich Sorge, ". . . the ruling classes had the 'Commune' on the brains . . . it was time to teach the 'dangerous classes,' that the 'order' could and would be saved also in the United States." The first exposures of Tweed Ring fraud on July 8 put added pressure on the Tammany response to events on the 12th. A suspect Tammany would have to prove it could act in the best interests of the business community.[66]

The Loyal Order of Orange's early July request for a police permit to hold their annual parade prompted a groan of disapproval from Irish Catholic organizations. Then, on July 10, with Tweed and Connolly's blessing, Police Superintendent James J. Kelso issued "General Order No. 57" forbidding the Orangemen from parading in honor of "William of Glorious Memory." Catholic Archbishop McClosky and Bishop O'Donovan applauded Kelso's order. But now a chorus of complaint issued from Wall Street. For over two hours on the 11th, indignant Protestant businessmen queued outside the Produce Exchange to sign a petition denouncing Order 57. The New York *Times* joined Wall Street merchants in condemning what it portrayed as the city authorities' cowardly surrender to the threats of a Catholic mob. Here, the Protestant upper classes anticipations of a home-grown Commune were blended with a much older fear for Protestant liberties menaced by Catholic-based political power. Arriving from Albany on the 11th, Governor Hoffman conferred with Tweed, Hall, and Connolly on the crisis. Mindful of his prospects as a Democratic presidential candidate, Hoffman rescinded General Order No. 57 and ordered that the parade take place. So great was the anticipation of violence after Hoffman's announcement that the *Herald* actually sent a reporter to interview the city's street pavers, quarrymen, and longshoremen about their plans for the 12th. In these near-exclusively Irish Catholic trades, blood was up and memories of the draft riots rekindled.

When asked whether the authorities would prevent the violence, one boulevarder confidently predicted, ". . . the city government . . . will do as they did in 1863—sit at home and drink punches and smoke cigars comfortably after dinner."[67]

Blood was spilled on schedule on the 12th. That morning, a little band of Orangemen guarded by five military regiments and an imposing body of police assembled on Eighth Avenue near Twenty-ninth Street in preparation for the march. Promptly at noon, throngs of Irish Catholic laborers halted work "along the wharves and unfinished buildings," and in dozens of strikers' parades, closed down the downtown waterfront and many uptown work sites. The "quarrymen employed in the Nineteenth Ward" and the "Mount Morris pipe men" were prominent in leading processions down to the Orangemen's assembly area on Eighth Avenue. "Armed Hibernians," one paper reported, "were marching up Eighth Avenue to intimidate the laborers on the Boulevard and make them quit work." Meanwhile, the same uptown police who battled the quarrymen in early May now lined Forty-second Street from Seventh Avenue to Eighth Avenue in an attempt to intercept the laborers on their march downtown. No matter. By two o'clock, the appointed hour of the parade, hundreds of laborers and their families had managed to elude police pickets. A hostile multitude of men, women, and children crowded around the Protestant assemblage at Twenty-ninth Street. When stoned by the Catholic spectators, the troops fired into the crowd, causing the death of thirty-seven and injuring sixty-seven others.[68]

Though many Protestant "best men" agreed with George Templeton Strong that by rescinding Kelso's Order Governor Hoffman had successfully repudiated the Ring and left Hall "floundering on his back in the mud," neither Hoffman nor Hall, Tweed, and company survived the repercussions of July 12. Reflecting the growing indignation of the upper classes, a late summer pamphlet entitled *Civil Rights. The Hibernian Riot and the "Insurrection of the Capitalists"* attacked Tammany as the author of both the million-dollar frauds against the New York City taxpayer and the July 12 violence. Quoting the *Tribune,* the pamphlet cited the recent "Tammany Riot" as evidence "again, as in 1863, [that] the criminal weakness and vacillation of the authorities have caused the peace of the city to be broken and its streets to be sprinkled with blood." Finally, the anonymous writer appealed to the city's Germans to throw off the tax burden of Tweed rule and declare complete independence from Tammany. Governor Hoffman, the early hero of the Protestant business community, soon watched the erosion of his political support from Protestants and Catholics alike. Peter Barr Sweeny wrote Hoffman on July 21, "I think the Republican Press have made such a bugbear of the Tammany ring that the People would not be likely to elect as President a man supposed to be in sympathy with them." The Irish were, of course, "permanently disaffected" toward Hoffman. By September, Samuel J. Tilden had begun to organize Democrats independently of both Hoffman and the

Ring in an attempt to save the state and national Democratic Party from the taint of insurrection and corruption. A coalition of Republicans, independent Democrats, and German leaders met in early September to form a "Committee of Seventy" to restore order to city government.[69]

The same fear of insurrection that had created the possibility for Tammany rule during the last years of the Civil War now bound Tammany leaders to the Irish Catholic community and insured Tammany's alienation of the "better classes." If Kelso's Order 57 had not been issued, Sweeny told Hoffman on the 21st, "our Police and our military would have been but infants against the armed host which would have scattered the Orangemen and Grant would have been put in possession of the City and would be there now. Some action was demanded by the almost universal sentiment of the people." Hoffman, Sweeny believed, should have left Kelso's Order alone and attempted merely by his personal presence in the city to stand "on both sides of the question." Years later, Tammany leader George Washington Plunkitt recalled Mayor Hall sending him into the crowd at the height of the riot on the 12th "with a small white flag." Plunkitt then "ordered Jim Fisk and the Eighth Regiment off the field of battle," presumably hoping that the departure of the military would have a calming effect. In July 1871, Sweeny, Hall, and Plunkitt still invoked the draft-riot strategy of quelling crowds with personal appearances, speeches, and white flags and rallying white New Yorkers of all classes, nationalities, and religious persuasions against federal military intervention in local affairs. But in 1871, the old Tammany strategy no longer worked. The mood of the city had changed. For one thing, the 1871 violence never evolved into a race riot: unlike so many of its predecessors, this challenge to the Protestant elite never led to attacks against that elite's black cultural and political clientele. Now, with the Civil War over, the threat of Republican military intervention in the metropolis had somewhat diminished. More than Republican martial law, the Committee of Seventy feared Tammany's version of "people's rule" based on indiscriminate hand-outs of public money and capitulation to the influence of immigrant and Catholic workers. That summer, there was much truth both to the *Times*'s charge that Tammany had become the instrument of the working classes and to Sweeny's defense that Tammany's Order 57 addressed "the almost universal sentiment of the people." The middle and upper classes who repudiated Tammany after the Orange riot abandoned the conservative style of political rule confirmed during and after the draft riots and turned back an unprecedented assertion of popular power.[70]

Reflecting in 1924 on the significance of the 1871 Orange riot, an elderly George Washington Plunkitt observed, ". . . there has been no religious riot in New York since that day." Though Plunkitt's recollection was not entirely accurate (witness Hoe Printing Press workers' 1902 assault on a Jewish funeral parade), the events of the Orange riot evinced a style of social relations rapidly declining in the metropolis. By contrast, the September 13, 1871, stonecutters' procession augured the rise of new forms of organization and resistance. The final limitation upon the Tweed political arrangement was the emerging power of the organized working class.[71]

The draft riots had erected or reinforced many barriers to working-class racial cooperation. Though there were scattered instances of journeymen in the building trades aiding black riot victims, the majority of the city's workers, inside construction and out, were content to allow the race riot of July 1863 to run its course. The scars were deepest among black workingmen and -women. The most serious and immediate problem black men faced in the years after the riots was finding a steady job. Especially in common laboring, the predominant source of black male employment, racist white contractors and work gangs collaborated to keep black wage earners away from docks, pits, and quarries and terrorize those actually hired. The first resolution of the black "New York City Labor Council" formed in October 1869 was thus "that it shall be the duty of each member 'to exercise vigilance and perseverance in securing employment and business to each member in whatever department of labor he or she may be engaged, and to advocate the cause of an equal right to labor with all other classes of our fellow citizens.' " A final declaration observed "that the exclusion of colored persons in this city from the right to labor in almost every department of industry, is a strong evidence of the power which the spirit of slavery and caste still holds over the minds of our white fellow citizens"; it cited blacks' citizenship and service during American wars as grounds for their claim to the right of employment. The wording of this manifesto had a distinctly cooperative flavor: only through the collective efforts of black workingmen and -women would the condition of black free labor in the city and the nation be improved. But while the black labor assembly elected delegates to the National Labor Convention in Washington in fall 1869, it as yet had no relationship with the all-white Joint Committee of the Workingmen's Union and Arbeiter Union. The German Section 1 of the International Workingmen's Association was the only citywide white labor body to promote actively the organization of black workers during the 1860s. Only after the changes in racial attitudes of the organized white working class in the wake of the ratification of the Fifteenth Amendment was an interracial working-class politics a possibility in New York City.[72]

The stonecutters' procession was the first evidence of a dramatic rejection of Tammany's brand of white supremacist popular rule in the building trades. At least eight thousand workingmen gathered in the rain on the Upper East Side and marched across to Broadway and down to City Hall to express sympathy with the stonecutters' eight-hour strike, demand general enforcement of the state eight-hour law, and denounce Tammany rule. While the building-trades associations of New York and Brooklyn led the procession—bricklayers, carpenters, stone masons, plasterers, painters, cabinetmakers, and construction laborers—they were joined by tailors, printers, and other artisan organizations. A company of Frenchmen waved a banner inscribed *"Comité International"* and German journeymen carried the red, white, and black flag of the United Fatherland alongside the Stars and Stripes. The presence of two other groups in the parade attracted the most notice: the International Workingmen's Association and a committee of black trade unionists. The marchers finally retired to a mass meeting at the Cooper Institute, where "it was difficult to ascertain which was the more important theme—the 'Eight-Hour' demonstration or the Corporation frauds and the fight between Mayor Hall and Controller Connolly." A unanimous vote to "throw off all allegiance to the Democratic party in the Fall elections" was followed by speeches in favor of enforcement of the eight-hour law by Workingmen's Union president William J. Jessup, Alexander Troup of the Printers and the National Labor Union, Richard Matthews of the Bricklayers, and John McDermott of the Plasterers. Resolutions also included the abolition of convict labor, the hiring of only skilled labor on public works, and the nationalization of all coal mines, railroads, gas works, canals, and telegraphs to increase the availability of steady employment. One speaker attacked the reckless metropolitan development that forced workingmen on the periphery of the city "to live in tenement rookeries in order to be near their work." So impressive was the display of organized cooperation across lines of nationality and race on the 13th that the *Herald* fancied the demonstration to be "a fraternization of the laboring classes of this city with the great Internationale of Europe. . . ."[73]

In a sense the stonecutters' demonstration of September 1871 was a working-class rejoinder to the Union League Club's presentation of colors to the black Twentieth Regiment and Broadway procession in March 1864. In the earlier episode, the Republican Club wives presented the soldiers of New York City's first black army regiment with their standard in a Union Square ceremony, and their husbands marched the new recruits down Broadway to their point of embarkment. The men and women of the Union League Club saw themselves reclaiming the public spaces of the city from the draft rioters and their traitorous upper-class allies, and instead associating the urban landscape with elite culture, loyal nationalism, and paternalist protection of the black community. The stonecutters' procession of September 1871, it seems, was also a symbolic racial repudiation of the draft riots by workers in many of the same trades that eight years before had taken to the streets. But the themes of September 13 were the interracial cooperation of labor, the power of organization, and, for many working people, independent politics. The Union League Club and stonecutters' processions highlighted a political transformation that had occurred between the last years of the Civil War and the first years of the next decade. By September 1871, the Republican upper classes had fully yielded the radical project of reshaping the metropolitan social and political order to the working classes. With the solutions of Tweed's Tammany discredited and the city's middle and upper classes in political disarray, the initiatives of working people now became the directing force behind city politics.[74]

What would a new assertion of working-class power look like? Troubled as much by revolutionary events in Paris as by local affairs in spring and summer 1871, middle- and upper-class New Yorkers began to conjure up visions of an "American Commune." The images recurring most were those of July 1863. In a June 1871 editorial entitled "Communists in New York," the New York *Times* reflected that

> only once in the history of the city did this terrible *proletaire* class show its revolutionary head. For a few days in 1863, New York seemed like Paris, under the Reds in 1870. . . . The cruelties inflicted by our "Reds" on the unhappy negroes quite equalled in atrocity anything that the French Reds perpetrated on their priests. Our "communists" had already begun to move toward the houses of the rich, and the cry of war to property was already heard, when the spirited assistance of the United States soldiery enabled the better classes to put down the disturbance. But had the rioters been able to hold their own a week longer, had they plundered the banks and begun to enjoy the luxuries of the rich, and been permitted to arm and organize themselves, we should have seen a communistic explosion in New York which would probably have left this city in ashes and blood. Every great city has within it the communistic elements of a revolution.[75]

The *Times* concluded that a working-class seizure of power resembling the Paris Commune was unlikely to occur on American shores. Reform was one deterrent: ". . . the industrial school and the children's charities are transforming the youthful Communists into industrious, law-abiding, property-earning citizens." Ultimately, however, the *Times* rested its hopes for public order in the acquisitive strain of American culture. "The American 'ouvrier,' " the paper reassured its readers, "always hopes to be capitalist." But as the working classes moved to political center stage in summer 1871, there was little evidence to confirm the *Times*'s pronouncement.[76]

The political arrangement of the Tweed regime, then, can be understood only in the context of the explosive social and ideological contests of the Civil War era. Tweed rule revealed the political centrality of the loyal urban working class during the 1860s and early 1870s. Tweed would remain in power only so long as he preserved civil peace and prevented the recurrence of treasonous 1863-style insurrections, on the one hand, and kept radical Republican martial law armies out of the city, on the other.

Tammany was able to clear a middle path between insurrectionary disloyalty and radical intervention because it successfully promoted the power, wealth, and physical growth of New York City. Tweed created an array of new political possibilities in the sixties based on the expansion of a New York City "interest." He coordinated the influence of city-based money markets, railroads, builders, and property owners and ultimately, an entire state government tied to a city political organization. His genius was to create a kind of party government that seemed to "solve" the problem of class conflict and social incoherence in the industrializing metropolis. Here was a "new" party that promised what the "old" parties of the fifties had failed to provide—an authentic majority rule. Indeed, with its reform of the city charter and its commitment to empowerment of the Irish, home rule, municipal expansion, preservation of an all-white polity, the eight-hour day, and even workers' right to organize and strike, Tammany was for a time offering the city's immigrant poor much of what they had been seeking. Many of the workers outraged at Republican government in 1863 could point to Tweed's policies in the late sixties with a sense of some accomplishment.

The Tweed Democracy was thus allowed to go its own way for six years. But Tweed's promise of loyal government free of popular disturbances was only as good as his ability to expand a metropolitan interest, preserve the *modus vivendi* among elites, and retain the political allegiance of the loyal white working class. Beginning with the bricklayers' strike of summer 1868, New York City workers became increasingly critical of the quality and pace of a frenetic Tammany-style development. Artisans' growing distaste for Tammany's racial position after 1869, combined with Irish Catholic laborers' challenge to Tammany's ethnic settlement and the withdrawal of business support in summer 1871, brought the Tweed Democracy to its final crisis. Tweed was no longer able to contain the volatile conflicts exposed by the draft riots. The loyal white working class that brought the Tweed regime to power ultimately presided over its demise.

CHAPTER 7

1872

The massive strikes for an eight-hour workday in spring 1872 were the sober denouement to the draft riots. The "winners" and "losers" of the crisis of 1863 now became clear and the unresolved conflicts of that summer were decided.

By early 1872 most of the groups that had seen Tweed's Tammany Hall as a respite from the corrupt party machinery of the past began to search elsewhere for solutions. In many ways, the contours of social and political conflict had much changed since the draft riots nine years earlier. The crisis of 1872 occurred not during a civil war but at a time when New Yorkers of the "better classes" were preoccupied with restoring civil peace. Merchant capitalists who had only a few years earlier battled each other over the issues of the sectional crisis now agreed that Reconstruction had ended and commercial and social stability had to be preserved both in the North and South. Reform-minded industrialists and manufacturers were now hardened by a decade of conflict with a powerful organized trade union movement. The middle- and upper-class New Yorkers who stood over the corpse of the Tweed Ring were both more unified and perhaps a little more frightened than they had been nine years before.

For all of these changes since 1863, certain features of metropolitan social life remained the same. Class relations were volatile during summer 1863 in part because of sharp ideological divisions between elite groups in the city. In spring 1872 one of those explosive conflicts still remained—that between employers who sympathized with the workers' movement toward organization and employers who did not. The eight-hour movement of May and June reopened a deep rift between those employers who rejected workers' social and political initiatives as coercive and destructive of class harmony and those who accepted them for their emphasis on discipline, sobriety, and moral and intellectual development. In a way, the conflict of 1872 was a redefined version of the divisions of 1863. In the earlier year industrialists, who cultivated loyal non-rioting workers and repudiated the riot in all its forms, opposed manufacturers—mostly small master artisans in building and woodworking—who approved or abided the limited demonstration on July 13. Between 1863 and 1872, manufacturers who had assented to the riot in its early phase were a dominant political voice in the city. They tolerated or endorsed wage earners' independent organizations. They formed one of the cornerstones of the Tammany post-riot political edifice. In the early stages of the eight-hour uprising of spring 1872, these "sympathetic" manufacturers still dominated employers' councils and guided the employers' response to the strike movement. But by the sixth week of the strike, in mid-June, the balance of power shifted to the industrialists. These employers, led by the militant Board of Iron Founders but now including many builders as well, branded all "eight-hour men" and trade unionists as disloyal and communistic and redefined "loyal" workmen as those who repudiated both the eight-hour work rule and the organizations that enforced it. These new definitions of "good" and "bad" workmen were endorsed by an Employers' Central Executive Committee formed June 18, which quickly won the support of most employers in the city for a police-aided suppression of the eight-hour strikes. By the end of that month employers once deeply divided over responses to independent organization among workers now united to suppress a politicized and independent eight-hour movement.

The draft-riot rift between an organized mass of workers who patrolled the streets in committees closing down factories and the vast swirling crowds that engaged in spontaneous acts of violence and arson was also evident in 1872. But now authority rested almost entirely with the organized majority of workers. The eight-hour movement was a test of the moral fiber of the "respectable" working class for workers as much as it was for employers. The very principle of the eight-hour day was grounded in workers' own ideas about respectability, self-discipline, and moral cultivation. Like the draft riots, the strikes of May and June 1872 occurred in two phases. The first phase began in early May and continued through June 10 and was led by artisans in the building and woodworking trades. The second phase was led by industrial workers in the metal trades, began in the second week of June, and continued, with ebbing energy, through the summer. In contrast to 1863, both phases of the 1872 revolt attempted to establish the organizations of a loyal working class as the foundation for a politicized and independent domain.

For New Yorkers in search of social and cultural consolidation, then, 1872 proved a time of reevaluation. The strikes of that year became a grand referendum on the question of 1850—could a competitive urban society be bound together by politically independent structures? But now, unlike in 1850, those structures of consolidation were entirely managed by workingmen. In the Industrial Congress of 1850 and afterward, the idea that wage earners should lead the project of consolidation was controversial but tolerable to many employers and even attractive to some. By the sixth week of the 1872 strike, the prospect of working-class leadership became real, perhaps for the first time, to many employers "sympathetic" to the eight-hour movement. In a moment of class reevaluation, these employers allied themselves with the industrialists, who were adamantly opposed to independent working-class organization on any terms. By the sixth week of the strike movement of 1872, middle- and upper-class New Yorkers finally agreed that the competitive individualism of urban society could not and should not be contained by the independent institutions of workingmen. Industrialists emerged as the ultimate winners of the 1863 crisis, and their acquisitive individualism and repudiation of working-class rule were now triumphant; workers in the draft riot trades met their final moment of defeat. A reunified middle and upper class still entertained hopes of distancing government from the ineffectual apparatus of the parties. But now employers did not defer to their working-class neighbors' independent political projects. Liberalism, New York's dominant bourgeois form of independent politics in the 1870s, would endorse a new and unyielding elitism.

A Tragic End for Horace Greeley

One sign that wage earners had wholly taken over the project of consolidation from middle-class reformers and politicians in spring 1872 was the absence of Horace Greeley from the councils of labor. Through the mid-sixties Greeley remained popular in some precincts of the labor movement. In the shipyard eight-hour strike of April 1866, Greeley was a favored speaker at support rallies, where he argued that the eight-hour day would contribute to workers' efficiency and capacity to help themselves. But Greeley went on to warn workers that "legislation was able to do very little for labor," and "though he was willing to see this experiment tried," cooperative production and not the eight-hour day promised workers the only true emancipation from the wages system. During the strike movement of 1871–72, Greeley's name did not appear in the *Tribune*'s labor columns but in its national campaign headlines. Greeley, champion of independence from the machinery of politics in 1850, was in 1872 the presidential candidate of the Democratic Party and a discredited Tammany Hall.[1]

Greeley had difficulty choosing among the ideological alternatives posed by the revolutionary upheavals of the sixties. Never entirely comfortable with the coercion required for a Republican Civil War victory, he secretly corresponded with Copperhead leaders Clement C. Vallandigham and Fernando Wood in an attempt to end the struggle by foreign mediation. More controversial still was Greeley's 1867 endorsement of general amnesty for ex-Confederates. On May 13 of that year he signed the bond for Jefferson Davis's release in Richmond. Immediately afterward, the *Tribune*'s circulation dropped by half, and thousands of subscribers to Greeley's Civil War history *The American Conflict* canceled orders. Ironically, Greeley's decision to run for the presidency on an independent Republican and Democratic ticket grew from his belief that the Grant administration was corrupt, indifferent to civil service reform, and illiberal toward the South. In a campaign distinguished for its scurrility, Greeley was assailed as a traitor, a crank, and a tool of Tweed Ring or Southern interests. Even the aging reformer Peter Cooper, an admirer of Greeley, was skeptical of the candidacy. Cooper wrote John A. Dix in fall 1872, "Our friend may find himself, with his best intentions, entirely powerless to control a government made up of men, who believe that 'the black man has no rights that the white man is bound to respect.' " The two Jacksonians, Cooper and Dix, were now members of the Republican Party; each, in his own way, had accepted the revolutionary consequences of the war era. Greeley, still seeking *rapprochement* with the old-line Democracy, lost to Grant in a landslide. Meanwhile the editor's life was crumbling around him. He came home from an exhausting election-eve campaign tour to find his wife Mary on her deathbed. After a long vigil, she died on October 30. Now returning to a neglected *Tribune,* Greeley learned that Whitelaw Reid had seized control of the newspaper during the campaign tour. Broken in spirit and half-insane, the journalist died three weeks later, in many ways a victim of his inability to make the revolutionary commitments of the era.[2]

Accusations that Greeley's presidential candidacy represented a continuation of Tweed rule may have contained an element of truth. Like Tweed, Greeley had tried to build a bridge of compromise across the chasms of sectional and class conflict. In order to attempt compromise, both men were compelled to embrace the machinery of the political parties, Tweed eagerly, Greeley with some reluctance. In 1872, the *idée fixe* of 1850, that the machinery of the political parties was unequal to the task of consolidation, had gained new momentum. In a meeting at the Fifth Avenue Hotel, "Liberal Democratic and Republican citizens" including Oswald Ottendorfer, Simon Sterne, and Parke Godwin of New York rejected both the major parties' candidates, Grant and Greeley, and nominated William S. Groesbeck of Ohio for President and Frederick Law Olmsted of New York for Vice President. The gathering resolved "that undue devotion to party has already greatly damaged the Republic, and we now engage ourselves to discountenance in every way possible the despotism of party organization and the abject submission of voters to the dictates of party politicians." "Horace Greeley," the meeting emphasized, "does not represent these principles but has been a life-long opponent of the most essential of them." By spring 1872, Horace Greeley, like Tweed, had been jettisoned by the representatives of New York's "independent" middle and upper class.[3]

The workers who led the eight-hour movement that spring were equally disenchanted with Greeley and other major party candidates. So discredited was the "corrupt" and "treasonous" Tweed and Greeley Democracy that leaders of the eight-hour movement almost universally agreed that the major parties and their politicians belonged outside the workers' domain. Siegfried Meyer observed in April 1872 that among the workers' associations of the city, the term "politician" mostly meant "the practitioner of a shady business." Even Democratic Governor John T. Hoffman, who many workers had believed was free of Tweed-Ring taint and would enforce the state eight-hour law and lend prestige to their movement, was on June 3 implicated in the Tammany frauds. By June 1872, Greeley, Tweed, and the whole political mechanism of class compromise they represented had been repudiated by large segments of both the middle and upper classes and the labor movement.[4]

The International Workingmen's Association

Like Greeley's disappearance from the labor movement, the prominence of the International Workingmen's Association in the New York City labor movement was evidence that workers sought to distance themselves from the major political parties in 1872. Independence from those parties and their leaders was the first principle of the IWA, a principle that drew to the organization a range of personalities, from old land reformers like John Commerford, George Henry Evans, Lewis Masquerier, and William West, to the German-born Marxists Friedrich Bolte and Friedrich Sorge.[5]

In many ways, the debates that led to the split among the New York City (and American) sections of the IWA resembled those that had divided the Industrial Congress of 1850. First, in 1871 and 1872, the leadership of the controversial Section 12—middle-class reformers such as Commerford, Evans, and West—was composed of the same figures that dominated the Congress twenty-two years earlier. These men and women were attracted to the broad agenda of the IWA, as it allowed them to introduce an array of anti-monopolist concerns. Commerford and his circle attacked landlords, clique politicians who controlled the machinery of government, and, in Masquerier's phrase, "Godology," or the clergyman's monopoly of "saving grace." By contrast, Sections 1 and 6 were controlled by wage earners, though not exclusively limited to a working-class membership. Here, a small group of Marxist leaders including Bolte, Sorge, Meyer, Vogt, and Karl Speyer of the Furniture Workers' League had much influence over a membership that included a proportionately large number of Germans in the woodworking trades. The celebrated late 1871 split in the IWA, it has been demonstrated, grew out of this division between sections controlled by the middle-class land reformers and those dominated by wage earners. Friedrich Sorge and his supporters sought to restrict the International to only those sections claiming a two-thirds' membership of wage earners. The French and American sections that seceded called Sorge's rule an act of "bigotry." Sorge denounced the middle-class reformers as Northern "carpetbaggers"; William

West retorted that "the bourgeoisie possess and acquire the experience and the intelligence which the movement needs." The vague and swirling debates of the Industrial Congress of 1850 over inclusion of middle-class reformers were by 1872 sharply focused—so much so as to fracture the First International in New York City.[6]

The IWA controversy of 1872 recalled the 1850 McCloskey-Barr debates over middle-class participation, with one critical difference. The group of delegates at the 1850 Congress led by James Bassett and Benjamin Price—who envisioned a politicized trades assembly drafting laws and defending working-class interests in the legislature—had no exact counterpart in the IWA. It was left to the organizations of the eight-hour movement, and not the IWA, to take up Benjamin Price's 1850 call for the working-class associations to "turn their own lawyers."

Bassett and Price's vision of a politicized trades assembly excluding politicians but tolerating sympathetic employers and reformers found many more working-class adherents in New York City than either Sorge's notion of an independent politics restricted to wage earners or the land reformers' nostalgic dream of an America free of all concentrated power. The difficulties of the New York City IWA, Siegfried Meyer observed in a rambling August 1871 letter to Karl Marx, could be blamed upon the peculiar social structure of the metropolis:

> The workers' movement has fallen off in the last five years in spite of the great clamor for an independent workers' party and in spite of the spread of labor unions throughout the land. And what accounts for this? The *kleinbürgerlichen* elements, which are sloughed off by the trade unions out of revolutionary instinct . . . are here drawn in [into the IWA] and in this way the systems, the hobby-horses, the idealizing rhetoric, etc. appear. . . . the American sections maintain no more than one or two workers, and while one German section after another is formed, their members neglect everything in order to strengthen the trades unions, in fact several try to destroy [the IWA] consciously. Precisely because the proletariat here comes into so much contact with the middle strata, one must keep the movement as pure as possible. . . . Actually, very large industry is more dominant in the New England states than in New York, where manufacturing shops with skilled workers are more general. If the bourgeois types turn up in the *Vereinen,* confusion and vagueness begin. . . .[7]

Here Meyer was acknowledging that intermediate territory of skilled workers and small employers that had been so controversial for the artisan associations and Industrial Congress of the early fifties. The IWA was unable to accommodate the "vagueness and confusion" attendant to the diverse political aspirations of New York City's sprawling lower middle class. As Meyer very nearly predicted, the IWA split over the issue of the participation of middle-class "outsiders" in the working-class domain. The great strength of the eight-hour movement in New York City was its ability to channel the social and political energy of the city's diverse working class and sympathetic *"kleinbürgerlichen* elements," under wage earners' leadership, into a movement capable of enforcing state labor legislation and contesting the employing classes for power.[8]

The Eight-Hour Movement

By the late sixties and early seventies, the movement for an eight-hour day had come to dominate workers' thinking about politics. The movement in New York had historical roots extending back to the deliberations of the Industrial Congress of 1850. But in New York as in so many other Northern cities it gained momentum only during the last years of the Civil War. The renewed demand for a law mandating a shorter workday in 1864 and 1865 was of a piece with the revived interest in housing and health legislation: artisans were again, as in 1850, defining consolidation as the use of government to regulate exploitative economic behavior. Along with this interest in legislative action went a virtual obsession with organization. One should here be careful not to place too much weight on the draft riots as a means of explaining post-war phenomena. But the riots did stigmatize a rough, saloon-frequenting, and insurrectionary poor and place a premium on the virtues of loyal and respectable workers—not only from the perspective of the middle and upper classes but also in the minds of many wage earners. The post-war eight-hour movement became a workshop in organization and discipline for artisans. The artisan leaders of the eight-hour movement became ambassadors of this respectable working class. As IWA activist Robert Blissert put it at an eight-hour rally in 1872, ". . . the workingmen had two enemies with which to contend—whiskey and the oppression of heartless capitalists." Only an orderly and respectable working class could answer Benjamin Price's call for the people to turn "their own lawyers, exercising their rights on all proper occasions." The outcome of the 1872 strikes—and the ultimate "winners" and "losers" of the July 1863 crisis—would turn on the character of this respectable workingman.[9]

How was it that the climactic political contest of this era was fought not in the legislature or at the polling place but on the shop floor and in the form of a strike? Of course, this shop-floor conflict had been building for two decades as industrialists endeavored to cast the "good workman" in an individualistic mold and organized workingmen tried, with considerable success, to define working-class respectability as part of a system of cooperative values. The workplace conflict intensified in spring 1872 as the now-powerful trade union movement mounted a grand effort to enforce the state eight-hour law on its own. Under these circumstances it was natural that industrialists and their allies would do all they could to challenge "combinations" among workmen and establish the dominance of their own definition of respectability. But the debate over consolidation was also resolved in the workplace because wage earners, in all their efforts to exercise political influence during the post-war years, never ventured too far from the trade union and the point of production as their locus of political authority and the strike as their political weapon.

The initial task of the eight-hour movement in 1866 and 1867 was the passage of a state eight-hour law. Most labor leaders agreed that the Workingmen's Union—the citywide trades assembly that coordinated union activities by spring 1864—was somehow central to the political project of enacting and enforcing such a law. John Woodruff of the Cigarmakers' Union argued in early 1864 that workingmen would obtain the shorter workday only if they voted for candidates favoring passage of an eight-hour bill. One Mr. McCleod of the Engineers' Union believed the Workingmen's Union should instead petition Congress for eight-hour legislation. Delegates to the Workingmen's Union from Carpenters' Union No. 2 and the Painters proposed a third alternative: that the eight-hour movement restrict its activities to strengthening the rule-enforcing power of the trade unions and launch the campaign through local activities of the Workingmen's Union. Messrs. Gridley and Fisher, the Carpenters' and Painters' delegates, expressed the fears of many of the city's trade unionists that politicians—even "independent" politicians who replied satisfactorily to interrogation by the Workingmen's Union—would divert the working-class eight-hour movement for their own uses. Accordingly, the Workingmen's Union relied mostly on its own working-class resources in the campaign for a state eight-hour law: it staged a two-day "eight-hour" picnic in late May 1865 reportedly attended by fifty thousand and a series of "eight-hour" mass meetings in December 1865 and April 1866. Such public pressure undoubtedly helped pass the law the following year. Here was James Bassett and Benjamin Price's project of 1850 realized: the trade organizations of the city assuming a quasi-political function as a legislative drafting and lobbying institution.[10]

The next phase of the eight-hour agitation in New York City required the use of citywide trade associations to compel enforcement of the new labor statute. During the first week of May 1867 the powerful Workingmen's Union already foresaw that city trade unions would have to take responsibility for the "ultimate adoption of the Eight Hour rule." Massive contests between labor and the combined forces of manufacturing and merchant capital such as the bricklayers' strike of 1868 educated artisans in the ways of organization and discipline. Through their trade unions, bricklayers, plasterers, and painters had intermittent success enforcing the eight-hour law through the late sixties. A Workingmen's Union and German Association of United Workingmen vitalized by the daily problem of enforcing work rules also served as the foundation for the independent labor candidacies of fall 1869, resolving to organize in "respective districts for the purpose of sending representatives direct from the ranks of labor to represent workingmen in the councils of the nation and State." The Workingmen's Union's quasi-political task of enforcing the eight-hour law had led to a well-defined plan for wage earners to intervene in the political arena as their own representatives, on their own terms. Though Tammany defeated the independent labor candidates of 1869, Governor Hoffman was, in the wake of that election, forced to make at least verbal commitment to the enforcement of the eight-hour statute.[11]

There were attempts to launch labor reform parties in New York City again in February 1870 and October 1871—each nurturing the hope that independent workingmen in the legislature might compel enforcement of the eight-hour law. The effort of winter 1870, led by Richard Trevellick, president of the National Labor Union, was quickly quashed by the Workingmen's Union, which denounced the plan for the "formation of a political party having for its object labor reform." By August 1870, the Workingmen's Union had come to regard the National Labor Congress as a "political dodge" and chose not to send representatives to its annual convention. The October 1871 attempt by Alexander Troup, Robert Blissert, and Dennis Griffin to form the "Workingmen's Party of New York City" gathered some support among delegates to the Workingmen's Union, but Plasterers' leader E. P. McDermott spoke for many other delegates when he complained that most of the new party's leaders were neither trade unionists nor mechanics. Like so many of the local workingmen's parties of the sixties and early seventies, the Workingmen's Party of 1871 failed to elect any of its nominees and soon disappeared. Of course, Tammany resistance to independent labor politics helped to account for these defeats. But the Workingmen's Union also seemed to undermine its own labor reform ventures. It was almost as if the city trade unionists were too uncomfortable with the ways and means of party politics to build an independent political organization. It would not be until 1886 that New York City workers launched a labor party with its own separate political apparatus, the United Labor Party. Through the early seventies, they relied entirely on the citywide trades assembly—and by extension on the trade unions themselves—to nominate "labor" candidates.[12]

After election day failures, the leaders of the Workingmen's Union often returned to the idea of cooperative production. The notion that workers could set up their own shops and establish their own domain apart from the wages system had captivated city workers since the early fifties. Following the crushing defeat of the workingmen's candidates in the 1869 election, Richard Trevellick and Nelson Young, enthusiastic proponents of a labor reform party, organized the "People's Cooperative Association," conceived as a national parent organization for wage earners' cooperatives. The officers of the Cooperative Association included Robert Blissert and John Browning, both leaders of the recent attempt to form a local labor party. In addition to promoting producers' cooperatives, the Cooperative Association also proposed to erect a working-class industrial institute housing reading rooms, lyceum halls, and cooperative stores: all in all, a working-class answer to the Cooper Union. The People's Cooperative Association and the many producers' cooperatives formed in 1870 were short-lived; the most successful of the cooperatives were formed by the Crispins in the boot and shoe trade. But it is important to note that the same trade unionists who were most active in attempts to build independent labor parties also led efforts to create a network of producers' cooperatives. The producers' cooperative offered these trade unionists the promise of a new organization of labor which simply ignored the political realm. Blissert and Browning may have had to resort to cooperative production by default after the electoral defeat of 1869. But these trade unionists seem to have believed that the ultimate object of a labor party was the creation of an independent working-class domain—what Robert Blissert referred to as the "glorious community and republic of labor"—and the protection of that domain from grasping employers and political party personnel. Perhaps these trade unionists regarded cooperative production not as a retreat from politics but as a more direct and less dangerous path toward the goal of establishing an independent working-class social realm free of the interference of "capitalists" and "politicians."[13]

In the end, the strike and not the independent political party or the producers' cooperative proved wage earners' most effective weapon during the middle decades. If the draft riots held any lesson for workers, it was the political power of the strike: had not the rioters' "strike" successfully nullified the draft in New York and kept a centralizing Republican government at bay? The successes of bricklayers, plasterers, and painters in implementing the eight-hour law through strikes in the late sixties confirmed the lesson. When its moment of opportunity arrived in spring 1872, the eight-hour movement relied upon the striking power of centralized trades organizations. The rule-making authority of wage earners on the shop floor, enforced by trades unions, was the foundation for workers' efforts at independent political influence in this era. It was on the shop floor and through the strike that workers' attempts at consolidation would either be vindicated or defeated.

With the fall of the Tweed Ring in late 1871, two movements quickly emerged to impose order on the city at a moment when all previous political solutions seemed to have failed. The first of these was the Committee of Seventy, an extra-partisan group of businessmen led by Democrats Henry G. Stebbins, William F. Havemeyer, and Oswald Ottendorfer and Republican James Brown. As a representative body of "the taxpayers and citizens of New York," the Committee authorized itself to examine the city ledgers and restore financial order. Though never more than a loose coalition of groups committed to the fiscal accountability of the city, the Committee of Seventy did much to keep the government running after the fall of Tweed.

The organizations of the eight-hour movement—Ira Steward's Eight Hour League, the Building Trades' League, the Furniture Workers' League, and the Iron and Metal Workers' League—offered their own interpretation of citizenship and accountable government. The exact form of that working-class citizenship would develop over the months of May and June in a series of strikes that may have involved as many as one hundred thousand workers. Working-class associations now intervened in the public life of the city to establish Robert Blissert's "glorious community and republic of labor."[14]

The eight-hour strikes of 1872 were not as spontaneous as the 1863 draft riot with its ad hoc worker committees and swirling crowds; nor were they orchestrated by one citywide labor organization. One of the great laments of William J. Jessup, leader of the city's Workingmen's Union, was the difficulty of imposing a centralized structure on the trades. Even the anti-coolie movement, Jessup confided hopefully to Siegfried Meyer in July 1870, had promise as a vehicle for centralizing control of the labor movement. Though leaders like Jessup and Robert Blissert would have liked to conduct the spring 1872 strikes through a central labor organization, such concerted action would not be possible until the rise of the Central Labor Union in the 1880s. The eight-hour strikes of 1872 developed, then, as a series of uncoordinated and improvised movements. Building-trades workers provided the overture, woodworkers entered with variations on the original theme, and then metal-trades workers delivered the tragic finale.[15]

With stonecutters, plasterers, and painters already working an eight-hour day, carpenters, bricklayers, plumbers, and housesmiths went on strike in early May 1872 to make the shorter working day general in the building trades. On May 20, the master masons and carpenters, in separate meetings, conceded the eight-hour rule to their employees, the boss carpenters' meeting doing so "under its most solemn protest." The strikes in the building trades succeeded, according to the *Times,* because "the small carpenters, who make up the great bulk of the Master Carpenters' Association, . . . will benefit by the eight-hour movement." The primary opposition to the eight-hour rule, the *Times* added, came from "those capitalist carpenters, who employ machinery largely, and who saw the prospect of heavy losses." According to this interpretation of May's events, the journeymen's victory on the 20th was in large measure a result of small bosses' sympathy for or coerced tolerance of the eight-hour law.[16]

There was much truth to the *Times*'s analysis of the events of May. David Montgomery has rightly argued that "employers who actually favored an eight-hour day for their own works were scarce as hens' teeth."[17] But many New York City small employers did tolerate the shorter workday: the competitive pressures on small bosses and subcontractors to continue production on an eight-hour basis in order to complete their contracts must have been immense. The broad intermediate stratum of part-time subcontractors in carpentering and bricklaying may have been decisive in winning the strike for the journeymen. These were the sort of men whose behavior the journeymen carpenters had with much success controlled through the Civil War epoch. There were some boss carpenters who even sympathized with the general aims of the eight-hour strike and were willing to adopt the rule as long as it was uniform through the trade. One Mr. Smith proclaimed at the May 20 employers' meeting, "I am for the eight-hour rule. It's better for the bosses. The plasterers, the bricklayers and the painters—men who do not require anything like the brains the journeyman carpenter requires—are working at eight hours; and it's not to our interest to make intelligent workmen serve at harder service than men of those other trades." Small bosses in building who sympathized with the eight-hour rule seemed to have believed, like Mr. Smith, that the rule was an appropriate one for "intelligent workmen."[18]

If Mr. Smith's opinion carried the day at the May 20 concession meeting of the boss carpenters, "Mr. Hume's" ideas also had much support. Hume announced that he was on the "ten-hours side":

> The men who are getting up this strike are the rag, tag and bobtail of the trade, and they are trying to get the good, capable men of the business into their ranks to give them calibre. They are acting in bad faith. . . . Talk about brains into carpenters. I have been in quest of brains into carpenters for a long time. There ain't any very intelligent or trusting men among these strikers.

The difference between Smith's and Hume's views hinged on their different interpretations of the "good workman." Smith accepted the workers' claims that the eight-hour movement would enhance the quality of their citizenship, their intellect, and their moral capacity. Hume replied that the strike by its very nature undermined the eight-hour men's claims to the status of good workmen. For Hume, the important point was that the strikers, almost regardless of their intellectual fiber, were "acting in bad faith," and could no longer be regarded as "trusting men."[19]

The building trades' eight-hour movement was sophisticated in its attempts to appeal to small bosses like Smith and other non-wage earners sympathetic to the eight-hour project. The Amalgamated Carpenters' Union does not seem to have included employers though it did count foremen among its members. On May 20, the journeymen debated whether, after having now won the eight-hour day, they should "refuse to work in any shop where non-Union men, be they foremen or others, are employed." After a furious debate, the majority of the carpenters voted to postpone the question. This decision may have reflected many carpenters' desire to preserve their ties with those journeymen who as part-time subcontractors were not eligible for union membership but whose support was critical to the eight-hour movement's success. The union sought to adopt a calm, at times even conciliatory tone with bosses who violated the short workday rule. "A committee of fifteen" was ordered on the 20th to "go to Yorkville, visit places where men were at work, but were particularly cautioned to use only moral suasion in inducing them to strike." Responding to early instances of policemen clubbing strikers, a carpenters' committee called upon lawyer Charles Spencer to assist in the prosecution of an offending officer, but only after ascertaining that "the lawyer [was] a Unionist, and heart and hand with them in their struggle." Albeit unofficially, the carpenters were even willing to bend the old injunction against lawyers' participation in order better to defend the movement against the coercive arm of the state. Finally, the building trades planned their own parade around the city to dramatize their cause to immigrants and non-workingmen; two weeks later, in concert with other trades in the Eight Hour League, they joined a boycott of all "storekeepers, bakers, butchers, tailors, undertakers, dry goods dealers and sewing-machine manufacturers" who "accumulate fortunes by the patronage of the working class" without supporting the strike. During a massive eight-hour parade to be held June 10, "every storekeeper" in the city and vicinity would be required to "place a card inscribed in large letters 'Eight hours for a day's work.' " That way, the Eight Hour League resolution concluded, "we may know our friends from our enemies." In this style, mixing two parts persuasion with one part coercion, the well-organized Building Trades' League sought to command the support of the workers and middle classes outside their organizations. The boycott, in particular, was an appropriate class strategy in a working-class community so intimately intertwined with a middle class of *petit* proprietors.[20]

The building trades' May uprising, then, combined an insistence that wage earners control the movement with a willingness to allow "privileged" sympathetic non-workers to participate. The battle lines between wage earners and employers were drawn more clearly in 1872 than ever before in the city's history. Indeed, journeymen carpenters learned that the bosses had yielded to the eight-hour rule from a worker "spy" who assumed the "character" of an employer and attended the master carpenters' meeting under cover. The rioters' highly charged definitions of loyalty in 1863—identifying who belonged and who did not belong to the "crowd"—were now applied by workers to the conflict between wage earners and employers. But the notion of some building-trades leaders in 1850—that certain middle-class sympathizers might be "privileged to membership" in the working-class domain—was ascendant in 1872. These foremen and small bosses were usually not allowed into the trade unions, but their support for the Eight Hour Leagues seems to have been critical to the latter institutions' success.

A final word on the complex ideology of the eight-hour movement is necessary. Some of the eight-hour agitators' ideas were attractive to some employers, especially the movement's emphasis on what one historian has called "citizenship, lectures, and concerts." But Ira Steward and other movement leaders could also verge dangerously close to a critique of the rights of property owners when they insisted on the inalienable right of workers to a portion of their time apart from the dictates of the wages system and the labor market. Labor, many eight-hour agitators insisted, could not be reduced to a commodity determined by a fixed standard. The eight-hour law moreover required the coercive power of the strike for enforcement. The respectable and sober artisan who contemplated coercion and threats of violence as a means of enforcing a work rule that verged on a critique of the rights of property owners—this artisan was a complicated animal indeed. He was in reality neither a "good" worker nor a "bad" worker in any of the bourgeois senses of those terms.[21]

Such complexities seem to have served the interests of wage earners more than they did those of employers. Small building contractors like Mr. Smith, who commended the intellectual qualities of the eight-hour men, very often blinded themselves to the class-conscious potentiality of the "intelligent workmen" they praised. The great power of the eight-hour movement was its ability to manipulate the "face" of working-class respectability that the eight-hour program presented to sympathetic employers. Workers could "sell" their respectability to sympathetic employers eager to "purchase" a cultivated and disciplined working class. This cultural "exchange" may ironically have provided the eight-hour movement with sufficient support among the middle classes to allow it to develop its class-conscious organization. Workers could take advantage of the anonymity of the large building site and the anonymity of a vast metropolis like New York to play the role of the "good workman" without an employer being much the wiser. It was unlikely by 1872 that Mr. Smith followed his "intelligent workmen" into their saloons, trade union halls, and homes to find out what they were "really" like. The "intelligent workman" role open to workers in the eight-hour movement undoubtedly helped to explain the success of that movement where the IWA failed. For many employers, a worker's IWA affiliation was, *prima facie,* disrespectable. By contrast, some employers seem to have regarded participation in the eight-hour movement as a sign of a worker's loyalty and moral fitness. Such employers' confidence in the faithfulness of the "good workman" was undermined by the events of June 1872, when the disciplined and loyal working class took to the streets to defend the eight-hour day against the assaults of organized industrialists. It was only during a grand *prise de conscience* like that of June 1872 that the employers and "loyal" workers of New York City truly came to know each other.[22]

Unlike the building-trades strikes of May, the woodworkers' movement beginning in the third week of May quickly gave lie to some employers' insistence that eight-hour men could be good workers. In New York City's cabinetmaking shops, class conflict may have been even more intense than in the building industry. Boss cabinetmakers had formed a strong association as early as 1860; by 1872 they were organized in a powerful Furniture

Manufacturers and Dealers Association. Journeymen cabinetmakers, mostly German, were among the best-organized artisans in the city through the Civil War epoch and were leaders of independent political experiments since the days of the *Arbeiterbund.* By 1872, the Furniture Workers' League was the base of operations for two of the city's most militant artisan leaders, Fred Homrighausen and Karl Speyer. In mid-century New York, labor radicalism often accompanied the smell of wood.

The Furniture Workers' League timed the eight-hour strikes in woodworking to begin at the height of the conflict in the building trades, May 20. In a communication to Governor Hoffman, the furniture manufacturers summarized what happened on the 20th: "On that day the factories were surrounded by an organized gang of strikers, under the lead of a committee from this League; and without seeking permission to enter—without asking the foreman—they plunged into the factory, invaded each floor, and by threats of violence compelled the workmen to desist from work, leave the shop, and join them in their movement. . . ." The manufacturers may have exaggerated the intimidation involved—by the first week in June seven thousand cabinet and furniture makers who belonged to the Eight Hour League had gone out on strike of their own volition. The sophisticated Furniture Workers' League understood that many of these German artisans were new to trade unionism—it led the effort to stage a massive eight-hour parade on June 10 to dramatize labor's cause. The woodworkers' leaders also invented the idea of boycotting those storeowners who did not display signs in support of the eight-hour cause.[23]

In the week before the parade, violence escalated around the city's largest uptown furniture and piano factories. Piano manufacture was the most degraded quarter of the city's cabinetmaking trade. Steinway and Sons and other larger employers were leaders in efforts to subdivide and mechanize the cabinetmakers' craft. On June 5 a decision by some of the workmen at Steinway to accept an increase in wages at ten hours was bitterly denounced in a mass meeting of piano workers. A procession of German cabinetmakers and piano workers marched on Steinway's Fifty-second Street factory, joined along the way by two hundred quarrymen. The crowd outside the factory attempted to dissuade the returning workmen from entering the building, "using threats of violence freely." Only the appearance of Police Inspector Walling prevented a confrontation between the strikers and the ten-hour men. Work now continued at Steinway under the protection of "the imposing front" of two hundred fifty policemen. The highly organized German cabinetmakers were clearly willing to use physical force to maintain the eight-hour rule. The slogan that occasionally appeared at city rallies—"Eight Hours—Peaceably If We Can, Forcibly If We Must"—well described the attitudes of many members of the Furniture Workers' League. These disciplined "intellectual workmen" were more willing than their building-trades brethren to entertain violence as a means of creating the "republic of labor." With their love of organization and their tolerance of violence, German woodworkers' activities in early June would force a sympathetic middle class to reevaluate its definition of the "good workman."[24]

That reevaluation was triggered by the grand eight-hour parade the Furniture Workers' League had scheduled for June 10. Five days before the parade, Theodore Banks, a Painters' leader influential in the anti-Marxist sections of the IWA and the Workingmen's Union, issued a circular calling for the use of arson by workers if capitalists refused to yield to the strike movement. Meanwhile, incidents of violence and confrontation between strikers' committees and scabs increased throughout the week. On the 6th one striker was arrested for trying to murder a scab, and another for threatening to burn down a factory where workers continued at ten hours. Banks was loudly denounced by the Eight Hour League and repudiated by several sections of the International. Meanwhile, thousands of workers—notably in the building trades—deserted the ranks of the planned parade on the 10th. When that day's demonstration finally took place, it was composed largely of furniture workers along with a contingent of metalworkers: cabinetmakers, upholsterers, polishers and varnishers, piano-makers, the Singer Sewing Machine Company employees, and two sections of the IWA. The banner, "Eight Hours—Peaceably If We Can, Forcibly If We Must"—flew under the auspices of the International. When Banks arrived at the assembly point, though, he was shoved to the rear. The disciplined woodworkers who dominated the ranks of the June 10 parade did not condone Banks's open endorsement of arson; they did, however, acknowledge physical coercion as a proper last resort of the eight-hour movement.[25]

Now, in the week after the parade, metal-trades workers launched their strike for the eight-hour day. The Iron and Metal Workers' League was the most poorly organized of the three umbrella eight-hour leagues in the city. Metal-trades workers hotly debated among themselves the appropriate way to conduct the strike. At a meeting of carriage blacksmiths, finishers, and helpers on the 7th, a suggestion of organizing "each branch" of the carriage-making trade into a separate union was fiercely attacked. When one worker asked "if there was any organization," cries of "Put him out" filled Germania Hall. The observation of another journeyman that they needed a "constitution and by-laws," since "at present they were an unorganized mob" finally led the gathering to appoint a committee to investigate the question. Through the second week of June, crowds of blacksmiths threatened J. B. Brewster's Carriage Factory and the Singer Sewing Machine plant, bastions of the ten-hour day in metallurgy. Meanwhile, another group of metalworkers sent a committee to Peter Cooper to give their approval to a Citizens' Association plan to end the strike through outside mediation. The metal-working trades were deeply divided over the question of coercion, which preoccupied the city after the June 10 parade.[26]

After June 10, the mood among employers had begun to change, as employers' organizations in all the trades stiffened their resistance. Strikes in progress on June 10 or beginning thereafter were doomed.[27] The most critical shift in mood occurred in the building trades. On June 17, a group of master masons held a meeting which revealed the city's largest builders' new determination to resist the eight-hour day and other journeymen's rules. The council proposed to "confer upon the results of the Eight-hour system and to consider the conduct of the men under its workings." This builders' meeting witnessed nothing less than a massive frontal assault on the idea that an eight-hour man could be a "good workman." John T. Conover, the city's largest builder and the celebrated target of the 1868 bricklayers' strike, presided, and Marc Eidlitz, Robert Darragh, and a few other of the most highly capitalized local builders dominated the proceedings. Nearly every builder who spoke that night claimed the eight-hour men were not the good workmen they appeared to be, and contested the trade unions' distinctions between good and bad workmen. Foremost was the complaint that

> although the eight hours has been granted, some of the men do not really do six hours' work. Another, that while all men are paid the same wages, there are some workmen who are almost worthless, and this class are rapidly superseding the good mechanic and driving him from the City. . . . Another is that under a certain law of the Unions a man who is an able workman is not allowed to do more than a certain amount of work within a given time, in order that the poorer workmen may not show to a disadvantage, and instances have occurred in which workmen have been fined $25 by the Union to which they belong for doing more work than they were allowed to do under the rules. Another is that if one man is discharged, and the others do not approve of it, then all quit.

Builders found this last union rule governing firing practices most annoying. One builder complained that on one of his jobs "every man out of seventy quit, because one worthless man shirked his duty and had been discharged." Another employer noted that his workmen put their tools down when "one non-society man was at work on his job." At the end of the evening, the master masons attending agreed to "reorganize and hold monthly meetings." After a controversial three-week trial run of the eight-hour law in building, these employers were rejecting the idea of the eight-hour man as a good workman. For the first time in the city's building industry, employers formally replaced journeymen's notion of good and bad workmen with their taxonomy. Employers like Mr. Smith, who on May 20 had sympathized with the "intelligent workmen" of the eight-hour movement, disappeared from public view in the third week of June. Now the soil was prepared for a unification of builders and industrialists in the name of the social and political rule of the "best men."[28]

At a mass meeting of over four hundred employers the following evening, this new possibility for a reunification of New York City's middle and upper classes was realized. The gathering was dominated by the industrialist Iron Founders' Association and its leaders John Roach, Adam S. Cameron, George B. Billerwell, and Andrew J. Campbell. But also attending were bodies of brass finishers, furniture-makers, builders, and piano-makers. For the first time in the Civil War epoch, "the bosses" were combined into a new "Employers' Central Executive Committee." The employers' resolutions established the trade unions' "arbitrary" distinctions between good and bad workers as the question of the evening:

> . . . we will hereafter pay our workmen by the hour, and we will only employ such as are willing to work ten hours per day, and til we can employ workmen on this basis. And the "trades unions" "societies," and "leagues," so-called have, by their unreasonable and arbitrary demands, done much to disturb the relations between employer and employed by forming combinations to secure the same rate of compensation for inferior as for superior workmen. . . .

The keynote speakers of the meeting were John W. Britton, partner in the firm of Brewster & Company, and Adam S. Cameron, a metal-trades employer. By the 18th, the strike among the metalworkers at Brewster's had ended in defeat. But Britton believed the mere fact of the strike signaled the failure of the Brewster & Company Industrial Association and its promise "to bring the workmen and the employer together" through a 10 percent profit-sharing plan. Britton lamented that "he found men took 'dirty' advantage though he had made some friends among the workmen of his establishment. . . . those very men on whom he fancied he could have placed dependence, were those who brought him into the difficulty." The tragedy, Britton concluded, was that the good workmen he had trusted were the very men who "could not be relied on." Cameron further explained to the group the character of the movement led by these deceiving workmen: "It is unhealthy, and I think I can see behind it the outline of Communism which it is sought to engraft upon this country, and we should take it by the throat and put an end to it." Finally, John Roach attributed the loss of the iron ship-building industry in New York City to the journeymen's work rules and the recurrence of strikes. The gathering then agreed to provide "loyal" non-union men with police protection. Employers in all branches of industry now agreed that good workmen would have no part of independent working-class organizations and their rules; bad workmen were an imported class of agitators who deserved the same draconian treatment that disloyal Confederates (and, presumably, draft rioters) had received during the Civil War.[29]

Though the Furniture Workers' League continued to meet each day through the middle of July, the eight-hour movement had for most purposes been defeated at the employers' meeting of June 18. From then on, the press reported a new wave of police beatings of workers, as the police began actively to serve the interests of the Employers' Central Committee. The employers' meeting of the 18th was also the end of an era of class relations in New York City. Merchants, manufacturers, and industrialists, though in many ways still at odds, had arrived at a *rapprochement* on certain basic questions in the wake of the massive working-class challenge of the eight-hour strikes. The project of consolidation—of finding new structures to bind a chaotic competitive order—had found adherents among all these middle- and upper-class groups after 1850. But during spring 1872, that project of consolidation had passed entirely into the control of a politically independent organized working class. The class conflicts of that spring forced employers sympathetic to or tolerant of a project of consolidation under working-class auspices to reevaluate their taxonomy of the working class. By late June, most of the city's employers redefined "the good workman" to exclude traits of independence and commitment to organization and work rules. It was no accident that the emerging "liberalism" of the 1870s grounded its vision of political independence not in loyal workers' initiatives but in a new and unyielding elitism.[30]

Shifts in outlook among the leadership of both the Association for Improving the Condition of the Poor and the Union League Club similarly suggest that an era of social relations had ended in New York. The AICP's October 1872 Annual Report featured a newly aggressive attack on coercive trade unions and their work rules. Frontally assaulting the eight-hour movement, the AICP reformers dismissed as a "pleasant theory" the belief in the "elevating effects on the character and condition of the workingman" resulting from "four hours a day for self-improvement and the benefit of his family." In reality, the moral reformers argued, a shortened workday would inevitably lead the wage earner to "lounging about street corners, gossiping and drinking at liquor-shops, neither [of which] tends to his own elevation nor to the happiness of his household, but rather to thriftless, dissipated habits and domestic wretchedness. . . ." Though the poor-relief organization had long expressed interest in labor issues, the moral reformers now for the first time attacked trade unions which "despotically enforce an arbitrary set of rules upon the observance of others, on the penalty of personal violence." The eight-hour strikes' encroachment on "individual rights" and their overtones of IWA and "Paris Communist" influence, the Annual Report concluded, "should be resisted, if necessary by the entire power of the State." Finally, the Report reprinted the resolutions from the June 18 "meeting representing the general manufacturing interest in the City of New-York." The AICP became the standard-bearer for the new and unified employers' interest consecrated on June 18. The industrialists and their outlook were finally triumphant.[31]

Nation editor E. L. Godkin may have best articulated the emerging consensus among middle- and upper-class New Yorkers. After the downfall of the Tweed Ring in September 1871, Godkin proclaimed in an article entitled "Rich Men in City Politics" that "one of the lessons which is most clearly enforced by our New York crisis is, that in a commercial and manufacturing community, it is not possible to prevent the union of wealth and political power." The issue, Godkin insisted, was not whether wealthy men should have influence in city politics, but what sort of rich men should dominate. "Property will weigh heavily in politics in spite of all we can do; what we have to decide is whether it shall weigh openly, legitimately, and by fair and moral means, or secretly, illicitly, and through bribery and corruption; and whether the property which weighs heavily in politics, shall be property honestly earned in commerce and manufactures and lawful speculation, or property accumulated in cheating, stealing, and corruption. . . ." Samuel J. Tilden, Democrat of Young America antecedents and leader of the anti-Tweed reform movement, and Godkin, Republican and spokesman for the Union League set, agreed that the middle and upper classes needed above all to insure the predominance of the "best men"—rich men of "legitimate wealth," not freebooting popular politicians or independent workers—in city affairs.[32]

The following year, the high priest of elite nationalism, Henry Whitney Bellows, delivered a sermon entitled "The Battle of Civilization," in which he announced that "the kingdom of heaven is to be taken by violence." The battle Bellows envisioned was not the conflict between Young America and the Union League Club that had in 1863 obsessed Frederick Law Olmsted and George Templeton Strong. One of the enemies in this new battle, Bellows began, was the familiar foe of "ease, and wealth, and position and friends to be made." Worry over material prosperity had been a familiar subject of Bellows's sermons since the oration at Jonathan Goodhue's funeral in 1848. Now, in 1873, Bellows directed merchant anxiety against new bogeys, "drifters with the current, and men who cry Peace when there is no peace; against bad citizens and corrupters of the city and the country . . . . against all who have abandoned the hope of God's kingdom, because too cowardly to apply the noble violence with which alone it is to be won." But at the very time Bellows proclaimed a new "battle of civilization" against old-line Democrats and corrupt Tammanyites, the Union League Club began to drift out of political reform activities. In 1879, Bellows wrote and published for private distribution a *Historical Sketch of the Union League Club of New York,* in part to remind a slothful membership of the wartime crisis that occasioned the Club's founding. In 1877, the Club's Committee on Political Reform reported " 'the action of the Club' upon subjects falling within their sphere 'had been more limited than in past years.' " That year, "to show that at least a great majority of the Club were still faithful to old principles," the Club felt obliged to adopt a series of resolutions affirming that "this Club has always been identified with the principles of the Republican party"—as if now there were some doubt. In 1872, LeGrand B. Cannon, chairman of the Club executive committee, asked the membership whether a patriotic club originating at a time of national peril "could permanently outlive the emergencies which led to its creation, and continue to flourish as a mere social institution, without the animating motive of a high public object." By 1872, the urgent need to establish a "club of true American aristocracy" in New York City had passed. The Union League Club, in spite of Bellows's prodding, was no longer the cultural barracks of a militant intelligentsia. The Club had for most purposes become, in Cannon's words, "a mere social institution." The war was over.[33]

By 1872, after over two decades of highly politicized social conflict, the schisms between elites and between workers and elites that had helped make draft rioting possible had been mediated. The problem of imposing order upon New York's intensely competitive and individualistic version of urban capitalism had preoccupied residents of the Civil War-era city. In the seventies, the problem persisted, but its political lineaments changed. After spring 1872, new forms of antagonism would emerge, well removed from the volcanic political conflicts of July 1863.

Epilogue: The Draft Riots' Lost Significance

The draft riots left New York and America a complex and ambiguous legacy. The Republican wartime government successfully marshaled enough regiments from the fields of Pennsylvania to suppress the riots on July 16–17, 1863. Lincoln and his advisors believed, with good reason, that an extended federal military supervision of New York would fail to restore order to the metropolis and might lead to recurring outbreaks. By not declaring martial law, the Republicans ended the violence, sustained conscription and the legitimacy of Republican rule, obtained enough men to preserve the momentum of the Union Army, and went on to win the war and receive the credit. But to accomplish this, Lincoln had to defer to conservative elites in New York by appointing Democratic financier John A. Dix as Commander of the Department of the East and sacrifice the ambitions of radical Republicans who saw martial law as an opportunity to reconstruct New York City. Dix's appointment not only confirmed New York as a Democratic city but suggested that there were strict limits to Republicans' national authority even at this early juncture, little more than two years after they took power. This early rebuke to Republican authority helps to explain why a radical, or "pure" Republican government did not come to power in the nation until 1866, and then, only for about two years, with license to proceed with military reconstruction not in New York, but only in the South. The national Republican government's "victory" during the draft riots also revealed the limited future prospects of radical rule.

Similarly, the conservatives who used the post-riot situation to secure their political power in New York City might have wondered, on deeper reflection, whether the draft riots did not augur their decline. For the conservatives, as for the radicals, the draft riot was a tale of two capitals: New York, the national economic capital with vaulting political aspirations, and Washington, the national political capital. Before the Civil War, the Democratic Party, led by metropolitan businessmen such as Dix and Belmont, occupied both seats of authority. Events of the sixties forced these captains of commerce to relinquish power in Washington and make vast compromises. The 1860 national election was a defeat for Young America and its New York City enthusiasts. When after the Battle of Antietam General McClellan rejected his aides' advice and declined to lead a counterrevolutionary march on Washington, it became increasingly likely that Young America merchants would remain out of power on the national level for some time to come. Barlow worked frantically backstage to orchestrate a McClellan victory in the presidential election of 1864, and Belmont even resorted to making stump speeches, but in vain. The Civil War disrupted Young America's relations with the American South and Great Britain—tore these businessmen from their economic and cultural moorings—and brought on a formidable challenge from Bellows, Olmsted, and the Union League Club. No less troubling from the perspective of Belmont, Barlow, and Dix, the draft riots and their aftermath insured that conservatives would retain power in the city only if they acknowledged the claims of two groups: the "Ring" of Tammany machine politicians who sat on the Board of Supervisors and presided over the draft exemption fund of the "County Substitute and Relief Committee," and a rapidly organizing working class. Indeed, if the draft riots represented a victory for conservatives, it was a victory riddled with concession and compromise. By fall 1863, it was clear that the draft riots had resulted not in clear "winners" and "losers" but a thicket of unresolved conflicts.

The white wage earners sympathetic to the revolt against the draft did not gain any enduring or meaningful control over the city as a result of the July riots. But the white working classes emerged from the event in a position of surprising influence. They frustrated an effective federal draft in New York City, precluded even the possibility of Republican rule in the metropolis for years to come, and guaranteed that the political program of local organized workers would receive an attentive hearing from the major parties over the next decade. Black wage earners, the riots' unequivocal victims, shared little of this growing working-class authority, the paternalistic attentions of the Union League Club notwithstanding. But white and black workers alike were swept up in the rage of organization that characterized the better part of the sixties and early seventies.

Further, the draft riots highlighted an extended era of working-class political influence. In this epoch, the riots were not the first time, nor the last, that workers' independent political initiatives shaped the political thinking and practice of other groups in the city. The trades congresses of the early fifties, the movements of the unemployed in 1854–57, and the post-war Workingmen's Union, International Workingmen's Association, and Eight Hour League commanded the attention of reformers of diverse social backgrounds and political allegiances. Wage earners were also influential participants in the Citizens' Association and related movements to build public parks, draft health and housing legislation, and create industrial schools. The conditions of working-class life were the main concern of these reform movements, and it was only natural that workers should take interest in their activities. While there was no successful independent labor party in Civil War-era New York, it was not because local workers sought such a party and failed. Rather, many wage earners were unwilling to make rigid distinctions between "economic" and "political" action in this period and regarded quasi-political structures such as the Industrial Congress, the Workingmen's Union, the IWA—and, for a time, Peter Cooper's 1853 City Reform and post-war Citizens' Association—as the most promising means of deploying the power of their trade organizations and imposing coherence on a competitive urban society. Through the early 1870s, New York City workers showed little enthusiasm for a labor party with its own political structure. Instead they preferred to nominate "labor" candidates from their trade assemblies. Such a strategy allowed working-class voters with strong commitments to the major parties to vote for a "labor" candidate without formally crossing party lines and affiliating with another organization—hence avoiding a problem with which alternative parties wrestled mightily in the late nineteenth and twentieth century. In the Civil War epoch, workers' creative attempts to translate economic prerogative into political influence and their growing skill in (and desire for) attracting middle- and upper-class allies to their movements afforded them substantial authority.

Some New Yorkers were more interested than others in the quasi-political institutional experiments of the fifties and sixties. Different trades organizations made the distinction between economic and political action to varying degrees. Building-trades leaders attempted in various quasi-political ways to influence reformers, politicians, and legislatures, while the recurring cry of tailors' leaders after 1850 was "no politics." Among the "better classes," Union League Club merchants and professional men had a vision of influence that blended economic, political, and cultural activities, at a distance from the party system. Industrialists formed the Citizens' Association, a quasi-political venture that could both enact reform legislation and cultivate *vrais ouvriers*. By contrast, Young America merchants, builders, and real estate interests readily looked to the machinery of the major political parties for solutions to problems of social conflict. Whether they turned away from the political party apparatus or toward it, all of these groups gave new urgency in the middle decades to the task of disciplining an especially volatile urban society and polity.

The artistry of a Boss Tweed lay in his creation of a system of party rule that appeared to be able to "solve" the problem of class conflict and social incoherence in the industrializing metropolis. If many New Yorkers had rejected the effete "old" political parties, then here was a "new" party staking its leadership on its ability to reconcile "people's rule" and class compromise. But by the early 1870s, the conflicts imbedded within the

Tweed Democracy could no longer be ignored or contained. The most prolonged and perhaps the most dramatic social confrontation of the era, the eight-hour strikes of 1872, centered on a political issue—whether the state eight-hour law, on the statute books for five years, would be enforced in New York City. But the protagonists in this conflict were not the major political parties, both of which had by 1872 been discredited in New York City as potential instruments of social and political order. The eight-hour strikes saw an array of allied trade organizations challenge the city's most powerful employers' societies. The eight-hour strikes ended in late June 1872 with a victory for a new "Employers' Central Committee" and the industrialists' version of acquisitive individualism. June 1872 represented a moment of reevaluation for all groups who had participated in projects of consolidation, as New York City began a new era in which workers were divided from the middle and upper classes by a widening social chasm, and a competitive urban capitalism entered its nineteenth-century heyday. Though none acknowledged it as such, June 1872 also represented the draft rioters' final moment of defeat. Only after nearly a decade of conflict, then, did the winners and losers of July 1863 become clear.

The significance of the draft riots, then, was their situation at the center of this era of politicized social relations in New York City. The controversy over the fairness of the Conscription Act merged with a much broader debate over the definition of justice in mid-century New York. That debate often proved highly corrosive to political order. For a time that debate and its attending conflicts allowed popular definitions of justice considerable influence in urban life. As a consequence, the more recognizably modern city of the late nineteenth century was a place where social debate was not so sweeping nor fundamentally critical, where justice as defined by the "best men" had far greater power and legitimacy, and where popular democracy, for all its vitality, was vitiated from its former self.

The business leadership of the Gilded Age metropolis was more unified and coherent than that of mid-century; it was far from monolithic. John Roach and his colleagues in metallurgy had the satisfaction of seeing their version of individualism dominate after 1872, but their victory that year earned them a reprieve, nothing more. By 1911, when economist Edward Pratt surveyed the city's factory landscape, there had been "a considerable movement" of heavy industry from Manhattan, and iron foundries had "long since ceased to exist" there. The industrial elite did not remain a vital force in the city. By the turn of the century, all of the city's important enterprises—financial administration, clothing, food, publications, and housing—were, in a broad sense, consumer industries, and most were committed to the viability of the downtown, center-island financial and manufacturing district. Skyscrapers (first built after the fall of Tweed) and sweatshops (more important in the economy than ever before) were the twin symbols of this new New York. Still, the late-nineteenth-century elite remained fragmented and found it difficult, as David C. Hammack has put it, "to develop and support a single economic program." Indeed, the divided elite endured as a hallmark of twentieth-century New York. New York's business and political leadership remains fractured by commitments that are at once international and national, as well as intensely local. Much as it was for Boss Tweed, the challenge for the twentieth-century "power brokers" has been to hold together New York as national center for financial administration and New York as a local venue of real estate deals and street improvements. It is fair to say that, while New York continues to have its political and fiscal crises, the problem of keeping the city's business leadership together and accommodating its national and local commitments has seldom been as formidable as it was in July 1863.[1]

The more unified and coherent metropolitan elite of the 1870s and 1880s helped to formulate a new national politics and culture. It was not the politics and culture Olmsted and his Union League colleagues had imagined in 1863. During that one shining season of opportunity, the Union Leaguers had contemplated all at once a reformation of New York City, the South, and the nation. If Olmsted had had his way, a standard of "democratic excellence" enforced by an active government would have been established in all these domains. But such a thoroughgoing centralization of the nation's political and cultural life flew in the face of New York's local interests—which asserted themselves forcefully during the draft riots—and so was doomed. The Union Leaguers also drew back from an active government of their own accord, frightened by what they perceived as the excesses of popular rule in the reconstructed legislatures of the post-war South, the defalcations of Boss Tweed, and the demands of eight-hour demonstrators. The New York gentry would not have another comparable opportunity to centralize and expand the American state until the turn of the century, when Theodore Roosevelt and Elihu Root took up the task. Nonetheless the "best men" who emerged triumphant from the draft riots had an immediate and lasting influence in national affairs. Through their literary journals and reform organizations, the Godkins and Tildens invented and promulgated the laissez-faire liberalism that became the creed of the newly self-conscious bourgeoisie of the late nineteenth century.[2]

Of course, the draft riots and their complicated legacy were not available to twentieth-century New Yorkers or Americans as part of a nineteenth-century inheritance. Public discussion of the event did continue for a time, largely among the increasingly embattled Gilded Age middle and upper class. William Osborn Stoddard's 1886 history of the draft riots, *The Volcano under the City,* was written in the wake of the sentencing of Chicago's Haymarket anarchists. Stoddard's arresting title and his recommendation that New York fortify itself against an unseen working-class threat well reflected the heightened anxieties of the late-nineteenth-century bourgeoisie. Wealthy New Yorkers tended to recall the draft riots only at moments of great social fear; other times they preferred to bury the memory of the event and its troubling conflicts. As the war generation died out at the end of the century, even the crisis-time evocations of the riots disappeared.

It is hard to imagine that the New York City poor did not continue to ponder the significance of the draft riots in unrecorded conversations, in the privacy of kitchens and taverns. But the insurrection was not part of workers' usable past. The event called forth discomfiting memories of treasonous resistance to a victorious national cause and of internecine hatreds among workers. In the last third of the nineteenth century, the white working classes of the urban North became increasingly committed to organization, to a more rigid distinction between "economic" and "political" action, and to at least a minimal acceptance of the black wage earner as a necessary element of any definition of "citizen." By the end of the century, the draft riots simply may not have made sense to a reconstituted working class with very different ideas about organization and politics. Indeed, many of the "new" mid-century social and political conditions that had made draft rioting possible were themselves rapidly transformed. Because the world of July 1863 was short-lived, the riots' ugliest and most sanguinary forms of violence disappeared from the political repertoire of succeeding generations of workers. Sadly, something of the draft rioters' audacity—their readiness to solve, all at once, the problems of an industrializing city and a divided nation—may have been lost as well.

APPENDIX A

Uptown Social Geography, 1863

In 1865, the Citizens' Association of New York issued a ward-by-ward report on the sanitary condition of the city entitled *Report of the Council of Hygiene and Public Health of the Citizens' Association of New York upon the Sanitary Condition of the City* (New York, 1865). Not merely a commentary on citywide health conditions, the *Report* also discussed the social life of each ward in minute detail. The *Report*'s comments on the Eighteenth Ward bear out the conclusion that the uptown middle-island district was very different in social constituency and tone from the uptown waterfront neighborhoods.

The *Report*'s review of conditions in the "18th Inspection District" (the southern half of the Eighteenth Ward), divided the area into three sections: an upper class zone west of Third Avenue, a zone of "artisans and tradespeople" from Third Avenue to First Avenue, and a laboring class, east of First Avenue, along the waterfront. Of the waterfront zone, the *Report* observed, "As regards the nationality of the population of this district, nearly all east of First Avenue are Irish and of Irish descent, with the occasional admixture of a family of Germans. More of the latter are found in the middle subdivision than in any other part; in fact quite a respectable portion of the population of that section are Germans" (*Report,* 209–10). It should be noted that the rioters' barricades just east of Third Avenue during July 1863 included behind their boundaries most residents of the middle zone as well as the zone along the waterfront.

On the West Side, in the Twentieth Ward, the class distinctions between middle island and waterfront were still evident, but somewhat less pronounced. The *Report* observed that the middle-island area east of Ninth Avenue was "occupied principally by people of American birth, many of them engaged in commercial pursuits," with some black families also residing between Sixth and Seventh avenues. In the area west of Ninth Avenue, the population was "by a large majority of foreign birth, and principally Irish. They are tradesmen, mechanics and laborers." The reformers noted that "there is rather more intemperance [along the waterfront] . . . than there is in the eastern portion of the district, though there are no localities in this district very notorious for the prevalence of vice and immorality" (*Report,* 240).

APPENDIX B

Occupations of Yorkville Anti-Draft Committee, July 14, 1863

Michael Cuskley, liquors
Isaac DuBoise, tailor
John Falvey, grocer
James Gallagher, lamps
Martin Hannan, liquors
William Hitchman, clerk, City Hall*
Owen J. Kelley, cooper(?)
John Keynton, attorney
Bernard McCabe, tailor
William McManus, liquors
Sylvester Ryan, laborer(?)
John T. Stewart, clerk

Source: New York *Herald,* July 15, 1863.

* Hitchman later became, and in 1863 may have already been, an active member of Tammany Hall.

Note: Only 12 of the 29 members of the Yorkville Committee were traceable in the city directory.

APPENDIX C

Decline in Black Population of New York City, 1861-1865

Ward Number	*Black Population, 1860*	*Black Population, 1865*
1	111	78
2	67	34
3	24	45
4	67	48
5	1396	865
6	334	289
7	141	70
8	2918	2174
9	424	476
10	198	96
11	225	124
12	263	436
13	562	302
14	1075	683
15	778	962
16	629	721
17	308	253
18	404	302
19	563	295
20	1471	1224
21	368	258
22	146	208
TOTAL	12,472	9943

Sources: U.S. Census Office, *Population of the United States in 1860* (Wash., D.C., 1864); Franklin B. Hough, *Census of the State of New York for 1865* (Albany, 1867).

APPENDIX D

List of Merchant Subscribers to Belmont Farewell Dinner, August 9, 1853

G. T. Adee
J. L. Aspinwall
S. J. Beebe
G. W. Beebee
J. D. Beers
R. M. Blatchford
J. H. Brower
P. Burrowes
W. B. Burrowes
Geo. Christ
R. F. Carman
E. Center
L. C. Clark
W. Clark
C. A. Clinton
H. A. Coit
J. Colles
J. Conkling
E. Croswell
T. E. Davis
T. Dehon
V. deLaunay
L. Dennison
E. Dodge
W. B. Duncan
F. W. Edmonds
B. H. Field
C. L. Frost
J. Gallatin
F. C. Gebhardt
A. D. F. Grant
H. Grinnell
C. A. Heckscher
Wm. Hoge
W. G. Hunt
C. D. Hurd
W. R. Jones, Jr.
H. H. Jonson
J. B. Kitching
S. Knapp
G. B. Lamar
A. M. Lawrence
C. B. Lawrence
Joseph Lawrence
James Lee
J. S. Libbey
E. B. Little
Jacob Little
M. Livingston
S. Livingston
A. Mann, Jr.
A. H. Markle
D. S. Miller
Charles Moran
M. Morgan
J. J. Palmer
Alfred Pell
J. J. Phelps
Royal Phelps
Edwin Post
Rufus Prime
J. F. Purdy
M. O. Roberts
G. S. Robbins
C. W. Rockwell
J. F. A. Sandford
R. Schell
F. Schuchardt
W[atts] Sherman
O. Smith
J. T. Soutter
P. Spofford
H. A. Stone
W. W. Stone
Joseph Stuart
Moses Taylor
I. Townsend
J. Trotter
W. H. Vermilye
J. A. Westervelt
S. Whitney
L. M. Wiley
W. E. Wilmerding
R. H. Winslow
B. R. Winthrop
R. Withers
N. H. Wolf
Wm. Wood
G. A. Worth

Source: August Belmont, *Letters, Speeches and Addresses* (New York, 1890), 3–4.

APPENDIX E

The Pine Street Meeting, December 15, 1860

Businessmen Who Signed Pine Street Resolutions

William H. Aspinwall
George E. Baldwin
John M. Barbour
James W. Beekman
J. H. Brower
Stewart Brown
T. W. Clerke
Edward Cooper
Edwin Croswell
Charles A. Davis
John A. Dix
Gerard Hallock
Elias J. Higgins
Wilson G. Hunt
A. S. Jarvis
John Kelly
G. Kemble
Richard Lathers
Thomas W. Ludlow
John McKeon
Charles O'Conor
John L. O'Sullivan
Edwards Pierrepont
Royal Phelps
Stephen P. Russel
Watts Sherman
Gustavus W. Smith
James T. Soutter
Samuel J. Tilden

Businessmen Reported To Have Attended Meeting Without Signing Resolutions

August Belmont
Erastus Corning
Henry Grinnell
William F. Havemeyer
Hiram Ketchum
A. A. Low
Robert B. Minturn
Alexander T. Stewart
Thomas Tileston

Source: Morgan Dix, *Memoirs of John Adams Dix* (New York, 1883), I:360.

APPENDIX F

Partial List of Businessmen and Business Concerns Endorsing People's Union Movement, November 1857

Horatio Allen
John H. Briggs
Gustavus A. Conover
Peter Cooper
Samuel Delamater
Frederic DePeyster
Benjamin H. Field
Hickson W. Field
Goodhue & Co.
C. Godfrey Gunther
Wm. Halsey & Co.
J[ames] Aug. Hamilton
William F. Havemeyer
Abram S. Hewitt
Wilson G. Hunt
S. Ingersoll & Field
Shepherd Knapp
Makesson & Robbins
Charles H. Marshall
Edwin D. Morgan
H. M. Schieffelin
C. Schwarzwaelder
John Stephenson
George Templeton Strong
Gerard Stuyvesant

Source: New York *Journal of Commerce,* Nov. 21, 1857.

APPENDIX G

List of Macready Petitioners, May 9, 1849

David Austen
Wilham C. Barrett
John R. Bartlett, Jr.
George Breuer
James Brooks
David C. Colden
James Colles
Edward Curtis
Francis Cutting
Charles A. Davis
J. E. Dekay
Robert L. Dillon
Simeon Draper
Denning Duer
Evert A. Duyckinck
Ogden P. Edwards
Hickson W. Field
J. Beekman Finlay
James Foster, Jr.
John W. Francis
Edward S. Gould
David Graham
Moses H. Grinnell
J. Prescott Hall
Willis Hall
Howard Henderson
Ogden Hoffman
Pierre M. Irving
Washington Irving
Ambrose L. Jordan
William Kent
Jacob Little
Ralph Lockwood
Cornelius Matthews
Herman Melville
Matthew Morgan
M. M. Noah
Duncan C. Pell
W. M. Prichard
Henry J. Raymond
Samuel B. Ruggles
Edward Sandford
Benjamin D. Silliman
Wessell S. Smith
Henry A. Stone
F. R. Tillou
Joseph L. White
Richard Grant White

Source: New York *Courier and Enquirer,* May 9, 1849.

APPENDIX H

Membership of the Union League Club of New York, 1863

John H. Abeel
A. G. Agnew
Cornelius R. Agnew
G. Albinola
George F. Allen
Benjamin G. Arnold
John L. Aspinwall
William H. Aspinwall
Daniel G. Bacon
Isaac H. Bailey
Latimer Bailey
George Bancroft
John C. Banon
William H. L. Barnes
Henry Bedlow
Gilbert L. Beekman
James W. Beekman
Henry W. Bellows
J. Philip Benkard
Josiah L. Bennett
Robert Benson
George F. Betts
Albert Bierstadt
O. William Bird
George Bliss, Jr.
Robert Bliss
William T. Blodgett
Edmund Blunt
George W. Blunt
Benjamin W. Bonney
William A. Booth
William Borden
Richard H. Bowne
Benjamin F. Breeden
Henry Brewster
Samuel W. Bridgham
John Crosby Brown
Thomas E. Brown
William C. Bryant
William A. Budd
J. E. Bulkley
Isaac M. Bull
William E. Bunker
A. M. Burr
Benjamin F. Butler, Jr.
Charles Butler
Charles E. Butler
Theron R. Butler
William Allen Butler
J. F. Butterworth
LeGrand B. Cannon
Nathaniel D. Carlisle
Charles Carow
James C. Carter
W. F. Cary
John B. Cecil
Nathan Chandler
Henry Chauncey, Jr.
T. M. Cheeseman
Everett Clapp
Henry E. Clark
W. Cockroft

(*Appendix H, continued*)

Thomas B. Coddington
C. J. Coggill
F. W. Coggill
Henry A. Coit
Joshua Coit
B. Collins
Charles Collins
James M. Constable
William S. Constant
Israel Corse
E. C. Cowdin
A. M. Cozzens
A. W. Craven
George William Curtis
Francis B. Cutting
Walter L. Cutting
John C. Dalton, Jr.
E. F. Davison
George B. De Forrest
James G. De Forrest
W. W. De Forrest
Henry Delafield
Edward Delano
Franklin H. Delano
John T. Denny
Thomas Denny
Christian E. Detmold
John A. Dix
William E. Dodge
William E. Dodge, Jr.
Henry C. Dorr
Levi A. Dowley
Davis Dows
William H. Draper
Henry Drisler
Abram Dubois
Denning Duer
Edwin J. Dunning
Dorman B. Eaton
Jonathan Edgar
Robert W. Edgar
Alfred L. Edwards
Richard S. Emmet
Walton W. Evans
William M. Evarts
Thomas Hall Faile
Thomas H. Faile, Jr.
Horace J. Fairchild
Haliburton Fales
C. B. Farnsworth
Charles N. Fearing
H. S. Fearing
Cyrus W. Field
Hamilton Fish
William H. Fogg
P. S. Forbes
Hobart Ford
John R. Ford
S. Conant Foster
Dudley B. Fuller
Joseph Gaillard
Sheppard Gandy
John E. Gavit
Sydney Howard Gay
E. T. Gerry
George Gibbs
Wolcott Gibbs
E. T. H. Gibson
James F. Gilbert
Parke Godwin
Richard Goodman
Charles C. Goodhue
John H. Gourlie
Joseph Grafton
H. W. Gray
J. F. Gray
O. De Forrest Grant
John C. Greene
Moses Grinnell
B. W. Griswold
George Griswold
J. N. A. Griswold
Egbert G. Guernsey
Edward Haight
William A. Hall
James Harper
George Griswold Haven
J. Woodward Hayden
Nathaniel Hayden
William Heath
Charles A. Heckscher
Henry F. Hitch

(*Appendix H, continued*)

Thomas Hitchcock
David Hoadley
Murray Hoffman
Philetus H. Holt
George T. Hope
Hamilton Hoppin
William J. Hoppin
Pendleton N. Hosack
Frank E. Howe
Joseph Howland
Mark Hoyt
H. W. Hubbell
Richard M. Hunt
Benjamin H. Hutton
Stephen Hyatt
Pierre M. Irving
John Jay
Charles Jenkins
Morris K. Jessup
James Boorman Johnston
William T. Johnston
George Jones
John Q. Jones
Charles A. Joy
Friedrich Kapp
James L. Kennedy
Robert Lennox Kennedy
John T. Kensett
Edgar Ketchum
Charles King
John A. King
Oliver Kane King
William G. King
Ambrose C. Kingsland
Charles P. Kirkland
Gideon Lee Knapp
Sheppard Knapp
William K. Knapp
Nehemiah Knight
George W. Land
Josiah Lane
John D. Lawson
Gideon Lee
William H. Lee
W. Creighton Lee
Francis Lieber
A. A. Livermore
Charles F. Livermore
George De Forrest Lord
J. Couper Lord
George Lorrillard
C. H. Ludington
Albon P. Man
Isaac P. Martin
John C. Martin
Charles H. Marshall
Charles H. Marshall, Jr.
Albert Matthews
Robert H. McCurdy
James McKaye
J. King Merrit
Edward Minturn
John W. Minturn
Robert B. Minturn
R. B. Minturn, Jr.
Samuel C. Moore
Hector Morrison
John H. Mortimer
Levi P. Morton
Benjamin Nathan
George Newbold
Adam Norrie
William Curtis Noyes
W. F. Oakey
Frederick Law Olmsted
George Palen
Francis A. Palmer
Henry Parish
Willard Parker
Eleazer Parmly
I. Green Pearson
Alfred Pell
Alfred Pell, Jr.
James K. Pell
Edmund Penfold
George A. Peters
J. C. Peters
Daniel Lewis Pettee
George D. Phelps
John J. Phelps
Henry L. Pierson
J. H. Pinkney

(*Appendix H, continued*)

Henry Varnum Poor
Alfred C. Post
Howard Potter
John Priestley
Frederick Prime
John W. Quincy
George S. Rainsford
J. H. Ransom
Silas Rawson
Henry J. Raymond
Robert G. Remsen
Philip Reynolds
S. W. Reynolds
P. Richards
George S. Robbins
Christopher R. Robert
Jones Rogers
James A. Roosevelt
Robert B. Roosevelt
Theodore Roosevelt
S. Weir Roosevelt
Austin L. Sands
Henry B. Sands
Mahlon D. Sands
George B. Satterlee
George C. Satterlee
R. S. Satterlee
Samuel B. Scheffelin
James L. Scheiffelin
Sidney Scheiffelin
J. S. Schultz
William Scott
Henry D. Sedgwick
Isaac Seymour
Thomas W. Shannon
Francis George Shaw
Isaac Sherman
E. F. Shonnard
Francis Skiddy
Samuel T. Skidmore
Augustus F. Smith
Charles D. Smith
N. Denton Smith
Samuel S. Smith
William A. Smith
Henry A. Smythe
Andrew S. Snelling
Rossell Stebbins, Jr.
G. S. Stephenson
Paul Spofford
Charles A. Stagg
John A. Stevens
Alexander T. Stewart
George E. Stone
John O. Stone
William Oliver Stone
Edwin W. Stoughton
Francis A. Stout
George Templeton Strong
P. Remsen Strong
Edward Sturges
Jonathan Sturges
Robert L. Stuart
Charles Sullivan
Nahum Sullivan
D. S. Suydam
James T. Suydam
Frederick G. Swan
Otis D. Swan
William H. Swan
James T. Swift
Henry A. Tailer
Charles N. Talbot
Moses Taylor
Henry G. Thompson
R. J. Thorne
Charles L. Tiffany
Thomas Tileston
Charles H. Trask
R. S. Tucker
Charles K. Tuckerman
Ernst Tuckerman
Henry T. Tuckerman
Joseph Tuckerman
Lucius Tuckerman
George E. Underhill
Robert Usher, Jr.
Henry F. Vail
A. A. Valentine
D. Van Nostrand
Alexander Van Rensselaer
Edgar S. Van Winkle

(*Appendix H, continued*)

Calvert Vaux
Washington R. Vermilye
Elias Wade, Jr.
George Cabot Ward
L. B. Ward
L. T. Warner
Ridley Watts
William H. Webb
John A. Weeks
Henry Weston
R. W. Weston
Sullivan H. Weston
Samuel Wetmore
Charles E. Whitehead
Thomas A. Whittaker
John E. Williams
R. Storrs Willis
W. B. Winterton
Henry R. Winthrop
Frederick H. Wolcott
John D. Wolfe
J. Butler Wright
E. M. Young

Source: Union League Club, *Articles of Association, By-Laws, Officers and Members of the Union League Club, 1863* (New York, 1863).

APPENDIX I

Social Composition of the Association for Improving the Condition of the Poor

The social composition of the New York Association for Improving the Condition of the Poor has to be analyzed on three distinct levels: executive leadership, supervisory ward councils, and contributing members. I place much emphasis on the participation in the AICP of James Brown (AICP president, Wall Street financier), Horatio Allen (AICP vice president, engine builder, and construction engineer of the nation's earliest railroads), and Thomas B. Stillman (master machinist, and through the 1840s and early 1850s, member of the Eleventh Ward Supervisory Council): all involved in the founding, financing, or management of the Novelty Iron Works, the city's largest industrial employer in the Civil War era. Howard Potter, banker, who would assume the presidency of the AICP after Brown's death and continue in that role through the eighties, was listed in the city directories of 1857–60 as "Secretary" of the Novelty Iron Works.

The managing officials of the AICP included a broader social constituency, with, predictably, a strong representation of certain groups: merchants, brokers, physicians, and, not surprisingly, iron hardware merchants and proprietors of metal and machine-building works. Of the ninety traceable members of the AICP "Supervisory Council" in 1863, there were 12 merchants of various kinds, 8 physicians, 8 iron founders and metal manufacturers, 6 brokers, 5 "agents," 4 hardware merchants, 4 tailors, 4 drug merchants or manufacturers, 3 builders or masons (presumably, master masons), 2 ministers, 2 connected with "real estate," 2 bankers, 2 painters, 2 paint manufacturers, 2 bakers, 2 boot manufacturers, and 1 each of book dealer, carriage manufacturer, cooper, "crockery," "express," feed salesman, "furnishing goods," "hangings," hat manufacturer, jeweler, lawyer, "leather and findings," marble manufacturer, plumber, policeman, saddler, stage proprietor, surveyor, "turner," "varnish," and "watches." Many of these members of the supervisory council also served as the organization's visitors.

Finally, and to my mind, most significant, is the appearance of many of the city's industrial employers and many of its largest employers among the AICP

"Members," which I take to mean those who contributed financially to the organization. The number of AICP "Members" was by 1863 well over a thousand. On the 1863 "Members" list appear, from an informal survey, the following large employers, listed either by proprietor or firm name: Horatio Allen, engine builder; George B. Billerwell, founder; William D. Andrews, metals manufacturer; C. W. Delamater, engine builder; Peter Morris, iron founder; Noah Worrall and "Worrall & Co.," iron founders; John Stephenson, car and carriage manufacturer; "Novelty Iron Works"; both Robert Hoe and "R. Hoe and Co.," printing press and machine manufacturers; "Scovill Manufacturing & Co.," photographic materials, metals and buttons; Mitchell, Vance & Co., chandeliers; L. H. Mace, Refrigerator Manufacturer; Manning Merrill, iron hardware manufacturer; Peter Lorillard, tobacco and snuff manufacturer; William Menzies, lumberyard proprietor; E. A. Quintard, iron ship and engine builder; William H. Webb, ship builder; to name only a small number.

It should be noted, as well, that of this small sample of AICP employers, Mitchell, Vance, & Co., L. H. Mace, and George H. Billerwell were leaders of the movement to form New York's first "Employers' Central Committee" in June 1872. New York *Tribune,* June 19, 1872.

Source: New York Association for Improving the Condition of the Poor, *Twentieth Annual Report . . . for the Year 1863* (New York, 1863), 5–8, 49–55.

APPENDIX J

Names and Occupations of Members of Uptown Property Associations, 1866-1871

East River Improvement Association

Harvey C. Calkins	John Roach
John Dunon	C. P. Rogers
George R. Jackson	Thomas T. Rowland
Benjamin F. Kreischer	T. F. Secor
H. W. Lowber	Erastus N. Smith
Christian Metzger	James R. Taylor
Ephriam Miller, Jr.	D. D. Wright
Andrew Mills	

Occupations (of 11 traceable): 6 engine builders and iron founders, 1 merchant, 1 joiner, 1 piano manufacturer, 1 hardware merchant, 1 builder.

East Side Association

H. P. Allen	J. I. Menges
James M. Boyd	C. E. Quackenbush
George W. Browne	Edward Roberts
J. E. Brush	Stephen Roberts
E. B. Bulkley	R. W. Roby
Charles Crary	Thomas Rutter
J. T. Cumming	Samuel Thompson
F. Geiger	J. M. Thorp
Joseph Hillenbrand	Robert Ward
John Holmes	J. H. Welch
P. G. Hubert	Noah Wheaton
E. C. Korner	W. A. Whitbeck

Occupations (of 16 traceable): 3 lawyers, 2 brewers, 2 hardware merchants, and 1 each of mason, grocer, surveyor, shades manufacturer, card manufacturer, eating house proprietor, fancy goods retailer, doors and sashes manufacturer, smith.

West Side Association

R. H. Arkenburgh	William R. Martin
Benjamin F. Beekman	James Monteith
William T. Blodgett	Courtlandt Palmer
Lewis B. Brown	John W. Pirsson
Andrew Carrigan	Marshall O. Roberts
Cyrus Clark	James F. Ruggles
John T. Daly	Charles Sanford
Jonathan Edgar	V. K. Stevenson
John A. C. Gray	Daniel F. Tiemann
Roswell D. Hatch	W. C. H. Waddell
A. S. Jarvis	Philip G. Weaver
John Q. Jones	Charles S. Webb
James T. King	William A. Whitbeck
David H. Knapp	(see East Side Association)
Charles F. Livermore	Fernando Wood

Occupations (of 18 traceable): 6 merchants, 5 lawyers, 1 each of teacher, "police," Commissioner of Castle Garden, "strawgoods," varnish manufacturer, tobacco merchant, banker.

Sources: Real Estate Record and Builders' Guide, May 9, 1868; West Side Association, *Proceedings of Six Public Meetings* (New York, 1871); East Side Association, *To the Friends of Rapid City Transit* (New York, 1871).

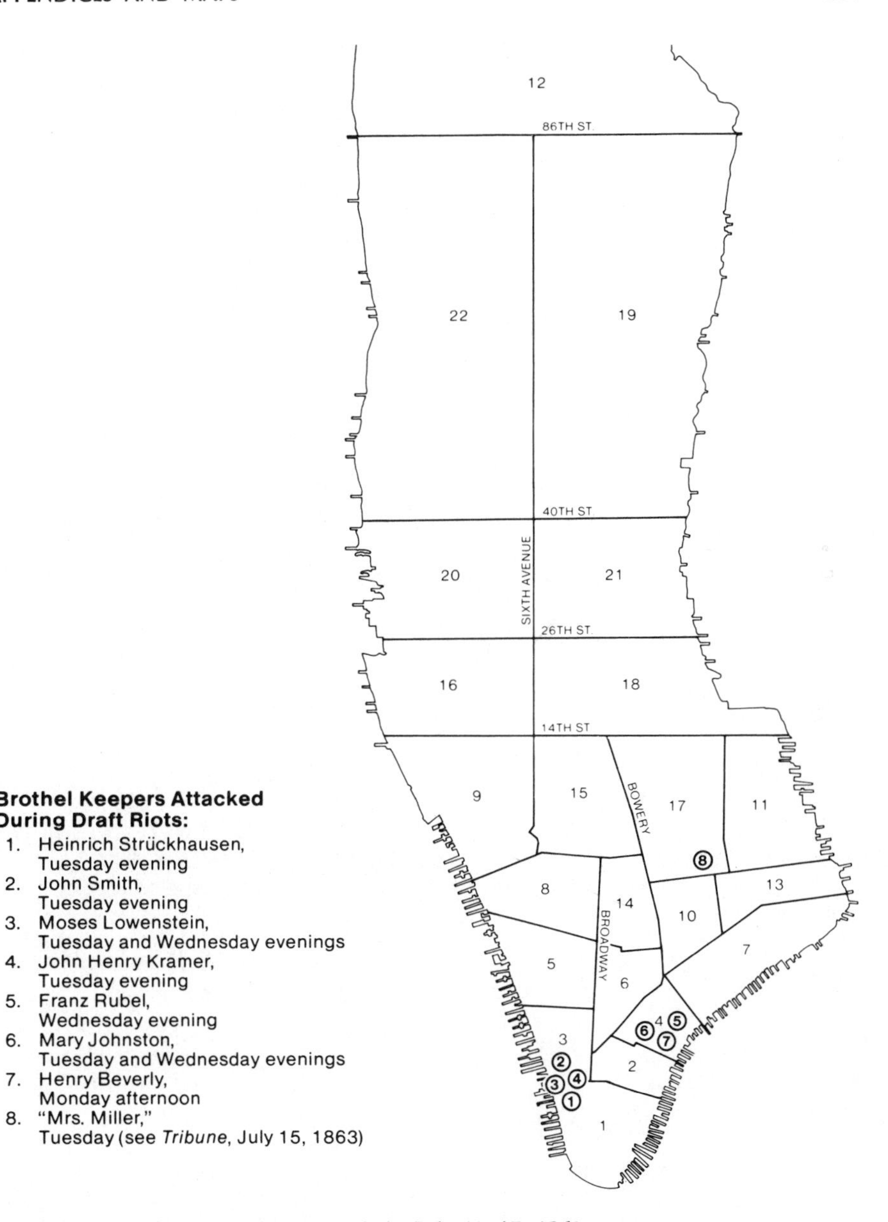

MAP 1. Map of the Attacks on Brothels, July 13–17, 1863
(*Source:* See Chapter 1, notes 118–124.)

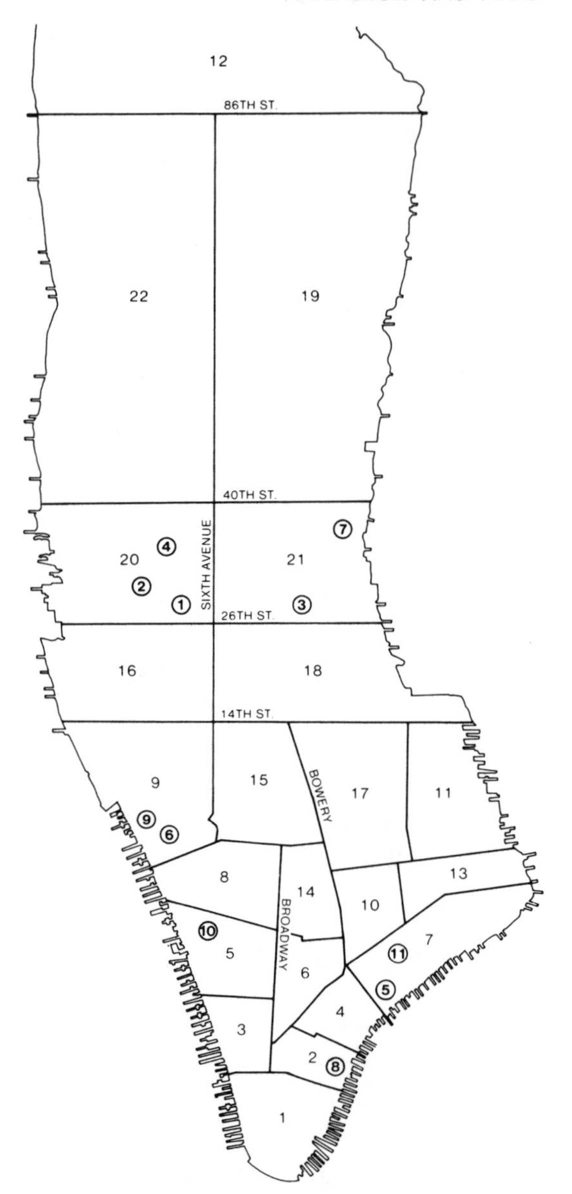

Black People Murdered By Draft Rioters:

1. Abraham Franklin, Wednesday morning
2. James Costello, Wednesday morning
3. William H. Nichols, Joseph Reed, Wednesday
4. Augustus Stuart, Wednesday evening
5. Jeremiah Robinson, (after Monday?)
6. William Jones, Monday afternoon
7. Joseph Jackson, Wednesday
8. Samuel Johnson, Tuesday night
9. William Williams, Tuesday morning
10. Ann Derrickson (white woman defending her mulatto son from beating), Tuesday evening
11. Peter Heuston, Monday afternoon or evening

MAP 2. The Geography of Racial Murders, July 13–July 17
(*Source:* David M. Barnes, *The Draft Riots in New York, July, 1863* (New York, 1863), 113–16.)

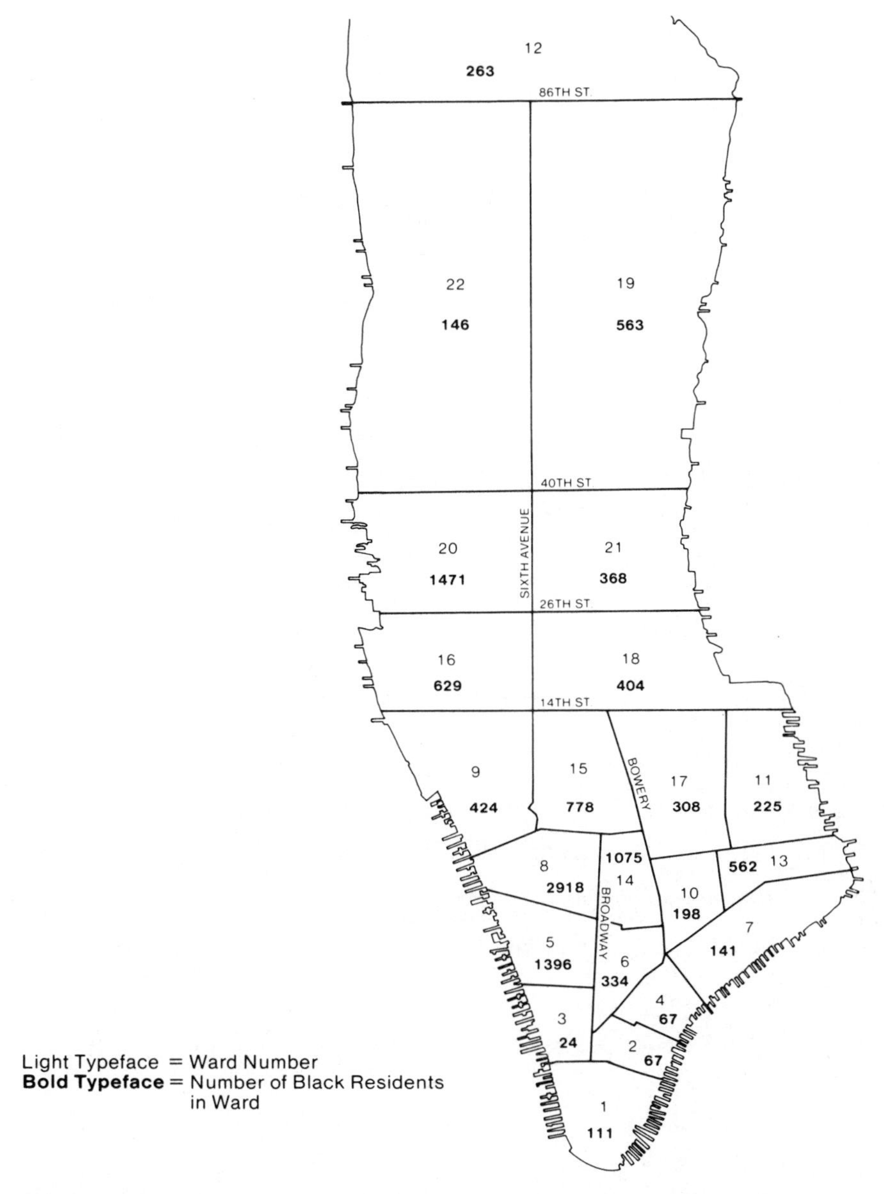

MAP 3. Distribution of Black Population in New York City, by Ward, 1860 (*Source:* U.S. Census Office, *Population of the United States in 1860* (Washington, D.C., 1864).)

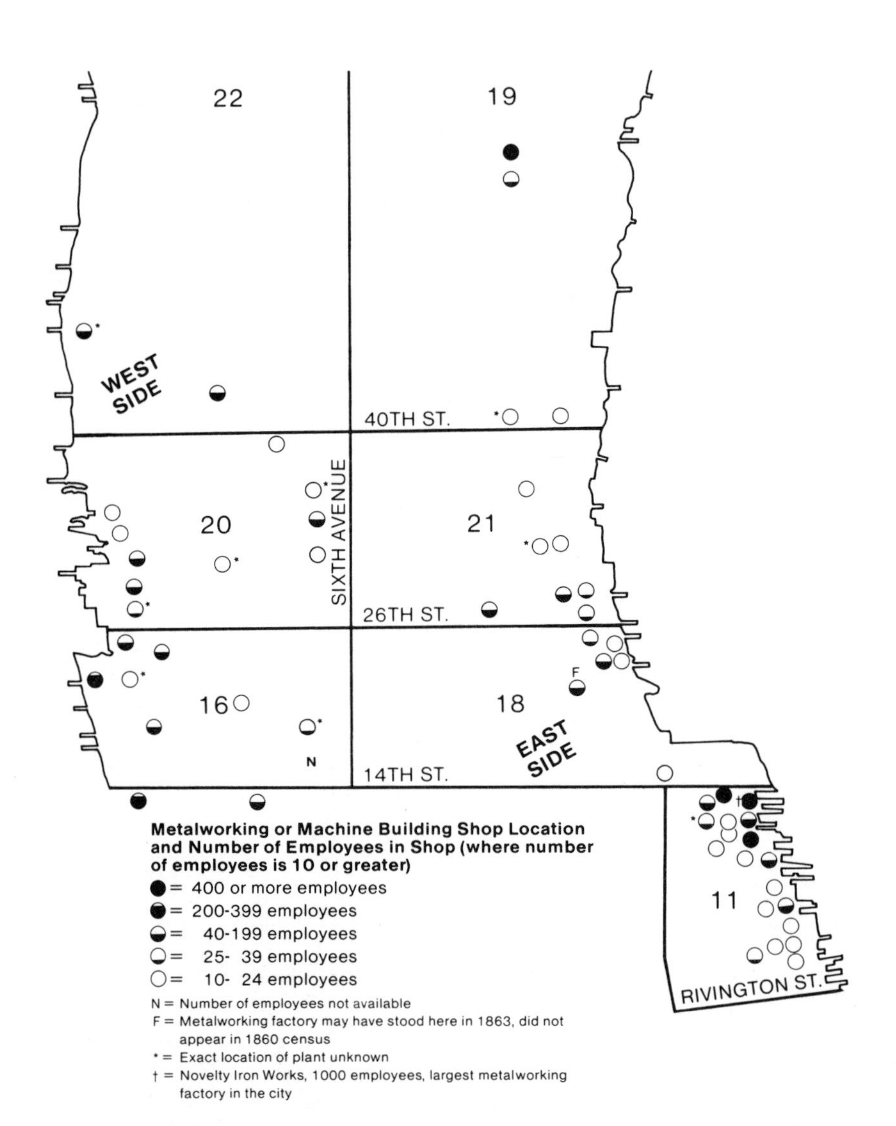

MAP 4. Location and Size of Metalworking and Machine Building Shops, Uptown Wards, 1860
(*Source:* United States Census Office, *U.S. Eighth Federal Census,* Industrial Manuscript Schedules, New York County, New York State Library, Albany, New York. Wards 11, 16, 18, 19, 20, 21, 22.)

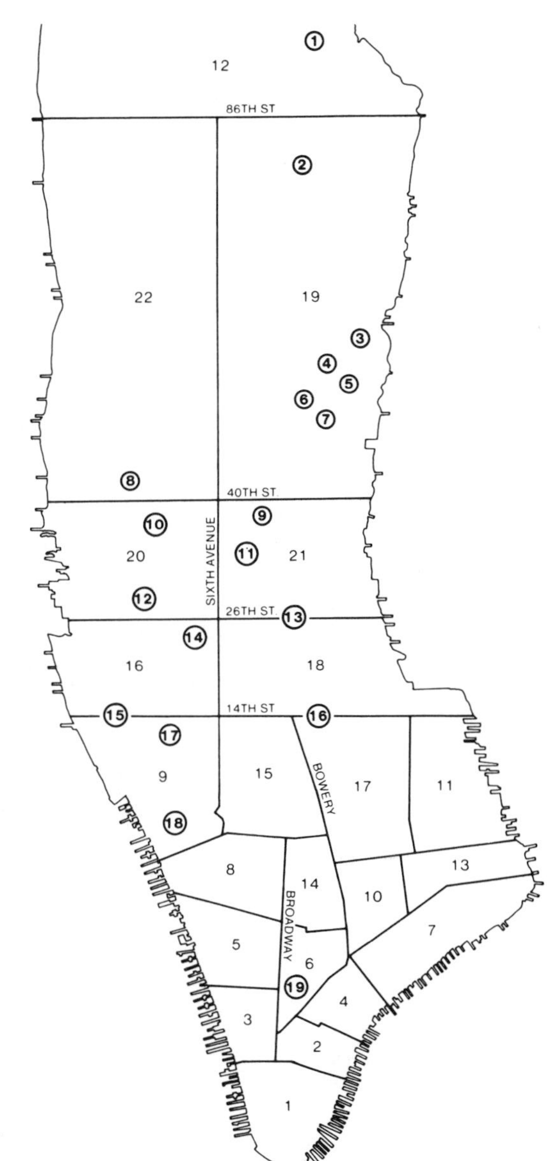

Tammany Contractors
Names and Places of Residence:

1. Patrick Slevin
2. Jeremiah D. Moore
3. Thomas Dunnelly
4. Michael Tracy
5. William Baird
6. Patrick Farley
7. G.C. Hegeman
8. Jeremiah Crowley
9. C.L. Purdy
10. William Jardines
11. J.W. and A.J. Pettigrew
12. John Martine
13. John Kinsley
14. Arthur Ahmenty
15. John Duffy
16. Christopher Keyes
17. Edward Guidley
18. Andrew J. McCool
19. C.G. Waterbury

Contractors listed are those whose addresses are provided on the Tammany Finance Committee List. For most, home and business addresses were the same or in the same locality.

MAP 5. Names and Geographic Distribution of Tammany Contractors, 1863 (*Source:* "Miscellaneous Contractors," in "Finance Committee, Democratic Republican General Committee, 1863," Misc. Mss., Manuscript Division, New York Public Library.)

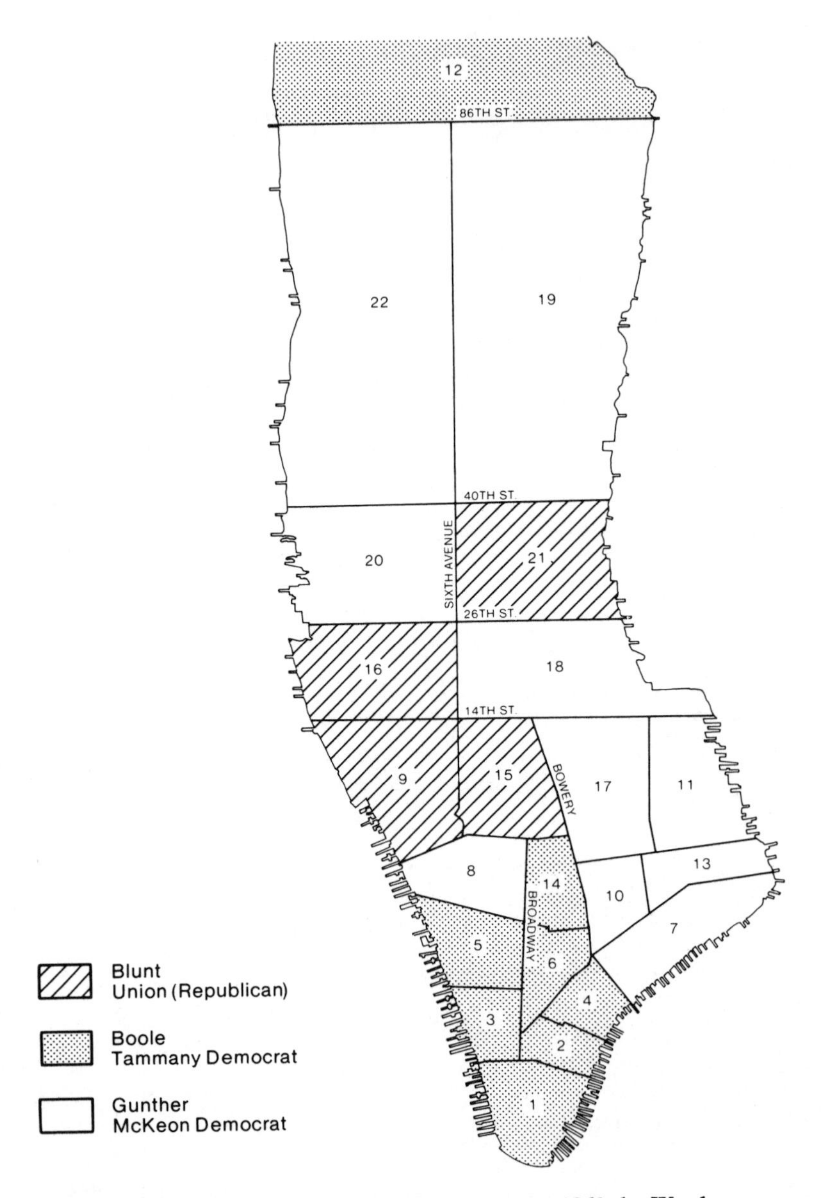

MAP 6. Election Returns, New York City Mayoral Race, 1863, by Ward (*Source: The Tribune Almanac and Political Register for 1864* (New York, 1864), 57.)

Notes

PREFACE

1. I borrow and edit for my own purposes a phrase used by Richard J. Evans in *Death in Hamburg: Society and Politics in the Cholera Years, 1830–1910* (Oxford, 1987), 567. Evans describes the cholera epidemic in late-nineteenth-century Hamburg as "one of those events that, as Lenin once put it, may perhaps be ultimately insignificant in themselves, but nevertheless, as in a flash of lightning, illuminate a whole historical landscape, throwing even the obscurest features into sharp and dramatic relief."

While the draft riots did indeed expose a contentious world, I readily acknowledge that mid-century society had quieter aspects that an event of this sort was less able to illuminate.

INTRODUCTION

1. Rioting was a common and important mode of cultural expression for antebellum urban Americans. On antebellum crowd violence, see Steven Novak, *The Rights of Youth: American Colleges and Student Revolt, 1798–1815* (Cambridge, 1977); David Allmendinger, "Dangers of Antebellum Student Life," *Journal of Social History* 7 (Fall 1973), 75–85 (youth); David Grimsted, "Rioting in Its Jacksonian Setting," *AHR* 77 (April 1972), 361–97 (rioting and republican ideology); Leonard L. Richards, *"Gentlemen of Property and Standing": Anti-Abolitionist Mobs in Jacksonian America* (New York, 1970); Theodore Hammett, "Two Mobs of Jacksonian Boston: Ideology and Interest," *JAH* 62 (March 1976), 845–68 (anti-abolitionist mobs); David Montgomery, "The Shuttle and the Cross: Weavers and Artisans in the Kensington Riots of 1844," *Journal of Social History* 5 (1972), 411–46; Bruce Laurie, "Fire Companies and Gangs in Southwark: The 1840s," in Allen F. Davis and Mark H. Haller, eds., *The Peoples of Philadelphia* (Phila., 1973), 71–87; Paul O. Weinbaum, "Temperance, Politics and the New York City Riots of 1857," *NYHSQ* 59 (July 1975), 246–70 (political violence of ethnic groups).

2. For a Northerner's anticipation of European intervention see Robert A. Maxwell to Abraham Lincoln, Phila., July 15, 1863, Abraham Lincoln Papers, LC.

3. Irving Werstein, *July, 1863* (New York, 1957); James McCague, *The Second Rebellion* (New York, 1968); Adrian Cook, *The Armies of the Streets: The New York City Draft Riots of 1863* (Lexington, Ky., 1974).

Accounts of the riots written by contemporaries and observers include William O. Stoddard, *The Volcano under the City. By a Volunteer Special* (New York, 1887); Joel Tyler Headley, *The Great Riots of New York, 1712–1873* (New York, 1873); David Barnes, *The Draft Riots in New York, July 1863* (New York, 1863); Ellen Leonard, *Three Days' Reign of Terror, or the July Riots in 1863, in New York* (New York, 1867); *The Bloody Week! Riot, Murder, and Arson, Containing a Full Account of the Wholesale Outrage on Life and Property, Accurately Prepared from Official Sources, by Eye Witnesses* (New York, 1863).

Other useful accounts include Hermann Schlüter, *Lincoln, Labor and Slavery* (New York, 1913), 207–10; Basil Leo Lee, *Discontent in New York City, 1861–1865* (Wash., D.C., 1943).

4. The New York City police were the one group to recall the events of the riot week with pride. On collective memory and history, see Maurice Halbwachs, *The Collective Memory,* trans. Francis J. Ditter, Jr., and Vida Yazdi Ditter (New York, 1980).

5. Adrian Cook writes that "riot was endemic in the social process of mid-nineteenth century New York," and links the draft riots to antebellum traditions of popular violence, though he does concede the distinctively anti-statist nature of the revolt. Cook, *Armies of the Streets,* 19.

6. Pauline Maier, *From Resistance to Revolution: Colonial Radicals and the Development of American Opposition to Britain, 1765–1776* (New York, 1972), 3–26; Grimsted, "Rioting in Its Jacksonian Setting," 361–97. Also, see George Rudé, *The Crowd in History, 1750–1848* (New York, 1964); E. P. Thompson, "The Moral Economy of the English Crowd in the Eighteenth Century," *Past and Present* 50 (Feb. 1971), 76–136.

7. On Parisian workers' activities in June 1848, see Georges Duveau, *1848: The Making of a Revolution* (New York, 1967); Donald McKay, *The National Workshops* (Cambridge, 1965); Mark Traugott, *Armies of the Poor: Determinants of Working-Class Participation in the Parisian Insurrection of June 1848* (Princeton, 1985).

8. Adrian Cook's revision of inflated riot-week estimates of a thousand or more draft riot casualties seems well-reasoned; Cook puts the death toll at 105 (or 119, if all doubtful cases are included). Cook, *Armies of the Streets,* 193–95. The drownings of black men in the East and North (Hudson) rivers and the retrieval and private burial of bodies by working-class families make casualty estimates difficult. Even if casualties numbered between 100 and 150, the draft riots were by far the most violent civil disorder in nineteenth-century America.

Michael Feldberg, *The Philadelphia Riots of 1844: A Study of Ethnic Conflict* (Westport, Conn., 1975), 156 (Southwark riot); on the Astor Place riot, see Peter G. Buckley, "To the Opera House: Culture and Society in New York City, 1820–1860" (Ph.D. diss., State Univ. of New York at Stony Brook, 1984), and below, Chapter 4. Buckley revises downward an earlier estimate of 31 deaths in the Astor Place riot.

It is impossible to draw up a precise body count of so many rioters, so many police, so many soldiers, so many blacks killed during the draft riots. The information about the identity of the dead and the circumstances of death is at best sketchy. Working with Adrian Cook's casualty list (*Armies of the Streets,* 213–18), supplemented by my own research, it is evident that at least 23 (and probably many more) of the approximately 105 dead were killed by troops, at least 11 were black males killed by the rioters, at least 8 were soldiers, and at least 3 were policemen who died as a result of riot duty. The number of dead whose identity and full circumstances of death are known is too small to allow generalizations. In many cases all that survives is the name of the deceased and the date and location of death. At most, one can hazard a guess that the largest group of casualties was that of rioters and bystanders killed by the military.

9. See Sean Wilentz, *Chants Democratic: New York City & the Rise of the American Working Class, 1788–1850* (New York, 1984), 219–71 and 377–86, on artisan republicanism in the context of the strikes of the thirties and early fifties; on the significance of the New York City tailors' strike of 1850, see below, Chapter 3.

10. On antebellum violence against black longshoremen in Philadelphia, see Bruce Laurie, *Working People of Philadelphia, 1800–1850* (Phila., 1980), 64–66; on anti-black violence in New York City, see Albon P. Man, Jr., "Labor Competition and the New York City Draft Riots of 1863," *Journal of Negro History* 36 (Oct. 1951), 376–405.

11. See Headley, *Great Riots of New York City,* 24–25, on the so-called Negro Riots of 1712 and 1741; see Linda K. Kerber, "Abolitionists and Amalgamators: The New York City Race Riots of 1834," *NYH* 48 (1967), 28–39, and Wilentz, *Chants Democratic,* 264–66, on the 1834 riots; Man, "Labor Competition," 403, on waterfront racial conflict during the 1850s. For a discussion of antebellum "racial rioting," see Paul A. Gilje, *The Road to Mobocracy: Popular Disorder in New York City, 1763–1834* (Chapel Hill, 1987), 143–70.

12. Laurie, "Fire Companies and Gangs," 79.

13. See Montgomery, "The Shuttle and the Cross," 411–46; Weinbaum, "Temperance, Politics," 246–70; also, W. J. Rorabaugh, "Rising Democratic Spirits: Immigrants, Temperance and Tammany Hall, 1854–1860," *Civil War History* 22 (June 1976), 138–57.

14. On the enrollment for and contemplated enforcement of the fall 1862 New York state draft, see *Times,* Aug. 18, 19, 21, 26–29, 31, Sept. 17, Oct. 15, Nov. 6, 9, 1862. It is possible, too, that the election of anti-draft Democrat Horatio Seymour as Governor of New York on November 4, 1862, led the Republican government in Washington to delay state conscription in New York. On New York's attempts to avoid a state draft through volunteering campaigns in late summer and fall 1862, see E. D. Morgan to E. M. Stanton, Albany, Aug. 14, 1862, and Thurlow Weed to E. M. Stanton, Albany, Aug. 15, 1862, in *WOR,* ser. III, vol. II:385, 393; George Winston Smith, "The National War Committee of the Citizens of New York," *NYH* 28 (Oct. 1947), 440–57.

15. Joel H. Silbey, *A Respectable Minority: The Democratic Party in the Civil War Era, 1860–1868* (New York, 1977), 144.

16. *WOR,* ser. III, vol. III (entries for March 3–10, 1863); on the provisions and operation of the Civil War draft, see Eugene C. Murdock, *One Million Men: The Civil War Draft in the North* (Madison, 1971).

17. [Edmund Ruffin], *Anticipations of the Future* (Richmond, 1860).

18. H. W. Bellows to his sister, Dec. 12, 1860, quoted in George M. Fredrickson, *The Inner Civil War* (New York, 1965), 54.

19. AICP, *Eighteenth Annual Report* (New York, 1861), 16.

20. *Weekly Caucasian,* Feb. 14, March 28, 1863; *Times,* March 25, 1863 (workers discuss conscription and commutation); on the actual operation of the commutation clause, see Hugh G. Earnhart, "Commutation: Democratic or Undemocratic?" *Civil War History* 12 (June 1966), 132–42, and especially, James M. McPherson, *Battle Cry of Freedom: The Civil War Era* (New York, 1988), 602–9; on Democratic Party reaction to commutation, see Earnhart, "Commutation," 133, and McPherson, *Battle Cry of Freedom,* 602–3, 608–9. I have not found any evidence of New York City Republicans defending commutation as an equitable measure in the debate over conscription triggered by the draft riots.

21. Maria L. Daly, "Diary of Maria L. Daly," I, July 14, 1863, Charles Patrick Daly Papers, NYPL.

22. One such dispute over Customs House patronage during the Buchanan administration is described in Leonard Chalmers, "Tammany Hall, Fernando Wood, and the Struggle to Control New York City, 1857–1859," *NYHSQ* 53 (Jan. 1969), 7–9. For a detailed discussion of New York Democrats' power in antebellum Washington, D.C., see Chapter 4, below.

23. James B. Fry, *New York and the Conscription Act of 1863: A Chapter in the History of the Civil War* (New York, 1885), 9, 12; *Herald,* Nov. 4, 1878 ("A Reminiscence of the Draft," by Horatio Seymour); John Jay to Edwin M. Stanton, NYC, July 18, 1863, *WOR,* ser. III, vol. III:540–41.

On the fairness of New York City and State conscription quotas see E. D. Morgan to Edwin M. Stanton, NYC, July 13, 1863, in *WOR,* ser. III, vol. III: 486 (Republicans battle widespread perception that New York draft quota was inequitable); Edwin Stanton to E. D. Morgan, Wash., D.C., July 17, 1863, Misc. Seymour, NYHS (Stanton informs Morgan that Seymour had every opportunity to review draft quotas); Richard Delafield to Col. J. B. Fry, NYC, Aug. 3, 1863, Delafield Letter-Press Book, NYHS (Republican official suggests that Nugent over-enrolled downtown business district "commuter" wards); and Horatio Seymour to John A. Dix, Albany, Aug. 7, 1863, John A. Dix Papers, CU. Nelson J. Waterbury, Democratic State Judge Advocate General, also believed that federal officials had punished New York with unfair conscription quotas.

24. *WOR,* ser. III, vol. III:241, 243–47, 360, 368–69.

25. Horatio Seymour to Samuel Sloan, NYC, July 13, 1863, *WOR,* ser. III, vol: III:530–31; *Proceedings of the Great Peace Convention, Held in the City of New-York, June 3d, 1863* (New York, 1863); *Herald,* June 1, 5, 6, 1863; *Daily News,* June 16, 1863.

26. T. J. Barnett to S. L. M. Barlow, Wash., D.C., June 10, 1863, Box 45, Samuel L. M. Barlow Papers, HL.

27. Col. J. B. Fry to Col. Robt. Nugent, Wash., D.C., July 3, 1863, *WOR,* ser. III, vol. III:467 (letter sent to all Acting Assistant Provost Marshal Generals); Col. Robt. Nugent to Col. J. B. Fry, NYC, July 9, 1863, Entry 1360, R. G. 110, NA; Fry, *New York and the Conscription of 1863,* 14–15 (Nugent's background).

28. Florence E. Gibson, *The Attitudes of the New York Irish Toward State*

and National Affairs, 1848–1892 (New York, 1951), 151; *Times*, July 6, 1863. Seymour himself, it should be said, supported the war effort even as he opposed the draft law and other centralizing Republican policies.

29. We can guess that reports of the slaughter at Gettysburg—the grim lists of men killed and maimed that comprised front-page news during the week of July 6—hardly enhanced the appeal of conscription. For lists of Gettysburg killed and wounded, see, for example, the front page of the *Times*, July 6, 7, 8, 9, 10, 1863; also see the headline, "Every Charge Repulsed With Great Slaughter," *Times*, July 6, 1863. I have found no specific evidence of draft rioters discussing the loss of life at Gettysburg, though one might interpret the rioters' cry, "die at home," as an implicit reference to carnage at the front; see below, Chapter 1.

30. *Herald*, July 14, 1863; Cook, *Armies of the Streets*, 55, on "grumbling about the draft" in Irish saloons of the Twentieth Ward on Sunday; *Sun*, July 11, 1863; George Washington Walling, *Recollections of a New York Chief of Police* (Denver, 1890), 78; *Tribune*, July 16, 1863 ("Organization").

CHAPTER 1. A MULTIPLICITY OF GRIEVANCES

1. See Chapter 2, below, for a discussion of each of these "readings" of the riots.

2. *Evening Express*, July 17, 1863. Though Brooks's calendar ends on Wednesday, it was not until sometime Friday that all areas of the city were pacified and work was generally resumed.

3. *Tribune*, July 14, 1863; *Irish American*, July 18, 1863; *Fincher's Trades Review*, July 18, 1863; Stoddard, *Volcano*, 30–31 (Croton Reservoir laborers); Barnes, *Draft Riots*, 10 (street contractors' men).

4. McCague, *Second Rebellion*, 61; *Bloody Week!*, 2. To my knowledge, no record of the proceedings in Central Park survives.

5. McCague, *Second Rebellion*, 61.

6. *Ibid.*, 62.

7. Barnes, *Draft Riots*, 10–12; Case of Francis Cusick, Grand Jury Dismissals, Aug. 1863, Supreme Court Cases, Location 9571, MARC; Cook, *Armies of the Streets*, 58–59.

8. Barnes, *Draft Riots*, 11–12; Cook, *Armies of the Streets*, 57.

9. Case of Thomas Fitzsimmons, Grand Jury Dismissals, Aug. 1863, and Case of Thomas Sutherland, Indictments, Aug. 1863, MARC. These cases involve the only two groups of East Side rioters from the morning of July 13 that are described in the court records in any detail. We do not know for certain that Fitzsimmons's crowd ever made it to the Ninth District lottery, though the timing of their confrontation with James Jackson (9:00 to 10:00 a.m.), the location (East Twenty-eighth Street), and their claim that they were demonstrating against the draft suggest they were headed toward the 10:30 lottery on East Forty-seventh Street. We do know that the Sutherland party, which closed Allaire's, Novelty, and several other shops in the Eleventh Ward, was headed uptown and that they arrived in time to participate in the attack on the Bull's Head (or Allerton's) Hotel, Forty-third Street between Fourth and Fifth avenues. There they joined crowds that had come to Allerton's from the Ninth District Provost Marshal's Office. See *Bloody Week!*, 7.

10. Stoddard, *Volcano,* 38.

11. Cook, *Armies of the Streets,* 60; *Leader,* July 18, 1863.

12. Stoddard, *Volcano,* 50; *Bloody Week!,* 4; for a biography of "Andrews the Mob Leader," see *Bloody Week!,* 30–31.

13. *Bloody Week!,* 3; *Tribune,* July 14, 1863; Stoddard, *Volcano,* 50; for Decker's speech to the Third Avenue crowd, see Cook, *Armies of the Streets,* 61.

14. *Herald,* July 14, 1863; for the Monday morning observations of one elite bystander, see *Strong Diary,* III:335–36.

15. *Herald,* July 14, 1863.

16. Jeter Allen Isely, *Horace Greeley and the Republican Party, 1853–1861: A Study of the New York Tribune* (Princeton, 1947), 10–11; Case of James H. Whitten, Indictments, Aug. 1863, MARC (see official court reporter's account of Whitten's trial for the *Tribune*); Cook, *Armies of the Streets,* 88–89.

17. Case of Thomas Fitzsimmons, Grand Jury Dismissals, Aug. 1863, MARC.

18. *Ibid.* (deposition of Charles Clinch).

19. *Ibid.*

20. Stoddard, *Volcano,* 50.

21. Case of Patrick Merry, Indictments, July 1863, MARC; Cook, *Armies of the Streets,* 247.

22. Headley, *The Great Riots of New York,* 169; Claim of James Vincent, Draft Riot Claims, Comptroller's Office, New York County, "Bundles Civil War Related," MARC.

23. *Tribune,* July 14, 1863; Barnes, *Draft Riots,* 46 (Eighth Precinct Policemen attacked), 58–59 (Fifteenth Precinct Policemen attacked).

24. *Tribune,* July 14, 1863; *Evening Express,* July 14, 1863.

25. *Irish American,* July 14, 1863. Accounts vary as to whether these houses were attacked because they were owned by Republicans or because their owners sheltered policemen fleeing the mob. Most likely, the rioters had both reasons in mind. See John Rogers, Jr., to Sarah Ellen Derby Rogers, July 18, 1863, Misc. Mss. Rogers, NYHS. Rogers saw the rioters searching for sheltered policemen in the Lexington Avenue homes that were burned down; he also overheard some rioters, possibly the same group, "inquiring where the republicans lived."

26. *Tribune,* July 14, 1863.

27. *Ibid.* (Eighteenth Ward Police Station House); Cook, *Armies of the Streets,* 68–70 (Twenty-first Street Armory).

28. *Tribune,* July 14, 1863.

29. Case of James H. Whitten, Indictments, Aug. 1863, MARC; for another indictment related to the attack on the *Tribune* office, see Case of George Burrows, Indictments, Oct. 1863, MARC; Cook, *Armies of the Streets,* 89–90.

30. Case of Thomas Fitzsimmons, Grand Jury Dismissals, Aug. 1863, MARC (depositions of Charles Irving, James W. Trimble).

31. *Ibid.* (deposition of Richard Hennessey).

32. *Ibid.* (deposition of Charles Irving).

33. See below, 39.

34. *Daily News,* July 23, 1863.

35. *Tribune,* July 24, 1863.

36. *Daily News,* July 23, 1863.

37. *Leader,* July 18, 1863.

38. *Evening Express,* July 15, 1863.

39. See Cook, *Armies of the Streets,* 57, for denunciations of Lincoln and Greeley at the attack on the Ninth District Provost Marshal's Office Monday morning.

40. See David Montgomery, "Strikes in Nineteenth Century America," *Social Science History* 4 (Feb. 1980), 81–104; Carl Degler, "Labor in the Economy and Politics of New York City, 1850–1860: A Study in the Impact of Early Industrialism" (Unpubl. Ph.D. diss., Columbia Univ., 1952).

41. See Case of Daniel Conroy and Thomas Kiernan, Grand Jury Dismissals, Aug. 1863, MARC; Case of Charles Dennin, Indictments, Aug. 1863, MARC (deposition of Peter Fowler). Dennin may have been printer Charles O. Denning.

42. Stoddard, *Volcano,* 30–31.

43. This last observation is based in part on discussion of the commutation and substitution clauses of the March 3rd Act in trade union circles. See *Fincher's Trades Review,* July 18, July 25, 1863. The offensive three-hundred-dollar clause was condemned by all Monday rioters, wrote one reporter, regardless of "whether one liked or disliked the Conscription Act." *Evening Express,* July 14, 1863.

44. *Tribune,* July 14, 1863 (Third Avenue, and Harlem and New Haven Railroads); *Evening Express,* July 15, 1863 (Weehawken Ferry, Hudson River Railroad); *Tribune,* July 15, 1863 (uptown bridges); for an arrest related to such attacks, see Case of Joseph Canary, Grand Jury Dismissals, Aug. 1863, MARC.

45. *Herald,* July 14, 1863.

46. Here the comment of the rioters who were trying to keep the authorities from summoning troops from Albany is relevant (see note 45), as is the report that rioters who attacked the gas house in the Eleventh Ward later in the week "were heard to say that they would put the city in darkness." *Herald,* July 16, 1863.

The one instance of Luddite-style machine-breaking during the riots occurred Tuesday in *Kleindeutschland.* Rioters burned some patent street-cleaning carts, "which undoubtedly were regarded as depriving rioting people of their profession." *Tribune,* July 15, 1863; on the Luddites, see Malcolm I. Thomis, *The Luddites: Machine-Breaking in Regency England* (New York, 1972).

47. Gideon Welles, *Diary of Gideon Welles* (Boston, 1911), I, entry of July 14, 1863, for one of many Republican claims that the riots were fomented by rebel agents; August Belmont to William H. Seward, Newport, R.I., July 20, 1863, in August Belmont, *Letters, Speeches, and Addresses* (New York, 1890), for evidence that some Democrats reached the same conclusion as Welles.

48. German involvement in the Monday processions is mentioned in Maria L. Daly, "Diary of Maria L. Daly," I, July 23, 1863, Charles Patrick Daly Papers, NYPL. On German rioters in *Kleindeutschland* on Tuesday, see *Tribune,* July 15, 1863. Also see Maria Daly's comment that "the principal actors in the mob were boys and I think many were Americans," in Maria L. Daly, "Diary," I, July 23, 1863, Daly Papers, NYPL.

49. On the building trades' participation in the Monday morning anti-draft procession, see *Irish American,* July 18, 1863; *Fincher's Trades Review,* July 18, 1863; *Tribune,* July 14, 1863.

Two points must be made with regard to the observation that artisans in the building trades formed the backbone of the uptown fire companies in 1863. First, each fire company had three leadership positions (foreman, assistant, secretary). In 26 of the 44 uptown fire companies (59.1 percent), at least one of the three officers was in the building trades. In 9 of those 44 companies (20 percent), at least two of the three officers were in the building trades. Second, among the membership of these companies, 343 of the 1430 uptown firemen (23.9 percent) were in the building trades. All told, the building trades were represented heavily among the leadership and substantially among the membership of the uptown fire companies (uptown is defined here as north of Fourteenth Street). For the lists of the officers and members of the New York fire companies, with occupations and addresses noted, that were used to compute these figures, see Office of the Chief Engineer, *Annual Report of the Chief Engineer of the Fire Department, of the City of New York, Document 14* (New York, 1863), 33–34, 36–37, 40–41, 52–57, 62–64, 66–69, 71–72, 79–81, 83–88, 90–91, 103, 112, 120–21, 123, 125, 127, 129, 132–34, 136, 139–40, 143, 145, 147, 152–55, 160–61, Municipal Reference and Research Center, New York.

50. See below, Chapters 3 and 6.

51. Dock work and quarry work were almost exclusively Irish occupations.

52. *Times,* July 14, 1863.

53. *Herald,* July 16, 1863 ("half-grown boys").

54. *Tribune,* July 17, 1863.

55. See Chapter 3 (1850 tailors' strike) and Chapter 4 (anti-abolitionist riot of 1834, police riot of 1857).

56. *Tribune,* July 18, 1863; *Leader,* July 18, 1863.

57. In this neighborhood, see claims of black women Eugenia Brown, Electra Cox, Elizabeth Dixon, Hannah Spencer, "Draft Riot Claims," Comptroller's Office, New York County, MARC; also Case of William Cruise, James Best, Moses Breen, Indictments, Dec. 1863, MARC.

58. *Herald,* July 16, 18, 1863.

59. *Leader,* July 18, 1863.

60. See Christine Stansell, "The Origins of the Sweatshop: Women and Early Industrialization in New York City," in Michael H. Frisch and Daniel J. Walkowitz, eds., *Working Class America: Essays on Labor, Community and American Society* (Urbana, 1983), 78–103; Wilentz, *Chants Democratic,* 107–42.

61. Lucy Gibbons Morse, "Personal Recollections of the Draft Riot of 1863," 2, Knapp/Powell, Trunk 1, NYHS.

62. *Ibid.,* 1.

63. *Ibid.,* 6; Case of John Corrigan, Grand Jury Dismissals, 1863, MARC (Corrigan was one of the men on horseback).

64. Claim of Abby H. Gibbons, in Board of Supervisors, *Communication,* II:874.

65. *Ibid.*

66. *Ibid.* (affidavit of James S. Gibbons).

67. Mrs. E. B. Sedgwick to Julia Gibbons, July 18, 1863, in Sarah Hopper Emerson, ed., *The Life of Abby Hopper Gibbons Told Chiefly Through Her Correspondence* (New York, 1896), II:52–53; for a theft claim related to the Gibbons attack, see Case of Mary Kennedy, Indictments, Aug. 1863, MARC.

68. Case of Michael O'Brien and John Fitzherbert, Indictments, Oct. 1863, MARC; Cook, *Armies of the Streets,* 125–26.

69. On cheering for Seymour and Fernando Wood on the afternoon of July 14th, see *Tribune,* July 18, 1863.

70. *Daily News,* July 16, 1863.

71. Thomas Bartlett, "An End to Moral Economy: The Irish Militia Disturbances of 1793," *Past and Present* 99 (May 1983), 41–64, and esp. 57.

72. On the significance of Baltimore as a Southern-sympathizing Northern city, see Barbara Jeanne Fields, *Slavery and Freedom on the Middle Ground: Maryland during the Nineteenth Century* (New Haven, 1985), 40–62. See Joel Tyler Headley, *The Great Rebellion: A History of the Civil War in the United States* (Hartford, 1865), I:70–82, on Baltimore crowds' attacks on the Massachusetts Sixth Regiment on its way to the front in April 1861. Governor Horatio Seymour's anti-draft agitation in New York State in 1863 was in some ways reminiscent of the attempts of Maryland's Governor Thomas Hicks to prevent federal troops from passing through Baltimore in the early days of the war.

73. John A. Kennedy to E. M. Stanton, July 8, 1863, in *WOR,* ser. III, vol. III:473.

74. McCague, *Second Rebellion,* 56.

75. Committee of Merchants for the Relief of Colored People, Suffering from the Late Riots in the City of New York, *Report,* in James M. McPherson, ed., *Anti-Negro Riots in the North, 1863,* 25; Cook, *Armies of the Streets,* 78.

76. Committee of Merchants, *Report,* 16; *Tribune,* July 14, 1863; Case of John Nicholson, Indictments, Oct. 1863, MARC; Cook, *Armies of the Streets,* 82.

77. *Daily News,* July 17, 1863; Albon P. Man, Jr., "Labor Competition and the New York Draft Riots of 1863," *Journal of Negro History* 36 (Oct. 1951), 375–405.

78. Case of Frank Shandley, Ann Shandley, James Shandley, James Cassidy, Indictments, Aug. 1863, MARC (street paver); Case of William Rigby, Indictments, July 1863, MARC (hack driver); Case of Edward Canfield, James Lamb, William Butney, Indictments, Aug. 1863, MARC; attacks on Charles Jackson, William Johnson and son, Committee of Merchants, *Report,* 21, *Evening Express,* July 17, 1863, "A Negro's Throat Cut in the First Ward," (longshoremen); Case of Dennis Carey, *Tribune,* July 24, 1863 (cartman).

79. *Evening Express,* July 17, 1863 ("A Negro's Throat Cut in the First Ward").

80. Claims of Mary Johnston, Henry Beverly, William Taylor, in Board of Supervisors of the County of New York, *Communication from the Comptroller . . . Relative to Damage by Riots of 1863,* II, Doc. No. 13 (New York, 1868), 163–64, 797–800, 926–29; Cook, *Armies of the Streets,* 79–80.

81. Cook, *Armies of the Streets,* 79–80; *Evening Express,* July 14, 15, 1863.

82. Committee of Merchants, *Report,* 15; Cook, *Armies of the Streets,* 82–83. Women of color were more often stoned or chased away than beaten or killed. For an instance of a white woman attacked because of her association with a targeted black man, see the rioters' beating murder of Ann Derrickson when she tried to shield her son Alfred from attack, Case of William Cruise, James Best, Moses Breen, Indictments, Dec. 1863, MARC.

83. *Evening Express,* July 14, 1863 ("the negroes have completely disappeared from the streets"); and see below, on Wednesday's attacks on Chinese men in the Fourth Ward. Such a change in the racial targets of downtown rioters suggests that by midweek most black people who were able to leave the downtown district had done so.

84. Case of Edward Canfield, James Lamb, William Butney, Indictments, Aug. 1863, MARC; Coroner's Inquest at the Body of William Williams, Indictments, Aug. 1863, MARC.

85. Case of Canfield, Lamb, Butney, Indictments, Aug. 1863, MARC. The reactions of the white witnesses of the murder are recorded in the testimony for the case. Also, see Committee of Merchants, *Report,* 20; Cook, *Armies of the Streets,* 97–98.

86. Case of Matthew Zweick, Indictments, Oct. 1863, MARC; *Tribune,* July 27, 1863.

87. Case of Matthew Zweick, Indictments, Oct. 1863, MARC. Zweick was indicted for riot, offending public decency, and first-degree murder. The district attorney refused to prosecute on the grounds of inconclusive evidence.

88. *Ibid.* Note that of the 23 rioters killed in a confrontation with Colonel Mott's troops after the lynching murder of James Costello on the 15th, 19 had waterfront addresses, listed in the newspaper as "Eleventh Avenue." Of the 16 men killed, 8 were traceable, with some allowance for error, in the city directory. This group was composed of 3 laborers, 3 grocers, an "agent," and either an employee or proprietor of a stoneyard (listed merely as "stone"). See New York *Times,* July 16, 1863 ("Partial List of the Killed and Wounded on 32nd Street").

89. *Tribune,* July 24, 25, 28, 1863; Cook, *Armies of the Streets,* 143.

90. Case of Patrick Butler, Indictments, July 1863, MARC. In his answer to an initial charge of murder, Butler responded, "I am not guilty, all I done was to take hold of his private parts." He later pleaded guilty to the charge of offending public decency. Cook, *Armies of the Streets,* 235.

91. *Bloody Week!,* 25; *Tribune,* July 16, 1863.

92. See especially the Claim of Maria Prince, County of New York, *Communication from the Comptroller,* 274–76 (Testimony of Enoch W. Jacques).

93. Note the importance of similarly violent rituals of purification in the Catholic religious riots of early modern France. See Natalie Zemon Davis, "The Rites of Violence: Religious Riot in Sixteenth-Century France," *Past and Present* 59 (1973), 51–91.

94. On the nature of street violence in antebellum cities, see Introduction, above.

95. The rise of the Republican Party and the demise of slavery did provide the context for a new black social and political assertiveness in Northern cities. See discussion below, Chapter 6.

96. On butchers, street violence, and gang activity in antebellum New York, see Wilentz, *Chants Democratic,* 269–70.

97. George Opdyke, *Official Documents, Addresses, Etc., of George Opdyke, Mayor of the City of New York, During the Years 1862 and 1863* (New York, 1866), 289.

98. See Susan G. Davis, " 'Making Night Hideous': Christmas Revelry and

Public Order in Nineteenth Century Philadelphia," *American Quarterly* 34 (Summer 1982), 185–99, for another argument associating youth rioting with the extreme casual employment and vagrancy of working-class youth at this stage of the industrial revolution. Also, see Michael B. Katz, Michael J. Doucet, and Mark J. Stern, *The Social Organization of Early Industrial Capitalism* (Cambridge and London, 1982), 255–56, on this point.

99. On anti-amalgamationist riots in antebellum New York, see Kerber, "Abolitionists and Amalgamators," 28–39; Richards, *"Gentlemen of Property and Standing,"* 150–55; Wilentz, *Chants Democratic,* 264–66. On the social and political issues of miscegenation, see Joel Williamson, *New People: Miscegenation and Mulattoes in the United States* (New York, 1980), 61–110.

100. Case of William Cruise, James Best, Moses Breen, Indictments, Dec. 1863, MARC; Cook, *Armies of the Streets,* 134–36.

101. Case of William Cruise, James Best, Moses Breen, Indictments, Dec. 1863, MARC.

102. Case of Thomas Fitzgerald, Indictments, Aug. 1863, MARC.

103. Case of Patrick Henrady, Daniel McGovern, Thomas Cumiskie, Indictments, Aug. 1863, MARC.

104. Case of John Halligan, Martin Hart, and Adam Schlosshauer, Indictments, Aug. 1863, MARC. See especially the affidavit of storekeeper Marcellus E. Randall. Randall saw Martin Hart and his party solicit "Sherwood and Conners," Mrs. Brown's dry goods store, a lager beer saloon, Sheridan's tinsmith shop, Mitchell's grocery store, Mrs. Colman's dry goods store, and Levi Adams's factory.

105. Case of John Piper, Indictments, Nov. 1863, MARC; Cook, *Armies of the Streets,* 95.

106. Hart's gang was arrested before it had a chance to carry out its threats, but stores and dwellings were burned for the offense of refusing to treat (this was usually not the only offense). See Case of Henry Saulsman, James Galvin and Thomas Kelly, Indictments, Aug. 1863, MARC, for the burning of a liquor store that was closed up during the early morning of July 27, 1863, and the sparing of a saloon next door that opened up to treat this Yorkville gang.

107. Case of John Piper, Indictments, Nov. 1863, MARC.

108. Alfred Goldsborough Jones, "Diary," XIV, July 14, 1863, Alfred Goldsborough Jones Papers, NYPL.

109. On the evangelical reform of prostitution in New York City, see John Barkley Jentz, "Artisans, Evangelicals, and the City: A Social History of Abolition and Labor Reform in Jacksonian New York" (Unpubl. Ph.D. diss., City Univ. of New York, 1977), 96–97.

110. Case of Richard Lynch and Nicholas Duffy, Indictments, Aug. 1863, MARC.

111. A. H. Dupree and L. H. Fishel, Jr., eds., "Eyewitness Account of the New York Draft Riots, July, 1863," *MVHR* 47 (Dec. 1966), 576.

112. *Ibid.*

113. *Ibid.*

114. *Evening Express,* July 14, 1863; Cook, *Armies of the Streets,* 145.

115. Case of Matthew Powers, Patrick Kiernan, Bernard Clark, Frederick

Hammer, Bernard Fagan, Indictments, Aug. 1863, MARC; Cook, *Armies of the Streets,* 145.

116. Case of Michael McCabe, Indictments, Aug. 1863, MARC. McCabe listed his occupation as "workman" in his court affidavit.

117. *Tribune,* July 17, 1863.

118. *Daily News,* July 15, 1863.

119. Claim of Heinrich Strückhausen, Draft Riot Claims, Comptroller's Office, New York County, MARC.

120. *Ibid.*

121. Case of John Smith, Coroner's Inquisition, July 1863, Supreme Court Collection, Location 7915, MARC.

122. Claim of Franz Rubel, in Board of Supervisors, *Communication,* II:800 (rioters destroy bass viol), and Claim of Moses Lowenstein, II:370–72 (rioters destroy piano).

123. On working-class perceptions of the brothel as an institution of bondage, in the Parisian setting, see Judith Coffin, "Artisans of the Sidewalk," *Radical History Review* 26 (1982), 89–101.

124. See Map 1, "Map of the Attacks on Brothels, July 13–17, 1863," for evidence that attacks on brothels centered on the waterfront districts. There were brothels in other areas of the city; see Timothy J. Gilfoyle, "The Urban Geography of Commercial Sex: Prostitution in New York City, 1790–1860," *Journal of Urban History,* 13 (Aug. 1987), 371–93, and esp. 385, 387–388. See below, Chapter 3, for a discussion of how attacks on brothels fit into laborers' social outlook.

125. Some newspapers remarked that a number of the downtown brothel-keepers were Democrats. See the Monday night attack on a Vandewater Street boarding house of a black man named Lyons. Lyons was reportedly "a well-known Democrat." *Evening Express,* July 14, 1863.

126. *Herald,* July 16, 1863 (Tuesday evening attacks on Jewish clothing stores in the East Thirties); *Tribune,* July 16, 1863 (attacks on Jewish clothing stores of Grand Street).

127. *Tribune,* July 15, 1863; *Evening Express,* July 15, 1863; Claim of Jacob Brush, Draft Riots Claims, Comptroller's Office, New York County, MARC; Cook, *Armies of the Streets,* 131 (bartender at 474 Grand Street). Among the targets in *Kleindeutschland* were Lincoln House; the German Republican headquarters, on Allen Street; Mr. Hoechster's hardware store, Avenue B (Hoechster was a known Republican). Prospective targets mentioned in the newspapers included Held's Hotel, owned by the "well-known German Republican" Andreas Willman, and Steuben House, belonging to Sixtus L. Rapp.

128. Cook, *Armies of the Streets,* 131 (attack on Elias Silberstein).

129. Weapons and clothing were the most commonly stolen items. See the Tuesday attack on Brooks Brothers's Catherine Street store. The firm had a reputation as a government war contractor, but in this attack any anti-government animus quickly evaporated in the ensuing free-for-all in which rioters helped themselves to stacks of clothing before police arrived on the scene. Claim of Brooks Brothers, in Board of Supervisors, *Communication,* II:963–1044. Also see Headley, *Great Riots,* 215–18; Barnes, *Draft Riots,* 55–56.

130. *Tribune,* July 15 (gang threatens Germans), July 18, 1863 (repudiation of the riots by the Germans of the Democratic Seventeenth Ward).

131. *Herald,* July 16, 1863; *Evening Express,* July 15, 1863.

132. It may be that crowds in the waterfront Fourth Ward turned against these few Chinese after the black residents of that district had been driven away.

133. Indeed the draft in New York City and Brooklyn had been suspended by Provost Marshal General Fry two days earlier, on July 14. On the Fowler episode, *Evening Express,* July 17, 1863; on crowds' earlier threats to the Armory, see *Evening Express,* July 14, 1863.

134. In part, this conjecture is based on evidence that the Upper West Side waterfront and adjacent blocks were home to many longshoremen in the mid- and late-nineteenth century, many of them involved in the coastwise trade in lumber and local traffic in brick and coal. See Joseph Jennings, *The Frauds of New York and the Aristocrats Who Sustain Them* (New York, 1874), 38; Charles B. Barnes, *The Longshoremen* (New York, 1915), 48–51, 60–62; and U.S. Commission on Industrial Relations, *Final Report and Testimony,* Sen. Doc. No. 415, Vol. III (Wash., D.C., 1916), 2053–67 (on Chelsea piers in the early 20th century); and below, Chapter 3. Longshoremen were undoubtedly frequent participants in Upper West Side riot incidents through the week. See esp. Case of Edward Canfield, Indictments, Aug. 1863, and Case of William Patten, Indictments, Aug. 1863, MARC. The Upper East Side had no comparable longshoring population.

135. Barnes, *Draft Riots,* 87; McCague, *Second Rebellion,* 73–74. Though we have no supporting affidavits here, Dolan's guilty plea (Cook, *Armies of the Streets,* 238) allows one to argue with confidence that he was indeed a leader of the attack on Opdyke's residence.

136. Case of Edward Clary, Grand Jury Dismissals, 1863, MARC.

137. Case of John Leavy and John Leavy, Jr., Indictments, Oct. 1863, MARC.

138. *Ibid.*

139. See Map 2, "The Geography of Racial Murders, July 13–July 17."

140. See Map 3, "Distribution of Black Population in New York City, by Ward, 1860." Note that in the Fourth Ward, listed as having only 67 black residents in 1860, there were violent attacks on black people and black property through the riot week. By comparison, the factory-district Eleventh Ward, with 225 black residents, and Eighteenth Ward, with 404 black residents, had many more black targets from which to choose, but little racial violence.

141. *Herald,* July 18, 1863.

142. Headley, *Great Riots,* 357 (Report of Captain Richard L. Shelley on operations near Gramercy Park on Thursday morning); *Strong Diary,* III:338; *Herald,* July 18, 1863.

143. Cook, *Armies of the Streets,* 145.

144. *Ibid.,* 118; *Tribune,* July 15, 1863.

145. *Bloody Week!,* 32.

146. Headley, *Great Riots,* 199.

147. Cook, *Armies of the Streets,* 119.

148. David Montgomery, *Beyond Equality,* 105, suggests that policemen were targets for the rioters more frequently than soldiers. Soldiers received a share

of the violence, but policemen were the most common targets. See Case of Daniel Conroy and Thomas Kiernan, Grand Jury Dismissals, 1863, MARC, for an arrest related to the beatings of members of the U.S. Invalid Corps.

149. See Weinbaum, "Temperance, Politics, and the New York City Riots of 1857," 246–70, on immigrant views of the Metropolitan Police.

150. That the rioters knew who Kennedy was, singled him out for attack, and made much of his near-fatal beating afterward is made clear in Barnes, *Draft Riots,* 10–12, 22; on Kennedy's use of the authority of the provost marshal as a means of consolidating Republican political power, see James F. Richardson, *The New York Police: Colonial Times to 1901* (New York, 1970), 124–29. In November 1862, Kennedy stationed police at every polling place with a list of those who had sworn themselves to be aliens in order to escape the draft. Those who had so sworn themselves were subject to arrest if they tried to vote. Indeed, some Republican leaders thought that in this and other forms of harassment Kennedy had gone too far. He was criticized by the *Times* and, after one suit for false arrest, censured by the Police Commissioners at a departmental trial. So close was the relation between Kennedy and the Lincoln administration that Kennedy and his detectives were used by Republicans in Washington to investigate a rumored assassination plot against Lincoln in Baltimore in January 1861, on the eve of the inauguration.

151. *Tribune,* July 15, 1863; Opdyke, *Official Documents,* 271. The Mayor had a financial interest in the armory, a fact that was widely known.

152. Cook, *Armies of the Streets,* 101.

153. Barnes, *Draft Riots,* 65.

154. *Ibid.,* 87–88.

155. *Ibid.,* 88; *Tribune,* July 15, 1863.

156. Cook, *Armies of the Streets,* 102.

157. *Times,* Oct. 21, 1863 ("Fires during the Riots").

158. *Bloody Week!,* 28; *Tribune,* July 31, 1863; Cook, *Armies of the Streets,* 126–27; Walling, *Recollections of a New York Chief of Police,* 81–82, on the "Battle of the Barricades."

159. See Appendix A, "Uptown Social Geography, 1863." See Friedrich Engels, "The June Revolution," in Karl Marx and Friedrich Engels, *The Revolution of 1848–49: Articles from the Neue Rheinische Zeitung* (New York, 1972), 50–59, for a discussion of where Parisian workers drew their class boundaries when the barricades went up in June 1848.

160. *Herald,* July 16, 1863.

161. *Ibid.,* July 17, 1863.

162. While many industrial workers in the metal trades lived and worked on the Upper West Side, the Upper East Side could more properly be called an industrial factory district. See note 134, above, on the Upper West Side. According to one student of the nineteenth-century West Side, "only one large factory, the Higgins Carpet Factory, was in operation down to war time." Otho G. Cartwright, *The Middle West Side: A Historical Sketch* (New York, 1914), 33. There were, in fact, a fair number of metalworking and machine-building factories on the West Side above Twenty-sixth Street during the Civil War years. By contrast, the Eighteenth, Twenty-first, and particularly the Eleventh wards on the

Upper East Side were better characterized as industrial worker wards, where large factories devoted to metallurgy were concentrated. Quarrymen, street pavers, and other common laborers did live and work alongside the factory workers of the East Side (though, notably, few longshoremen). But more than the West Side, the Upper East Side featured clusters of large metallurgical establishments along its waterfront. The congregation of huge factories at the northeastern end of the Eleventh Ward, drawing on a vast labor pool of metalworkers residing in the low-income districts along the Upper East Side waterfront, had no counterpart on the West Side. See Map 4, "Location and Size of Metalworking and Machine Building Shops, Uptown Wards, 1860."

In those few instances where the map indicates "exact location of plant unknown" (that is, the address of the factory could not be found in the city directory), we do know the "division" or district of the ward where the factory was to be found. That, and the movement of the census taker between known addresses, often allows one to guess the location of the factory within a range of several blocks.

Metal-trades workers were frequent participants in Upper East Side riot incidents through the week. See esp. Case of John Leavy and John Leavy, Jr., Indictments, Oct. 1863, Case of Thomas Sutherland, Indictments, Aug. 1863, MARC, and discussion of Twenty-second Street prisoners, below.

163. Evidence that some East Side rioters regarded Catholic identity as grounds for acquitting individuals of whose political loyalties they were uncertain can be found in Leonard, *Three Days' Reign of Terror,* 19.

164. Upper East Side rioters tended to live within a few blocks of their targets. Soldiers later found stolen arms from the Union Steam Works in surrounding tenements on Twenty-second Street and belongings from Colonel Henry O'Brien's gutted Thirty-fourth Street home in the adjacent tenements on that block. See Alan N. Burstein, "Immigrants and Residential Mobility: The Irish and Germans in Philadelphia, 1850–1880," and other essays emphasizing the importance of the "journey to work" to the experience of nineteenth-century urban workers, in Theodore Hershberg, ed., *Philadelphia: Work, Space, Family and Group Experience in the Nineteenth Century* (New York, 1981). Burstein notes the high residential concentration of Irish and German industrial workers around the factories where they worked. See discussion of New York's factory districts, below, Chapter 3.

On women's use of the phrase "die at home" on the Upper East Side as "a favorite watch-word," see Leonard, *Three Days' Reign of Terror,* 7.

165. *Daily News,* July 17, 1863.

166. Barnes, *Draft Riots,* 56.

167. *Weekly Caucasian,* July 25, 1863; *Daily News,* July 17, 1863.

168. *Tribune,* July 17, 1863.

169. Cook, *Armies of the Streets,* 160–62; *Evening Express,* July 17, 1863.

170. Leonard, *Three Days' Reign of Terror,* 16–20.

171. Cook, *Armies of the Streets,* 162.

172. Letter of Henry M. Congdon to his Father (Charles Congdon, Esq.?), July 17, 1863, Misc. Mss., NYHS.

173. *Herald,* July 18, 1863.

174. *Daily News,* July 18, 1863.
175. *Ibid.,* July 17, 1863.
176. *Ibid.*
177. These were Thomas Lube, molder, residing at 131 E. 22nd St., and James Smith, blacksmith, 128 E. 22nd St. See *Times,* July 18, 1863 ("Rioters Arrested").
178. Employees on the Upper West Side had returned to work by 2 p.m. Thursday. East Side shipyards and factories remained closed at least until Friday morning. *Daily News,* July 17, 1863.
179. See Chapter 3, below, for a discussion of the labor movement of 1863.
180. *Tribune,* July 23, 1863; *Fincher's Trades Review,* July 25, 1863 (Keady's letter).
181. *Times,* July 14, 1863 (Germans organizing to protect property late Monday); *Tribune,* July 15, 1863 (some Germans rioting on Tuesday); *Herald,* July 17, 1863 ("Meeting in the Seventeenth Ward").
182. *Herald,* July 17, 1863 ("Germans of Division Street"); *Strong Diary,* III:343 (Germans of the Seventh Ward); *Tribune,* July 30, 1863 (*Schützenverein*); Montgomery, *Beyond Equality,* 106 (*Turnverein*).
183. *Daily News,* July 17, 1863 ("Peace Democrats of the Thirteenth Ward").

CHAPTER 2. THE TWO TEMPERS OF DRACO

1. Edwin D. Morgan to Adm. Hiram Paulding, [n.p.], Nov. 2, 1863, Edwin D. Morgan Papers, NYSL; Morgan Dix, *Memoirs of John Adams Dix* (New York, 1883), II:85.
2. Morgan was no *rara avis.* Hamilton Fish, William Aspinwall, James and Stewart Brown, George Law, and a number of other Republican merchants and bankers formed a powerful conservative Republican faction in New York City.
3. Opdyke to Edwin M. Stanton, N.Y., July 13, 1863; George Opdyke to Rear Adm. Hiram Paulding, N.Y., July 13, 1863, Opdyke Administration Letterbook, Mayors' Papers, MARC; Maj. Gen. John E. Wool to E. M. Stanton, N.Y., July 20, 1863; Col. Robt. Nugent to Col. J. B. Fry, N.Y., July 15, 1863, *WOR,* ser. I, vol. 27, part II:878–79, 899–901.
4. *Strong Diary,* III:336–37.
5. Opdyke, *Official Documents,* 265.
6. Headley, *Great Riots,* 149–51.
7. Opdyke, *Official Documents,* 265.
8. George Opdyke to Maj. Gen. John E. Wool, N.Y., July 13, 1863, Opdyke Administration Letterbook, Mayors' Papers, MARC.
9. Col. Robert Nugent to Col. J. B. Fry, N.Y., July 15, 1863, *WOR,* ser. I, vol. 27, part II:899.
10. Col. Robert Nugent to Col. J. B. Fry, N.Y., July 13, 1863, Letters Sent, Southern Division of New York, Acting Asst. PMG, Entry 1360, R.G. 110; Col. J. B. Fry to Maj. Gen. John E. Wool, Wash., D.C., July 13, 1863, Entry 1403, R.G. 393, NA.
11. Opdyke, *Official Documents,* 269.

12. Richard Moody, *The Astor Place Riot* (Bloomington, Ind., 1958); also see Buckley, "To the Opera House: Culture and Society in New York City, 1820–1860."

13. *Morning Express*, Nov. 4, 7, 9, 13, 1857; see William I. Garfinkel, "Guarding the Liberty Tree: New York City and the Panic of 1857" (Unpubl. Senior Essay, Yale Univ., 1977); Jack Drums Foner, "Some Social and Economic Aspects of the Crisis of 1857, with Special Reference to New York City" (M.A. thesis, Columbia Univ., 1933); also, see Chapter 4, below.

14. Man, "Labor Competition and the New York Draft Riots," 398–99.

15. Opdyke, *Official Documents*, 269.

16. *Ibid.*, 270.

17. Opdyke yielded to the Police Board of Commissioners' wish to consult Governor Seymour before arming the police with muskets on July 13. On the 14th, Seymour sanctioned the change. Opdyke, *Official Documents*, 272–73. It was not automatically assumed that live ammunition would be used by the military. See Gen. E. S. Sandford to E. M. Stanton, N.Y., July 14, 1863, *WOR*, ser. I, vol. 27, part II:888.

18. C. T. Christensen to Major Wainwright, N.Y., July 14, 1863, Entry 1394, R.G. 393, NA; *Herald*, July 14, 1863.

19. Opdyke, *Official Documents*, 270; *Herald*, July 14, 1863.

20. Opdyke discusses the strategy of sending police and military on "expeditions" in *Official Documents*, 273.

21. *Herald*, July 14, 1863.

22. *Times*, July 14, 1863.

23. Barnes, *Draft Riots*, 12–14; Headley, *Great Riots*, 171–74; *Times*, July 14, 1863.

24. Headley, *Great Riots*, 174. See pp. 24–25, above, for a discussion of the political geography of districts unsympathetic to the riots.

25. *Herald*, July 14, 1863; Opdyke, *Official Documents*, 271.

26. *Herald*, July 14, 1863.

27. *Ibid.*

28. *Ibid.*

29. See esp. John T. Hogeboom to Salmon P. Chase, N.Y., July 18, 1863, Salmon Portland Chase Papers, LC; also, Chapter 1, above.

30. Mattie Griffith to Mary Estlin, N.Y., July 27, 1863, Estlin Papers, Dr. Williams's Library, London, England. I thank Eric Foner for bringing this letter to my attention.

31. Morgan Dix, *Memoirs of John Adams Dix* (New York, 1883), II:74.

32. Claim of George Spriggs, Draft Riot Claims, Comptroller's Office, New York County, MARC.

33. *Strong Diary*, III:336–37; *Herald*, July 14, 1863.

34. *Herald*, July 14, 1863; August Belmont to W. H. Seward, Newport, R.I., July 20, 1863, in David Black, *The King of Fifth Avenue: The Fortunes of August Belmont* (New York, 1981), 231.

35. We have only the brief accounts of the St. Nicholas meeting in the *Herald*, July 14, 1863, and the *Strong Diary*. For evidence that Masterson and Farley's outlook had wider currency among Democratic leaders, see esp. the editorial of Congressman James Brooks's *Evening Express*, July 13, 1863, in

which Brooks laments that the riot was becoming so violent that it was passing "the bounds of all law, and of all justification" and inviting "the despotism in Washington" to declare martial law; on Belmont's sympathy for the draft, see A. Belmont to W. H. Seward, Newport, R.I., July 20, 1863, in Black, *The King of Fifth Avenue,* 231.

36. Opdyke, *Official Documents,* 273; this position was also that of Belmont, who predicted that martial law would antagonize "large and powerful organizations, which had not at all participated in the riots, [but] are prepared to withstand the draft at all hazards." Black, *The King of Fifth Avenue,* 231.

37. *Strong Diary,* III:337.

38. Welles, *Diary,* I:369; John Jay to E. M. Stanton, N.Y., July 18, 1863, *WOR,* ser. III, vol. III:541.

39. Howard Carroll, *Twelve Americans* (New York, 1883), and Stewart Mitchell, *Horatio Seymour of New York* (Cambridge, 1938), are both sympathetic to Seymour's career and response to the riot; Dix, *Memoirs of John Adams Dix,* II, James B. Fry, *New York and the Conscription of 1863,* and, more recently, Eugene C. Murdock, "Horatio Seymour and the 1863 Draft," *Civil War History* 11 (June 1965), 117–41, are hostile to Seymour.

40. Mitchell, *Horatio Seymour,* esp. 556–57.

41. On Seymour's perceptions of New York City, see Horatio Seymour to Samuel J. Tilden, Utica, Oct. 25, 1866, Samuel J. Tilden Papers, NYPL, and Felix Oldboy, *A Tour around New York* (New York, 1892), 294.

42. Whether Seymour actually greeted the crowd outside City Hall as his "friends" is a subject of some controversy. See Cook, *Armies of the Streets,* 104–06.

43. *Herald,* July 15, 1863; Mitchell, *Horatio Seymour,* 298–336; Cook, *Armies of the Streets,* 104.

44. Cook, *Armies of the Streets,* 104.

45. *Times,* July 15, 1863.

46. *Ibid.*

47. *Ibid.; Herald,* July 15, 1863; Denis Tilden Lynch, *Boss Tweed, The Story of a Grim Generation* (New York, 1927), on Tweed and Seymour's tour of the city.

48. *Times,* July 15, 1863.

49. See N. Hill Fowler's speech on the Upper West Side, above, pp. 34–35. On Barlow and McClellan as Seymour's advisors at the Saint Nicholas, see Samuel L. M. Barlow to J. H. Dillon, N.Y., July 15, 1863, Letterpress Vol. VIII, Samuel L. M. Barlow Papers, HL.

50. Accordingly Seymour's orators continued to announce to crowds that the draft had been temporarily suspended long after the suspension had been made public in the press on the 14th.

51. *Herald,* July 16, 1863, for the text of the proclamation; S. L. M. Barlow to J. H. Dillon, N.Y., July 15, 1863, Letterpress Vol. VIII, Barlow Papers, HL; *Evening Express,* July 14, 1863, for Democratic support of the proclamation as an alternative to martial law.

52. *Evening Express,* July 14, 1863.

53. *Herald,* July 15, 1863 (referred to as "Mr. Downing's House").

54. *Ibid.;* Cook, *Armies of the Streets,* 121. See Appendix B, "Occupations

of Yorkville Anti-Draft Committee, July 14, 1863." One-quarter of the traceable committee members were saloonkeepers.

55. *Tribune,* July 16, 1863 (Farley's and Long's votes in favor of the Tammany draft exemption fund).

56. In fact, the fund would be activated by the Board of Supervisors before the constitutionality of the draft was upheld in the state courts. The United States Supreme Court never had an opportunity to hear and decide the draft issue during the Civil War.

57. Carroll Smith Rosenberg, *Religion and the Rise of the American City: The New York City Mission Movement, 1812–1870* (Ithaca, 1971), 225–45, on mid-nineteenth-century American urban philanthropy; see also Gareth Stedman Jones, *Outcast London: A Study in the Relationship between Classes in Victorian Society* (Oxford, 1971), ch. 13, "The Deformation of the Gift: The Problem of the 1860s," on how an ontology of giving was central to the emerging London middle classes' self-perception; on this problem in the New York context, see Chapter 5, below.

58. On the efforts of Samuel J. Tilden, Edwin D. Morgan, George T. Curtis, and William B. Reed to raise the constitutionality of the draft in the courts, see G. T. Curtis to S. J. Tilden, Rockaway, N.Y., July 17, 1863; William B. Reed to S. J. Tilden, Phila., July 28, 1863, Box 7, Samuel J. Tilden Papers, NYPL; on Tilden and Morgan's July 20 audience with Navy Secretary Gideon Welles and President Lincoln, see Welles, *Diary,* I:380; Alexander Flick, *Samuel Jones Tilden: A Study in Political Sagacity* (New York, 1939), 144; James A. Rawley, *Edwin D. Morgan, 1811–1883, Merchant in Politics* (New York, 1955), 191–92.

59. *Daily News,* July 14, 1863; *Evening Express,* July 14, 1863.

60. *Daily News,* July 14, 1863.

61. New York *Leader,* July 18, 1863, quoting from its own Feb. 28, 1863 editorial, a propos of the draft riots.

62. Maria L. Daly, "Diary of Maria L. Daly," I, July 23, 1863, Charles Patrick Daly Papers, NYPL; also, Harold E. Hammond, ed., *Diary of a Union Lady, 1861–1865* (New York, 1962), 252; on Charles P. Daly, see Harold E. Hammond, *A Commoners' Judge: The Life and Times of Charles Patrick Daly* (Boston, 1954), 172–76.

63. William B. Reed to S. J. Tilden, Phila., July 28, 1863, Tilden Papers, NYPL.

64. C. T. Christensen, Orders, July 13, 1863, Dept. of the East, Letters Sent, Entry 1394, R.G. 393, NA.

65. Sandford concentrated his troops at the Seventh Avenue Armory to protect it from attack and resisted Police Commissioner Acton's appeals to send out detachments to engage rioters on the East Side. Cook, *Armies of the Streets,* 84–85, 102–3.

66. Opdyke, *Official Documents,* 274–75; George Opdyke to E. M. Stanton, N.Y., July 14, 1863, *WOR,* ser. I, vol. 27, part II: 916.

67. E. M. Stanton to George Opdyke, Wash., D.C., July 14, 1863, *WOR,* ser. I, vol. 27, part II:916.

68. Thomas T. Davis to E. M. Stanton, N.Y., July 14, 1863, *WOR,* ser. I, vol. 27, part II:917; "James and Briggs" to S. P. Chase, N.Y., July 16, 1863,

Salmon P. Chase Papers, LC, for two of the many critical assessments of Wool's riot-week command. Also, Cook, *Armies of the Streets,* 84–85.

69. Opdyke wrote three years later, "I counseled prudence and forbearance on all sides in regard to the political aspects of the riot, and withheld my sanction from whatever seemed to me calculated to create feelings of hostility between the opposing political parties." Opdyke, *Official Documents,* 282–83.

70. Maj. Gen. Robert C. Schenck to Maj. Gen. Henry W. Halleck, Baltimore, July 14, 1863, *WOR,* ser. I, vol. 27, part II:916; on speculation as to whether state troops would enforce martial law in a confrontation with Washington, see J. W. Alden to S. P. Chase, N.Y., July 24, 1863, Chase Papers, LC.

71. William Helfenstein to Edwin M. Stanton, N.Y., July 14, 1863, Letters Received by the Sec'y of War, Irr. Series, Microcopy 492, R.G. 94, NA.

72. *Tribune,* July 14, 1863 (editorial).

73. William Helfenstein to E. M. Stanton, N.Y., July 14, 1863, Letters Received by the Sec'y of War, Irr. Series, Microcopy 492, R.G. 94, NA.

74. *Times,* July 15, 1863; *Herald,* July 15, 1863.

75. David Dudley Field to Abraham Lincoln, N.Y., July 15, 1863, Abraham Lincoln Papers, LC.

76. William A. Hall to Montgomery Blair, N.Y., July 15, 1863, Reel 55, Abraham Lincoln Papers, LC; for Hall's business interests, see Case of James H. Whitten, Indictments, Aug. 1863, MARC.

77. On Butler in New Orleans, see Hans Louis Trefousse, *Ben Butler: The South Called Him Beast* (New York, 1957), 107–34; James M. McPherson, *Ordeal by Fire: The Civil War and Reconstruction* (New York, 1982), 380–81; on martial law as a remedy for wartime disloyalty and social disorder, see Grace Palladino, "The Poor Man's Fight: Draft Resistance and Labor Organization in Schuylkill County, Pennsylvania, 1860–1865" (Ph.D. diss., Univ. of Pittsburgh, 1983).

78. *Tribune,* July 14, 17, 1863. Greeley was satisfied with General Harvey Brown's handling of the insurrection.

79. See Chapter 1, above.

80. F. L. Olmsted to O. W. Gibbs, N.Y., Nov. 5, 1862, "Safe File," Mss., Club Library, Union League Club, N.Y. A heavily edited version of this letter can be found in Henry W. Bellows, *Historical Sketch of the Union League Club of New York* (New York, 1879), 11–16.

81. Union League Club, "List of Members," in *Articles of Association, By-laws, Officers and Members of the Union League Club, 1863* (New York, 1863), 14–17.

82. F. L. Olmsted to Edwin L. Godkin, Baltimore, July 19, 1863, Edwin L. Godkin Papers, HU.

83. F. L. Olmsted to Edwin L. Godkin, Frederick, Md., July 15, 1863, Edwin L. Godkin Papers, HU.

84. As early as October 1862, George Templeton Strong is contemplating the hanging of elite traitors. *Strong Diary,* III:268–69.

85. On the Club's understanding of "semi-loyalty," see Union League Club, *Report of the Executive Committee, June, 1864* (New York, 1864), 22–23.

86. *Times,* July 17, 1863 (editorial).

87. *Times,* July 19, 1863.

88. *Tribune,* July 21, 1863.

89. Committee of Merchants for the Relief of Colored People Suffering from the Late Riots in the City of New York, *Report* (New York, 1863), esp. 9–11, 31.

90. *Ibid.,* 9 (article reprinted from *Times,* [n.d.]).

91. *Tribune,* July 21, 1863.

92. *Strong Diary,* III:336, 343, for Strong's anti-Irish (though not anti-German) musings after the riot. On mercantile nativism as a political movement in New York, see Chapter 4.

93. Union League Club, *Banquet Given by the Members of the Union League Club of 1863 and 1864, to Commemorate the Departure for the Seat of War of the Twentieth Regiment of United States Colored Troops Raised by the Club* (New York, 1886), 24–25.

94. Union League Club, *Report of the Executive Committee, January 1866* (New York, 1866), 9–10; Paul Migliore, "The Business of Union: The New York Business Community and the Civil War" (Unpubl. Ph.D. diss., Columbia Univ., 1975), 3.

95. Union League Club, *Banquet,* 7.

96. *Herald,* July 16, 1863 (Boardman); Horatio Allen to Maj. Gen. John E. Wool, Novelty Iron Works, July 16, 1863, Letters Received, Dept. of the East, Entry 1403, R.G. 393, NA. Allen's request to Wool for 100 muskets and 400 rounds of ammunition was approved.

97. *Herald,* July 14, 1863.

98. *Ibid.*

99. *Tribune,* July 24, 1863 (Parker employees' letter of July 22).

100. *Times,* Jan. 11, 1887 (obituary of John Roach); by the mid-sixties the Parker shop was part of James Brewster's carriage manufacturing complex. See Chapter 5 on the managerial styles of Roach and Brewster.

101. *Tribune,* July 17, 1863; *Evening Express,* July 17, 1863; Cook, *Armies of the Streets,* 160–61.

102. *Tribune,* July 17, 1863.

103. *Ibid.; Daily News,* July 16, 1863.

104. *Herald,* July 18, 1863, on the release of Morris Boyle ("The First Habeas Corpus").

105. W. F. H. Garley, H. M. Hoxie, S. Silbey, W. A. Woodward, etc., to Abraham Lincoln, Des Moines, July 15, 1863, Abraham Lincoln Papers, LC.

106. Stewart Pearce to William H. Seward, Wilkes Barre, Pa., July —, 1863, *WOR,* ser. I, vol. 27, part II:934.

107. Robert A. Maxwell to Abraham Lincoln, Phila., July 15, 1863, Abraham Lincoln Papers, LC.

108. *Tribune,* July 17, 1863. Many radicals were unhappy with Lincoln's support for a moderate solution in New York City. See Geo. B. Cheever to S. P. Chase, Saratoga Springs, Aug. 7, 1863, Salmon P. Chase Papers, LC.

109. Georges Duveau, *1848: The Making of a Revolution* (New York, 1967), 180–81; Stewart Edwards, *The Paris Commune, 1871* (New York, 1971), 345–50; E. J. Hobsbawm, *The Age of Capital, 1848–1875* (New York, 1975), 12.

110. As it was, military desertion and fraternization with rioters were rare during the riots. The details of the one reported incident—the desertion of James

Ruttgers, a German soldier in the Seventeenth New York—were much debated (did he desert to the mob and call his comrades to follow, or merely run away?). See *Herald,* July 18, 1863. Certainly there was reason to fear that with a declaration of martial law and escalation of the violence, the number of desertions would have increased.

111. *Herald,* July 16, 1863.

112. George W. Walling, *Recollections of a New York City Chief of Police* (Denver, 1890), 86; Headley, *The Great Riots of New York,* 221.

113. Engels, "The June Revolution," for a discussion of the strategy of the military during the Parisian June days.

114. See Colonel Thaddeus Mott's battle on Twenty-second Street, Chapter 1, above.

115. See lynching of Abraham Franklin, Chapter 1, and riot activities of Colonel Mayer, Chapter 2, above.

116. Cook, *Armies of the Streets,* 165; McCague, *Second Rebellion,* 174.

117. *Tribune,* July 18, 1863; on Hughes, see Albon P. Man, Jr., "The Church and the New York Draft Riots of 1863," *Records of the American Catholic Historical Society of Philadelphia* 62 (March, 1951), 33–50.

118. Maj. Gen. John E. Wool to Brig. Gen. E. R. S. Canby, N.Y., July 17, 1863, Entry 1394, R.G. 393, NA.

119. John Jay Knox, *United States Notes* (New York, 1884), 76; E. M. Stanton to John A. Dix, Wash., D.C., March 16, 1861; John Jay to John A. Dix, Katonah, N.Y., Jan. 21, 1861, John Adams Dix Papers, CU.

120. T. J. Barnett to S. L. M. Barlow, Wash., D.C., July 23, 1863, Box 45, Barlow Papers, HL.

121. Of course, it would be wonderful to share Lincoln's actual thoughts during the riot week. The President does not seem to have left a recorded impression of the violence in New York. The closest we can come is T. J. Barnett's two letters to Samuel Barlow on Lincoln's views of New York and the draft, June 10 and July 23, in the Barlow Papers. Another point of entry into Lincoln's inner circle during the riots is Welles, *Diary,* I:372–73 (July 16, 1863), where Navy Secretary Gideon Welles attributed the appointment of Dix (and, in Welles's view, the unfortunate failure to declare martial law) to the influence of Seward and Stanton upon Lincoln. Stanton may have been particularly important here. The lawyer from Ohio was a former Democrat who had served briefly as Attorney General in the Buchanan administration and reversed his politics upon entering the Lincoln cabinet. Stanton and Dix were friends: Stanton had entrusted to Dix the task of selling off his stock investments in March 1860 when he began to fear the collapse of the Union was imminent. See Edwin M. Stanton to John A. Dix, Wash., D.C., March 16, 1861, John Adams Dix Papers, CU.

122. James R. Gilmore to A. Lincoln, N.Y., July 24, 1863; Sidney Howard Gay to A. Lincoln, N.Y., July 26, 1863, Abraham Lincoln Papers, LC; Lawrence Lader, "New York's Bloodiest Week," *American Heritage* 10 (June 1959), 98.

123. Cook, *Armies of the Streets,* 177–87; Hoffman was reelected to the Recorder post with bipartisan support in 1863. See Charles F. Wingate, "An

Episode in Municipal Government," *North American Review* 119 (Oct. 1874), 359–407.

124. See esp. Roy P. Basler, ed., *The Collected Works of Abraham Lincoln* (New Brunswick, 1953), VI:369–70.

125. Horatio Seymour to S. J. Tilden, Albany, Aug. 6, 1863, Folder 58, Samuel J. Tilden Papers, NYPL.

126. See John A. Dix to Horatio Seymour, N.Y., July 30, 1863; H. Seymour to J. A. Dix, Albany, July 31, 1863; J. A. Dix to H. Seymour, N.Y., Aug. 8, 1863; H. Seymour to J. A. Dix, Albany, Aug. 7, 1863; H. Seymour to J. A. Dix, Albany, Aug. 15, 1863; J. A. Dix to H. Seymour, N.Y., Aug. 18, 1863; H. Seymour to J. A. Dix, Albany, Aug. 20, 1863, John Adams Dix Papers, CU.

127. John A. Dix to Horatio Seymour, N.Y., Aug. 18, 1863, John Adams Dix Papers, CU.

128. E. M. Stanton to John A. Dix, Wash., D.C., Aug. 15, 1863, John Adams Dix Papers, CU.

129. *Ibid.*

130. Mason Whiting Tyler, *Recollections of the Civil War* (New York, 1912), 118.

131. *Ibid.*

132. Entry of Aug. 24, 1863, Provost Marshal Record of the Seventh Congressional District, N.Y., Item 1660, R.G. 110, NA.

133. *Ibid.*

134. William Dusinberre, *Civil War Issues in Philadelphia, 1856–1865* (Phila., 1965), 170; Nicholas B. Wainwright, ed., *A Philadelphia Perspective: The Diary of Sidney George Fisher Covering the Years 1834–1871* (Phila., 1967), 458; for a comprehensive discussion of why civil peace was maintained in wartime Philadelphia that takes into account a variety of social and political factors, see James Matthew Gallman, "Mastering Wartime: A Social History of Philadelphia during the Civil War" (Ph.D. diss., Brandeis Univ., 1986).

135. *Times,* Aug. 28, 1863.

136. Mayor Opdyke succeeded in lowering the exemption appropriation from $3 million to $2 million, but failed to block the bill despite a court order enjoining the City Corporation from acting under the ordinance. See Opdyke Administration Letterbook, Mayors' Papers, MARC.

137. *Sun,* Sept. 21, 1863.

138. *Sun,* Sept. 26, 1863.

139. *Sun,* Sept. 10, 1863.

140. *Sun,* Sept. 26, 1863.

141. *Ibid.*

142. *Sun,* Sept. 29, 1863.

143. See Lawrence Lader, "New York's Bloodiest Week," 98.

144. Cook, *Armies of the Streets,* 133; also, see Maria L. Daly, "Diary of Maria L. Daly," I, July 23, 1863, Charles Patrick Daly Papers, NYPL, on the psalm-singing of armed black families in the Fifteenth Ward; *Tribune,* July 22, 1863.

145. Cook, *Armies of the Streets,* 175.

146. *Ibid.*

147. For one instance, see *Tribune,* July 30, 1863.

148. See "Draft Riot Claims," Comptroller's Office, New York County, MARC, for examples.

149. See Appendix C, "Decline in Black Population of New York City, 1861–1865."

150. *Tribune,* Aug. 3, 1863.

151. Society for the Diffusion of Political Knowledge, *The Constitution. Addresses of Prof. Morse, Mr. G. T. Curtis and Mr. S. J. Tilden to the Organization* (New York, 1863); on the outlook and activities of the Society, see Chapter 4.

152. John Jay to Horatio Seymour, ca. Aug. 1863, John Jay Papers, CU.

153. *Strong Diary,* III:411.

154. Bellows, *Historical Sketch of the Union League Club of New York,* 187, for the text of the address.

155. T. C. Perry, ed., *Life and Letters of Francis Lieber* (Boston, 1882), 342.

156. Bellows, *Historical Sketch,* 187.

157. Union League Club, *Banquet,* 56.

158. *Herald,* March 9, 1864.

159. AICP, *Eighteenth Annual Report* (New York, 1861), 16.

160. AICP, *Twentieth Annual Report* (New York, 1863), 23–24, and Chapter 4, below.

161. *Ibid.,* 35.

162. *Ibid.,* 34.

163. *Ibid.*

164. AICP, *Twenty-ninth Annual Report* (New York, 1872), 53–54.

165. See Chapter 5 for a discussion of the formation of the Citizens' Association in December 1863.

166. AICP, *Twentieth Annual Report,* 45.

167. *Ibid.,* 31.

168. *Ibid.,* 22.

169. Opdyke, *Official Documents,* 302.

170. *Ibid.,* 293.

171. Herman Melville, "The House-top. A Night Piece," in Hennig Cohen, ed., *The Battle-Pieces of Herman Melville* (New York, 1964), 89–90.

172. Montgomery, *Beyond Equality,* 110–11, on the political defeat of New York City's Peace Democrats in 1864–66; also below, Chapter 6.

173. On Melville's view of the Civil War, see esp. Fredrickson, *The Inner Civil War,* 185–86.

CHAPTER 3. WORKERS AND CONSOLIDATION

1. See Margaret Shortreed, "The Anti-Slavery Radicals: From Crusade to Revolution," *Past and Present* 16 (Nov. 1959), 65–66, and Eric Foner, *Free Soil, Free Labor, Free Men: The Ideology of the Republican Party before the Civil War* (New York, 1970), for a discussion of the relation between the Republican middle classes and Civil War-era democratic nationalism; see Mont-

gomery, *Beyond Equality,* for one of the few works that does situate labor within the social and political debate of this period.

2. See John R. Commons et al., *History of Labour in the United States* (New York, 1918), I:575–613, on "the new trade unionism"; also Norman Ware, *The Industrial Worker, 1840–1860* (Boston, 1924), 227–40, on the "aggressive" labor movement of the 1850s.

3. Horace Greeley, *Industrial Association: An Address to the People of the United States* (Boston, 1850), 6–7; John Higham, *From Boundlessness to Consolidation: The Transformation of American Culture, 1848–1860* (Ann Arbor, 1969), 15–16, 26; John R. Commons, ed., *A Documentary History of American Industrial Society* (New York, 1938), VII:306–07; on Olmsted and Bellows, see Chapter 4, below.

4. *Tribune,* April 23, 1850 (on Gamble).

5. On the antebellum New York clothing industry, see *Herald,* June 7, 1853; Egal Feldman, *Fit for Men: A Study of New York's Clothing Trade* (Washington, 1960), 90 (18th-century tailor enlists family help during the rush season), 95–111 (changing structure of the antebellum trade); Jesse Eliphalet Pope, *The Clothing Industry in New York* (Columbia, Mo., 1905), 1–44; Christine Stansell, "The Origins of the Sweatshop: Women and Early Industrialization in New York City," in Michael H. Frisch and Daniel J. Walkowitz, eds., *Working Class America: Essays on Labor, Community and American Society* (Urbana, 1983), 78–103; Wilentz, *Chants Democratic,* 119–24; on social change in the shoe industry, see *Tribune,* Sept. 5, 9, 1845; May 27, 1853; Wilentz, *Chants Democratic,* 124–77; Wilhelm Weitling, *Garantien der Harmonie und Freiheit* (Berlin, 1955 edition), 289.

6. U.S. Senate Committee, *Report . . . upon the Relations between Labor and Capital* (Wash., D.C., 1885), I:420; the rise of the subcontracting journeymen was already well advanced by the fifties, contrary to the observation of Pope that subcontracting was not prevalent until the eighties, Pope, *The Clothing Industry,* 19.

7. E. J. Hobsbawm and Joan Wallach Scott, "Political Shoemakers," *Past and Present* 89 (Nov. 1980), 97–98; Peter Burke, *Popular Culture in Early Modern Europe* (London, 1978), 38–39 (shoemakers' culture); *Tribune,* April 26, 1853, *Herald,* Aug. 6, 1850 (conditions in tailoring and shoemaking and the need for shop meetings); on the problems of organization in tailoring and shoemaking in the fifties and sixties, see Iver Bernstein, "The New York City Draft Riots and Class Relations on the Eve of Industrial Capitalism" (Ph.D. diss., Yale Univ., 1985), 176–82.

8. John R. Commons et al., *A Documentary History of American Industrial Society,* VIII:297–99; *Tribune,* July 16, 1850, *Herald,* July 26, 1850 ("subbosses" as leaders); *Herald,* Aug. 5 (employer named Crawford guest speaker at tailors' rally), Aug. 6, 1850 (no unilateral dictation of prices); *Herald,* Aug. 9, 13, 1850 (the "honorable" boss).

For accounts of the tailors' strike of 1850, see *Tribune,* July 26, 29, Aug. 13, 1850; *Herald,* July 25, 26, Aug. 5, 6, 8, 13, 1850; Carl Wittke, *The Utopian Communist: A Biography of Wilhelm Weitling, Nineteenth Century Reformer* (Baton Rouge, 1950), 191–95; Degler, "Labor in the Economy and Politics of New York City," on "Strikes"; Wilentz, *Chants Democratic,* 377–83.

9. Daniel J. Walkowitz, "The Artisans and Builders of Nineteenth-Century New York: The Case of the 1834 Stonecutters' Riot" (paper in the author's possession); Wilentz, *Chants Democratic,* 132–34; Elizabeth S. Blackmar, "Housing and Property Relations in New York City, 1785–1850" (Ph.D. diss., Harvard Univ., 1980), 386–492.

10. Robert Ernst, *Immigrant Life in New York City, 1825–1863* (New York, 1949), 73–75; Wilentz, *Chants Democratic,* 132–34; see also Richard Schneirov and Thomas Suhrbur, "Origin of the Carpenters' Union in Chicago, 1863–1891" (Paper to the Conference Commemorating One Hundred Years of Organized Labor in Illinois, Oct. 9–10, 1981), on the multi-talented artisanry of pre-Civil War carpenters.

11. Ware, *The Industrial Worker,* 233; Wilentz, *Chants Democratic,* 132–34; Walkowitz, "Artisans and Builders," 11–12; also see Solomon Blum, "Trade Union Rules in the Building Trades," in Jacob H. Hollander, ed., *Studies in American Trade Unionism* (New York, 1906), 295–319, on the survival of the subcontracting journeyman and his prerogatives in the building industry of the early twentieth century.

12. For a year-by-year review of the rapid expansion of the metropolitan building industry between 1846 and 1850, see *Hunt's Merchants' Magazine and Commercial Review* 25 (July–Dec. 1851), 136, "Buildings Erected in New York"; see *Putnam's Monthly Magazine* I (Feb. 1853) and III (March 1854) for the continuation of this building boom into the mid-fifties; also, Charles Lockwood, *Manhattan Moves Uptown* (Boston, 1976), 82, 96, 264.

13. Sketch of Homer Morgan's Life in *Real Estate Record and Builders' Guide,* II (Sept. 19, 1868).

14. Blackmar, "Housing and Property Relations in New York," 386–492.

15. *Ibid.*

16. *Tribune,* May 9, 25, 1850.

17. *Tribune,* June 13 (Constitution and By-laws, Bricklayers' and Plasterers' Association), July 16 (Constitution and By-laws, House Carpenters' Association), May 15, 1850 (United Association of Coach Painters).

On the formalization of work rules as the inauguration of a new and "more advanced" phase of craft control in the mid-nineteenth century, see Benson Soffer, "A Theory of Trade Union Development: The Role of the 'Autonomous Workman,'" *Labor History* 1 (Spring 1960), 141–63; David Montgomery, *Workers' Control in America* (Cambridge, Eng., 1979), 15.

18. *Tribune,* June 13, 1850; *Constitution and By-laws of the City of New-York Bricklayers' Benevolent and Protective Association* (n.p., n.d., [1851]), Beekman Collection, NYHS.

19. On the Industrial Congress and the strikes of 1850, see Commons et al., *History of Labour in the United States,* I:551–62; Degler, "Labor in the Economy and Politics of New York City, 1850–1860"; Wilentz, *Chants Democratic,* 363–90. On the programs and politics of 1848, see Duveau, *1848: The Making of a Revolution,* Donald McKay, *The National Workshops* (Cambridge, 1938), and Bernard Moss, *The Origins of the French Labor Movement: The Socialism of Skilled Workers* (Berkeley, 1977), on events in France; see especially P. H. Noyes, *Organization and Revolution: Working-Class Associations in the German Revolutions of 1848–1849* (Princeton, 1966), on Germany. Greeley's

"Council of Delegates" was probably his own version of the French *Prud-hommes*—arbitration courts run by magistrates.

20. *Tribune,* May 3, June 1, 22, July 1, 1850 (ethnic cooperation in 1850); Ira M. Leonard, "The Rise and Fall of the American Republican Party in New York City, 1843–1845," *NYHSQ* 50 (April 1966), 150–92; Wilentz, *Chants Democratic,* 299–325; on ethnic composition of the work force, see Richard Briggs Stott, "The Worker in the Metropolis: New York City, 1820–1860" (Ph.D. diss., Cornell Univ., 1983), 59–60.

21. Wilentz, *Chants Democratic,* 364; *Tribune,* June 15, 1850.

22. *Herald,* July 2, 1850.

23. Wilentz, *Chants Democratic,* 353 (on L.U.B.A.).

24. *Tribune,* July 1, 1850 (Downey on benevolent associations); Wilentz, *Chants Democratic,* 370 (L.U.B.A. strike).

25. *Herald,* July 26, 1850.

26. *Ibid.* (eliminate state law against combinations); *Tribune,* Aug. 8, 1850 (abolition of contract system, minimum wage on public works, Inspector of Rents, District Surveyors, Price Committee report); *Tribune,* Sept. 25, 1850 (eight-hour law).

27. Horace Greeley to T. C. Wittenberg, N.Y., Sept. 28, 1850, Horace Greeley Papers, NYPL; *Herald,* July 16, 1850 (James Gordon Bennett editorial); see Chapter 4, below, on the *Democratic Review*'s efforts to rebut the Industrials' interpretation of political economy.

28. *Tribune,* July 25, Aug. 13, Sept. 18, 25, 1850.

29. See the discussion of the tailors' unconditional acceptance of the Democratic reformers during 1850 in Bernstein, "The New York City Draft Riots," 225–27.

30. *Tribune,* Dec. 2, 1850, Wilentz, *Chants Democratic,* 382–83 (the Industrial Congress and the election of 1850); Commons et al., *History of Labour in the United States,* I:560–61; Wilentz, *Chants Democratic,* 383–84 (the Tammany take-over of the Congress). To be fair, Walsh introduced the persona of the "workingman's friend" in America; the popular politician John Wilkes had experimented with many elements of this style in eighteenth-century England. See esp. John Brewer, *Party Ideology and Popular Politics at the Accession of George III* (Cambridge, 1976), ch. 9.

31. *Tribune,* Aug. 8, 1850. Indeed, the program of Price, Bassett, and company was especially ambitious in view of the general trend in American political economy at mid-century—state governments were then beginning to withdraw from public projects and encourage courts and corporations to guide the economy. See Harry N. Scheiber, "Government and the Economy: Studies of the 'Commonwealth' Policy in Nineteenth-Century America," *Journal of Interdisciplinary History* III (Summer 1972), 135–51.

32. Michael F. Holt, "The Politics of Impatience: The Origins of Know Nothingism," *JAH* 60 (Sept. 1973), 309–31; on the Common Council of 1852–53 and the use of bribery in metropolitan expansion, see Jerome Mushkat, *Tammany: The Evolution of a Political Machine, 1789–1865* (Syracuse, 1971), 275–76, Edward K. Spann, *The New Metropolis, New York City, 1840–1857* (New York, 1981), 299–305; on corruption in New York and other American cities during the 1850s, see Mark W. Summers, *The Plundering Generation: Corrup-*

tion and the Crisis of the Union, 1849–1861 (New York, 1987), 138–50; *Times,* Sept. 5, 1853 (Greeley at Vegetarian Festival).

33. *Times,* Sept. 5, 1853 (speech of Thomas Doyle). Almost nothing has been written about the Amalgamated Trades Convention of 1853. See the brief mention in Commons et al., *History of Labour in the United States,* I:608; also see the tantalizing collection of documents from the Convention reprinted in Commons et al., *A Documentary History of American Industrial Society,* VIII: 336–43.

34. *Times,* Sept. 5, 1853.

35. *Times,* Sept. 21, 28, Oct. 5, 12, 25, 1853; *Tribune,* Sept. 21, 1853.

36. *Times,* Sept. 5, 1853; *Tribune,* Sept. 14, 27, 1853.

37. Hermann Schlüter, *Die Anfänge der deutschen Arbeiterbewegung in Amerika* (Stuttgart, 1907), 143–44.

38. Schlüter, *Die Anfänge,* 140–41 (demands of the *Arbeiterbund*); *Times,* March 2, 1854, and Schlüter, *Lincoln, Labor and Slavery: A Chapter from the Social History of America* (New York, 1913), 75–77, on the *Arbeiterbund* and the Nebraska Act.

39. *Tribune,* Aug. 15, 22, 1850; Wilentz, *Chants Democratic,* 382.

40. *Times,* Aug. 4, 1854 (Walford on work rules and "The Advantages of Trade Associations"), Sept. 26, 1854 (Walford and Masterson on Nebraska controversy).

41. *Herald,* March 2, 4, 7, 1854; *Times,* March 4, 1854; Bruce C. Levine, "In the Heat of Two Revolutions: The Forging of German-American Radicalism," in Dirk Hoerder, " 'Struggle a Hard Battle': Essays on Working-Class Immigrants," (Dekalb, 1986), 35–38 (on the *Arbeiterbund*'s Anti-Nebraska demonstrations), 38-39 (on the Republican Party's electoral difficulties in *Kleindeutschland* relative to the German wards of other Northern cities). Levine also suggests that New York City German workers were reluctant to vote Republican because they feared Republican victory would lead to Southern secession and loss of Southern markets, with adverse effects upon the New York City economy.

42. *Times,* Dec. 22, 23, 1854, Jan. 6, 10, 11, 12, 16, 30, 1855; Amy Bridges, *A City in the Republic: Antebellum New York and the Origins of Machine Politics* (New York, 1984), 137–40, gives these workingmen's meetings brief treatment in a different context.

43. *Times,* Nov. 30, 1857; Degler, "Labor in the Economy and Politics of New York City," 314–21.

44. *Herald,* March 13, April 10, 1860, *Tribune,* March 28, 1860, Degler, "Labor in the Economy and Politics of New York City," 321–22 (on the Lynn strike sympathy meetings); *Herald,* Jan. 16, 1861, *Tribune,* Jan. 16, 1861 ("Anti-Coercion Mass Meeting"); Julia Blodgett Curtis, "The Organized Few: Labor in Philadelphia, 1857–1873" (Unpubl. Ph.D. diss., Bryn Mawr College, 1970) (on the Philadelphia "Committee of Thirty-four" and the workingmen's peace movement of the secession winter). It should be noted that the "Anti-Coercion Meeting" did resolve to petition the New York State legislature to convene a state convention to express "public sentiment" against the Republican Party.

45. *Tribune,* April 19, 1861 (Gulick Hose No. 12 organizes firemen), April 20, 1861 (workers marching to Union Square meeting by shop); Levine, "In the Heat of Two Revolutions," 39 (*Turnverein* organizes New York's Twentieth

(United Turner) Regiment); Basil Leo Lee, *Discontent in New York City, 1861–1865* (Wash., D.C., 1943), 175–76, 200–201, 209–10.

46. *Times,* March 25, 1863; *Weekly Caucasian,* March 28, 1863.

47. *Times,* March 25, 1863; *Weekly Caucasian,* May 29, June 13, 20, 27, July 11, 1863.

Davis's reform circle of the 1850s did not disappear permanently. The land and labor reformers would reemerge in the middle-class sections of the International Workingmen's Association in the early 1870s. See Montgomery, *Beyond Equality,* 416–17, on "sentimental reformers" John Commerford, William West, and Lewis Masquerier. Also see Thomas A. Devyr to Karl Marx, Greenpoint, N.Y., Jan. 16, 1872, Folder D1028, IISH.

48. *Fincher's Trades Review,* July 25, 1863.

49. *Weekly Caucasian,* Oct. 3, 10, 17, 1863; *Weekly Day Book,* Nov. 14, 1863.

50. *Sun,* Oct. 26, Nov. 4, 9, 10, 11, 22, Dec. 5, 8, 16, 1863; *Weekly Day Book,* Dec. 26, 1863; also see Lee, *Discontent in New York,* 209–10, for a brief description of the machinists' strike.

51. See George T. Strong's description of the area east of Second Avenue during the draft riots in *Strong Diary,* III:338; AICP, *Eleventh Annual Report* (New York, 1854), 25; *Times,* March 4, 1853 (Walks among the New York Poor); Jacob Abbott, "The Novelty Works," *Harper's New Monthly Magazine* II (May 1851), 721–22; Joel H. Ross, *What I Saw in New-York* (Auburn, N.Y., 1852), 200.

52. Allan Pred, "Manufacturing in the American Mercantile City: 1800–1840," *Annals of the Association of American Geographers* 56 (June 1966), 327; Robert Ernst, *Immigrant Life in New York City, 1825–1863* (Port Washington, N.Y., 1965), 82; John H. Morrison, *History of New York Ship Yards* (New York, 1909), 151–52.

53. Abbott, "The Novelty Works," 724. On industrial workers' need to live nearby the waterfront factories, the Citizens' Association observed of the industrial Eleventh Ward in 1865, "The shipyards, iron, lead, and copper works here, also give employment to many thousand hands to whom a residence near is a great necessity. Hence the excessive crowding in this locality." Citizens' Association of New York, *Report of the Council of Hygiene and Public Health of the Citizens' Association of New York upon the Sanitary Condition of the City* (New York, 1865), 174.

54. *Sun,* Nov. 9, 1863; Leonard, *Three Days' Reign of Terror,* 7, 12; Charles Stelzle vividly recalled the role saloonkeepers played as defenders of workers in the relations with the community in the 1880s: "I recall that as a young [machinists'] apprentice, when I was arrested . . . the first man to whom my friends turned was the saloon-keeper on the block. And he furnished bail gladly. . . . He loaned money without setting up the work basis of the Charity Organization Society and similar relief organizations." Charles Stelzle, *A Son of the Bowery: The Life Story of an East Side American* (New York, 1926), 48.

55. Paul O. Weinbaum, "Temperance, Politics and the New York City Riots of 1857," *NYHSQ* 59 (July 1975), 246–70.

56. *Herald,* Feb. 22, 1854; *Times,* April 7, 1854; Weinbaum, "Temperance, Politics," 246–70, on German and Irish support for Wood in the East Side fac-

tory district; S. H. Harlow and H. H. Boone, *Life Sketches of the State Officers, Senators and Members of the Assembly of the State of New York, in 1867* (Albany, 1867), 197–98 (John J. Blair).

57. Harlow and Boone, *Life Sketches,* 197–98.

58. See Raphael Samuel, "Workshop of the World: Steam Power and Hand Technology in Mid-Victorian Britain," *History Workshop Journal* 3 (Spring 1977), 6–72, for a discussion of the blending of hand and machine work in the new industrial occupations of mid-century Great Britain; for a similar view of American industrialization, see Dolores Greenberg, "Reassessing the Power Patterns of the Industrial Revolution: An Anglo-American Comparison," *AHR* 87 (Dec. 1982), 1237–61.

59. See Case of John Leavy and John Leavy, Jr., Indictments, Oct. 1863, MARC, for instance of journeyman blacksmith and helper among the draft rioters; see John H. Ashworth, *The Helper and American Trade Unions* (Baltimore, 1915), on journeymen employing helpers in the nineteenth-century metal trades.

60. *Sun,* Nov. 12, 1863 (Singer, piece rates); Case of Matthew Powers, Indictments, Box 82–1863, 1867, MARC (Powers was a brassworker at Brown's Scale Factory, which may have used inside contracting); see John Buttrick, "The Inside Contract System," *The Journal of Economic History* 12 (Summer 1952), 205–21; I have found no evidence of inside contracting at the large waterfront iron shipbuilding plants.

61. Stelzle, *A Son of the Bowery,* 44.

62. Frank Roney, *Frank Roney: Irish Rebel and California Labor Leader, an Autobiography* (Ira B. Cross, ed.) (Berkeley, 1931), 180–81, on New York City iron foundries' "open-shop" status in the 1860s.

63. See David Montgomery, "Strikes in Nineteenth-Century America," *Social Science History* 4 (Feb. 1980), 88. Montgomery observes that work rules, conditions, and hours "were seldom explicit strike issues" in the nineteenth century.

64. On infrequency of machinists' strikes before 1900 and frequency of molders' strikes, see David Montgomery, "Culture and Conflict in Machine-Building Enterprises: Machinists before Scientific Management," 3 (paper in author's possession).

65. Stelzle, *Son of the Bowery,* 44; Terence Powderly, *The Path I Trod* (Harry J. Carman, Henry David, and Paul N. Guthrie, eds.) (New York, 1940), 41–42.

66. On Allaire's shop, see Chapter 5, below; Harlow and Boone, *Life Sketches,* 197–98 (Blair); William F. Ford, *The Industrial Interests of Newark, N.J., Containing an Historical Sketch of the City* (New York, 1874), 61–64 (Seth Boyden); Stephen D. Tucker, *History of R. Hoe and Company New York,* R. Hoe and Company Collection, CU; Charles H. Haswell, *Engineer's and Mechanic's Pocket-book* (New York, 1903); "Charles Haynes Haswell," *Engineering News,* 57 (1907), pp. 509–11; on the nineteenth-century machinist, see Monte A. Calvert, *The Mechanical Engineer in America, 1830–1910* (Baltimore, 1967); on the many similarities and important differences between the attitudes of machinists and their employers, see David Montgomery, *The Fall of the House of Labor: The Workplace, the State, and American Labor Activism, 1865–1925* (Cambridge, Eng., 1987), 180–93.

67. Abbott, "The Novelty Works," 726–27.

68. Generalizations made for machinists here probably apply to other industrial worker "aristocrats" such as patternmakers and draughtsmen.

69. *Morning Express,* Nov. 1, 1853 (City Reform), Nov. 4, 1857 (election rioting); *Times,* Sept. 1, 1853 (Painters), Jan. 30, 1855 (Irving); *Tribune,* March 7, 1853 (City Reform), April 11, 1853 (machinists' strike).

70. *Tribune,* Feb. 7, 1863; *Weekly Caucasian,* Feb. 14, 1863.

71. *Weekly Caucasian,* Feb. 14, 1863.

72. John Talbot Smith, *The Catholic Church in New York: A History of the New York Diocese from Its Establishment in 1808 to the Present Time* (New York, 1905), I:267–69; for further discussion of the Republican industrialists and the report of an anti-Catholic plot, see below, Chapter 5.

73. Montgomery, *Beyond Equality,* 95.

74. Carl Wittke, *The Irish in America* (New York, 1956), 127, 132; on the attitudes of Irish Catholics toward abolitionism, the Republican Party, and the black community, see Man, "The Church and the New York Draft Riots of 1863," 33–36; Lee, *Discontent in New York City,* 125–64; Florence E. Gibson, *The Attitudes of the New York Irish toward State and National Affairs, 1848–1892* (New York, 1951), 135.

75. Karl Marx to S. Meyer and A. Vogt, London, April 9, 1870, in Karl Marx and Friedrich Engels, *Letters to Americans, 1848–1895* (New York, 1953), 79.

76. On laboring trades whose sense of civic importance was connected to medieval municipal privileges see Walter M. Stern, *The Porters of London* (New York, 1960); Graham Russell Hodges, *New York City Cartmen, 1667–1850* (New York, 1986).

77. Elizabeth Fox-Genovese and Eugene D. Genovese, *Fruits of Merchant Capital: Slavery and Bourgeois Property in the Rise and Expansion of Capitalism* (New York, 1983), 76.

78. William Hartman, "Customs House Patronage under Lincoln," *NYHSQ* 41 (Oct. 1957), 440–57; Joseph Jennings, *The Frauds of New York and the Aristocrats Who Sustain Them* (New York, 1874), 24; see Jerome Mushkat, *Tammany: The Evolution of a Political Machine, 1789–1865* (New York, 1971), 307, 312, 315, and Leonard Chalmers, "Tammany Hall, Fernando Wood, and the Struggle to Control New York City, 1857–59," *NYHSQ* 53 (Jan. 1969), 7–13, on the political struggles over Customs House patronage in the Buchanan administration.

79. Francis Schell, *Memoir of the Hon. Augustus Schell* (New York, 1885), 18, 58; James Buchanan to Augustus Schell, July 25, 1863, in J. B. Moore, ed., *The Works of James Buchanan* (Philadelphia, 1910), XI:341–42; *Strong Diary,* II:344, III:158; Hartman, "Customs House Patronage," 440–57.

80. Hartman, "Customs House Patronage," 440–57; Jennings, *The Frauds of New York,* 21.

81. *Tribune,* Oct. 16, 1852; Man, "Labor Competition," 392–93; Jennings, *Frauds of New York,* 23–24; see U.S. Commission on Industrial Relations, *Final Report and Testimony* (Sen. Doc. 415, Vol. III) (Wash., D.C., 1916), 2162, 2166, on the strength of longshoremen's unions before the 1874 strike.

82. *Times,* Jan. 19, 1855; *Tribune,* Jan. 19, 1855; Jennings, *Frauds of New York,* 22–23, 29, 45, 47, 50–52.

83. Jennings, *Frauds of New York,* 30; on subcontracting practices of nineteenth-century stevedores and the social structure of the waterfront, see Charles B. Barnes, *The Longshoremen* (New York, 1915), 13–75.

84. *Herald,* Oct. 12, 1852; *Tribune,* Oct. 13, 1852.

85. Barnes, *The Longshoremen,* 22; U.S. Commission on Industrial Relations, *Final Report,* 2119.

86. See Leonard P. Curry, *The Free Black in Urban America, 1800–1850* (Chicago, 1981), on the difficulty free black men had finding secure employment in antebellum Northern cities.

87. *Tribune,* Jan. 18, 1855; April 14, 1863; Man, "Labor Competition," 398–99.

88. *Tribune,* Feb. 15, 1855; *Daily News,* July 17, 1863; *Irish American,* March 23, 1860; U.S. Commission on Industrial Relations, *Final Report,* 2069; see *Daily News,* June 20, 1863, "Mass Meeting of the Journeymen Cartmen," for discussion of the Irish dirt cartmen's "all-white" exclusionary rule.

89. Case of William Patten, Indictments, Aug. 1863, MARC; *Trow's New York City Directory, 1863–64* (New York, 1864). Most of the addresses and occupations of Patten's defense witnesses are listed in the court record. In several instances it was necessary to use the address given in the court records to trace an occupation in the directory.

90. Jennings, *Frauds of New York,* 19, 38–39.

91. See Natalie Zemon Davis, *Society and Culture in Early Modern France* (Stanford, 1975), 97–123, on youth charivaris in sixteenth-century Lyon serving as moral regulators of marriages disapproved by the community.

92. *Weekly Day Book,* Nov. 28, 1863.

CHAPTER 4. MERCHANTS DIVIDED

1. Roland Barthes, *Mythologies* (New York, 1976), 138, and for discussion, 137–42.

2. Alfred D. Chandler, Jr., *The Visible Hand: The Managerial Revolution in American Business* (Cambridge, 1977), 17–19.

3. Robert G. Albion, *The Rise of New York Port* (New York, 1939); P. Glenn Porter and Harold Livesay, *Merchants and Manufacturers: Studies in the Changing Structure of Nineteenth-Century Marketing* (Baltimore, 1971), 19–20, 22; Chandler, *Visible Hand,* 26–27.

4. George Rogers Taylor, "The National Economy before and after the Civil War," and Alfred D. Chandler, Jr., "Organization of Manufacturing and Transportation," in David T. Gilchrist and W. David Lewis, eds., *Economic Change in the Civil War Era* (Greenville, Del., 1965), 10–22, 138, 157–58; George Rogers Taylor, *The Transportation Revolution, 1815–1860* (New York, 1962); Edward K. Spann, *The New Metropolis,* 4; New York Chamber of Commerce, *Sixth Annual Report* (New York, 1864), 37 (Perit's address); William E. Dodge, Jr., *Old New York: A Lecture* (New York, 1880), 51.

5. Amy Bridges, *A City in the Republic: Antebellum New York and the Origins of Machine Politics* (Cambridge, Eng., 1984), 70–72; see Paul O. Weinbaum, "Mobs and Demagogues" (Ph.D. diss., Univ. of Rochester, 1974), and Richards, *"Gentlemen of Property and Standing,"* on elite leadership of mobs in the 1830s.

6. Richard Lowitt, *A Merchant Prince of the Nineteenth Century: William E. Dodge* (New York, 1954), 202; Allan Nevins, ed., *The Diary of Philip Hone, 1828–1851* (New York, 1936), 925.

7. Chandler, *The Visible Hand,* 91–92; J. F. D. Lanier, *Sketch of the Life of J. F. D. Lanier* (New York, 1870), 43–62 (on innovations of Duncan, Sherman, and Winslow, Lanier); Fritz Redlich, *The Molding of American Banking: Men and Ideas* (New York, 1951), II:362 (on Cisco); on Democratic elite's intimacy with Buchanan, see esp. Albert V. House, "The Samuel Latham Mitchill Barlow Papers in the Huntington Library," *Huntington Library Quarterly* 28 (Aug. 1965), 347, Irving Katz, *August Belmont: A Political Biography* (New York, 1968), 50–57, and correspondence between Buchanan and John A. Dix, discussed below, pp. 141.

8. Frederick Law Olmsted to Wolcott Gibbs, N.Y., Nov. 5, 1862, in Safe File Box Mss., Union League Club, N.Y.; Thomas Bender, *New York Intellect: A History of Intellectual Life in New York City, from 1750 to the Beginnings of Our Own Time* (New York, 1987), 182; *Strong Diary,* III:303, 307; Robert Bowne Minturn, *Memoir of Robert Bowne Minturn* (New York, 1871); Mary W. Tileston, *Thomas Tileston, 1793–1864* (New York, 1925); George Wilson, *Portrait Gallery of the Chamber of Commerce of the State of New York* (New York, 1890), 48–52, 60, 133 (Perit, Sturges, Griswold family).

9. It should be noted here that the draft-riot divisions between pro- and anti-martial law merchants do *not* seem to have corresponded to various other economic distinctions between these businessmen, either in terms of the markets they served (western as opposed to southern traders, or non-local versus local market traders), the kind of specialization of enterprise (such as financier, importer, exporter, wholesaler), the degree of specialization of enterprise (generalist versus specialist), or the type of merchandise traded (dry-goods merchant versus produce merchant, or the like).

10. On Olmsted's hope that participation of socially established New Yorkers in the Union League Club would lend authority to his ideas about nationality, see F. L. Olmsted to Wolcott Gibbs, Nov. 5, 1862, Safe File Box Mss., Union League Club, N.Y.

11. See the list of subscribers to the Belmont farewell dinner, Appendix D; also, New York Merchants to August Belmont, N.Y., Aug. 9, 1853, in August Belmont, ed., *Letters, Speeches and Addresses* (New York, 1890), 3–4.

12. August Belmont to Merchants of New York, N.Y., Aug. 12, 1853, in Belmont, ed., *Letters, Speeches and Addresses,* 5–7.

13. See James P. Shenton, *Robert John Walker: A Politician from Jackson to Lincoln* (New York, 1961), 114–23, on the Polk administration's connections to the "eastern business constituency."

14. On the widespread merchant opposition to the high-duty Morrill tariff bill of 1861, see Philip Foner, *Business and Slavery: The New York Merchants and the Irrepressible Conflict* (Chapel Hill, 1941), 261–84.

15. *The United States Magazine and Democratic Review* 27 (Oct. 1850), 304.

16. Alvan F. Sanborn, ed., *Reminiscences of Richard Lathers* (New York, 1907), 5–7, 63–119, 226–28; Thomas Prentice Kettell, *Southern Wealth and Northern Profits, as Exhibited in Statistical Facts and Official Figures* (New York, 1860), 10.

17. Herbert D. A. Donovan, *The Barnburners: A Study of the Internal Movements in the Political History of New York State and of the Resulting Changes in Political Affiliation, 1830–1852* (New York, 1925), Marvin Meyers, *The Jacksonian Persuasion* (New York, 1960), 234–75, and Eric Foner, "Racial Attitudes of the New York Free Soilers," *NYH* 46 (Oct. 1965), 311–29 (on Barnburners); John Bigelow, ed., *The Writings and Speeches of Samuel J. Tilden* (New York, 1885), I:423; Samuel J. Tilden, "Address of the Governor at the Washington County Agricultural Fair," Undated, Samuel J. Tilden Papers, NYPL; John A. Dix, *Speeches and Occasional Addresses* (New York, 1864), II:360–82; Morgan Dix, *Memoirs of John Adams Dix* (New York, 1883), I:92; Samuel J. Tilden, Draft of Letter to Tammany Society, July 4, 1866, Samuel J. Tilden Papers, NYPL.

18. On the Belmont circle's attitudes toward race, see below, pp.. 146–47; on the Jeffersonian theme in the writings of Samuel Tilden, see Robert Kelley, *The Transatlantic Persuasion: The Liberal-Democratic Mind in the Age of Gladstone* (New York, 1969), 238–92; see Merrill D. Peterson, *The Jefferson Image in the American Mind* (New York, 1960), 190–94, for Free Soil Democrats' use of Jeffersonian themes.

For the best record of the Pine Street proceedings, see *Journal of Commerce,* Dec. 17, 1860; also Foner, *Business and Slavery,* 232–36; Sanborn, ed., *Reminiscences of Richard Lathers,* 74–119; Dix, *Memoirs of John Adams Dix,* I:347–60. See, too, the list of merchants known to have attended the Pine Street Meeting and list of those who signed the Pine Street Address, Appendix E. For an extended discussion of the significance of the Pine Street Meeting, see Bernstein, "The New York City Draft Riots," chapter 5.

19. Merle Curti, " 'Young America,' " *AHR* 32 (Oct. 1926), 34–55, is the best existing discussion of the Young America outlook and constituency; on the movement's literary and cultural underpinnings, see Perry Miller, *The Raven and the Whale* (New York, 1956).

20. On Belmont's involvement in Young America, see Curti, " 'Young America,' " 51–54, and in schemes to purchase Cuba, Robert E. May, *The Southern Dream of a Caribbean Empire, 1854–1861* (Baton Rouge, 1973), 141–42, 164–65; on Belmont's relationship with Douglas in the 1850s, Katz, *August Belmont,* 23–32, 62–91.

21. August Belmont to George N. Sanders, Aug. 1854, quoted in Curti, " 'Young America,' " 53–54.

22. Thomas Jefferson, *Notes on the State of Virginia* (New York, 1964), 154.

23. Bigelow, ed., *The Writings and Speeches of Samuel J. Tilden,* I:89–100 (Shakers), 279–83 (state temperance law); see Kelley, *The Transatlantic Persuasion,* 404–5, on Tilden and the secularist tradition.

24. Dix, *Speeches and Occasional Addresses,* II:275–77.

25. Dix, *Memoirs of John Adams Dix,* I:354; Society for the Diffusion of Political Knowledge, *The Constitution. Addresses of Prof. Morse, Mr. Geo. Ticknor Curtis and Mr. S. J. Tilden* (New York, 1864), 3; S. F. B. Morse, *Imminent Dangers to the Free Institutions of the United States Through Foreign Immigration and the Present State of the Naturalization Laws* (New York, 1835), 16, 23; on the views of Pius IX and the critique of secularism in the 1864 Syllabus of Errors, see E. J. Hobsbawm, *The Age of Capital, 1848–1875* (New York, 1975), 114–15.

26. AICP, *Fifteenth Annual Report* (New York, 1858), 22.

27. On Wood's October 22 message, see New York City Board of Aldermen, *Proceedings,* LXVIII (New York, 1857), 156–61, also reprinted in the *Times,* Oct. 23, 1857; on the debate over poor relief during the depression of 1857, see above, Chapter 3; also, Benjamin J. Klebaner, "Poor Relief and Public Works during the Depression of 1857," *The Historian* 22 (May 1960), 272–73; Leah H. Feder, *Unemployment Relief in Periods of Depression* (New York, 1936), 31–34; and William I. Garfinkel, "Guarding the Liberty Tree: New York Labor and Politics during the Panic of 1857" (Unpubl. Senior Essay, Yale Univ., 1977).

28. On Wood's early career and relationship with West Side stevedore gangs, see Samuel A. Pleasants, *Fernando Wood of New York* (New York, 1948), 12–14; on Wood's political style and relationship to the immigrant laborer community, see Matthew Hale Smith, *Sunshine and Shadow in New York* (Hartford, 1868), 268–75, and Leonard Chalmers, "Fernando Wood and Tammany Hall: The First Phase," *NYHSQ* 52 (Oct. 1968), 379–402; on Wood as mid-fifties reformer and property-holders' endorsement of his reelection, see *ibid.* and Harvey O'Connor, *The Astors* (New York, 1941), 101–4; on Wood and Central Park, see Ian R. Stewart, "Politics and the Park: The Fight for Central Park," *NYHSQ* 61 (July/Oct. 1977), 135.

29. *Tribune,* Oct. 15, 1857; *Journal of Commerce,* Nov. 7, 1857; *Times,* Oct. 23, 1857; George Francis Train, *Young America in Wall Street* (New York, 1857), 328; also see *Herald,* Oct. 23, 1857.

30. *Morning Express,* Nov. 4, 5, 6, 9, 12, 13, 1857; *Journal of Commerce,* Nov. 7, 9, 10, 12, 1857; Garfinkel, "Guarding the Liberty Tree," 6–26 (demonstrations and bread riot of early Nov. 1857).

31. *Times,* Oct. 27, Nov. 7, 12, 1857 (Raymond's increasing hostility to Wood); *Times,* Nov. 16, 1857, *Journal of Commerce,* Nov. 16, 21, 1857, Garfinkel, "Guarding the Liberty Tree," 31–34, and below, Appendix F, "Partial List of Businessmen Endorsing People's Union Movement, November 1857" (People's Union); Howard B. Furer, *William Frederick Havemeyer: A Political Biography* (New York, 1965); *Strong Diary,* II:373–75.

32. *Times,* Nov. 7, 1857, *Journal of Commerce,* Nov. 10, 1857 (reprint of Fernando Wood to Royal Phelps, New York, Nov. 6, 1857)—Phelps published the Wood letter in the *Journal of Commerce* with an introductory note allowing a strong inference that Phelps approved of Wood's ideas and wished to publicize them.

33. *Journal of Commerce,* Nov. 10, 1857; John A. Dix to President James A. Buchanan, N.Y., Nov. 10, 1857, John A. Dix Papers, CU.

34. John A. Dix to David R. Thomason, N.Y., Nov. 9, 1857; also, see David

R. Thomason to John A. Dix, N.Y., Nov. 7, 1857, John A. Dix Papers, CU; Dix's letter to Thomason is also reprinted in the *Journal of Commerce,* Nov. 12, 1857.

35. Daniel E. Sickles to James Buchanan, N.Y., Nov. 20, 1857; John A. Dix to James Buchanan, N.Y., April 20, 1858, James Buchanan Papers, Historical Society of Pennsylvania; August Belmont to John Slidell, N.Y., Dec. 8, 1857, quoted in Jerome Mushkat, *Tammany: The Evolution of a Political Machine, 1789–1865* (Syracuse, N.Y., 1971), 309–10; on Sickles's struggle against Augustus Schell and Fernando Wood for federal patronage, see Leonard Chalmers, "Tammany Hall, Fernando Wood, and the Struggle to Control New York City, 1857–1859," *NYHSQ* 53 (Jan. 1969), 7–13; also George N. Sanders to James Buchanan, N.Y., July 26, 1857, James Buchanan Papers, Historical Society of Pennsylvania.

36. *Journal of Commerce,* Dec. 17, 1860 (speech of Hiram Ketchum); "The United States and the United Kingdom," *The United States Review* 32 (May 1853), 396, 399.

37. Dix, *Memoirs of John Adams Dix,* I:343; James Buchanan to Royal Phelps, Wash., D.C., Dec. 22, 1860; James Buchanan to James Gordon Bennett, Wash., D.C., Dec. 20, 1860, in John Bassett Moore, ed., *The Works of James Buchanan,* XI:69–70, 73–74; on Fernando Wood's free city proposal, see Foner, *Business and Slavery,* 285–96; DeAlva S. Alexander, *A Political History of the State of New York* (New York, 1906), II:348; James J. Heslin, " 'Peaceful Compromise' in New York City, 1860–1861," *NYHSQ* 44 (Oct. 1960), 349–62; Emory Thomas, *The Confederate Nation, 1861–1865* (New York, 1979), 307 ff., for the Constitution of the Confederate States of America with its free-trade clause.

38. Foner, *Business and Slavery,* 310.

39. "Lieber to Peletiah Perit," in *The Merchants' Magazine and Commercial Review* [also *Hunt's*] XLV (Nov. 1861), 513–15; Foner, *Business and Slavery,* 322. Foner's notion that merchants perceived their economic "interest" as connected with national unity after Sumter has some validity. But one also needs to consider merchants' nationalism as an independent variable, apart from economic interest; see discussion of Belmont, below. See Daniel Hodas, *The Business Career of Moses Taylor: Merchant, Finance Capitalist, and Industrialist* (New York, 1976), 177, for the argument that western trade connections brought New York merchants to a pro-Union stance in 1861.

40. August Belmont to John Forsyth, N.Y., Dec. 19, 1860, in Belmont, ed., *Letters, Speeches and Addresses,* 36–39.

41. Dix, *Memoirs of John Adams Dix,* I:349–50 (Soutter); Coddington, "The Activities and Attitudes of a Confederate Businessman," 3–36 (Lamar); John L. O'Sullivan, *Peace the Sole Chance Now Left for Reunion* (London, 1863), 18; on O'Sullivan's career, see Julius W. Pratt, "John L. O'Sullivan and Manifest Destiny," *NYH* 14 (July 1933), 213–34; Sheldon H. Harris, "John Louis O'Sullivan and the Election of 1844 in New York," *NYH* 61 (July 1960), 278–98.

42. Dix, *Memoirs of John Adams Dix,* II:11–15 (on Dix and the Union Defense Committee—Democrats Royal Phelps, Richard M. Blatchford, and Ed-

wards Pierrepont were all Committee members); Hodas, *Business Career of Moses Taylor,* 178–88; Katz, *August Belmont,* 93–115 (New York City bankers and the financing of the war effort); Belmont, ed., *Letters, Speeches and Addresses,* 64–101 (Belmont's pro-Union correspondence with English capitalists and politicians).

43. George Winston Smith, "The National War Committee of the Citizens of New York," *NYH* 28 (Oct. 1947), 440–57.

44. *Evening Express,* Oct. 17, 31, Nov. 1, 1862; on the large pro-slavery nativist movement in New York City and the New York Know Nothing Party's campaign to commit the national Know Nothing organization to a pro-slavery platform, see *Tribune,* June 19, 1855; also, Thomas J. Curran, "Know Nothings of New York" (Unpubl. Ph.D. diss., Columbia Univ., 1963), 292–93, on the political path from the Know Nothing Party in the mid-fifties through the Constitutional Union movement of 1860–61 to the pro-slavery Democracy during the middle years of the war. For some New York merchants, political nativism was a rehearsal for the pro-slavery Democracy, not the Republican Party; O'Connor, *The Astors,* 294 (paraphrasing Chanler's speech).

45. August Belmont to S. J. Tilden, New York, Jan. 27, 1863, Samuel J. Tilden Papers, NYPL; *Strong Diary,* III:297; on the Society for the Diffusion of Political Knowledge (hereafter SDPK), see SDPK, *The Constitution* (New York, 1863); also, Katz, *August Belmont,* 120–21; Flick, *Samuel Jones Tilden,* 140–41.

On the Society's position on race and slavery, see SDPK, "The Letter of a Republican and Prof. Morse's Reply" (Paper 4), 8–11; "An Argument on the Ethical Position of Slavery in the Social System" (Paper 12), 16–17; "Speech of the Hon. James Brooks" (Paper 3), 7–8; "Emancipation and Its Results" (Paper 5), 26–27, all in SDPK, *Papers,* Butler Library Collection, Columbia University.

46. SDPK, "The Letter of a Republican and Prof. Morse's Reply" (Paper 4), 9.

47. Fox-Genovese and Genovese, *Fruits of Merchant Capital,* for a recent restatement and interpretation of this position.

48. Richard Moody, *The Astor Place Riot* (Bloomington, 1958), 101–75; Harry Brinton Henderson III, "Young America and the Astor Place Riot" (Unpubl. M.A. Thesis, Columbia Univ., 1963), 70–72, 83–86; Buckley, "To the Opera House: Culture and Society in New York City, 1820–1860"; the figure of 22 casualties is Buckley's revised figure. On the 1834 riot, see Linda K. Kerber, "Abolitionists and Amalgamators: The New York City Race Riots of 1834," *NYH* 48 (1967), 28–39; Wilentz, *Chants Democratic,* 264–66.

See Appendix G, "List of Macready Petitioners, May 9, 1849," for merchants who wished to see the military protect the British actor; New York *Courier and Enquirer,* May 9, 1849. Matthew Morgan, James Colles, Henry A. Stone and Jacob Little appeared both on the Macready petitioner list and the Belmont dinner list.

49. *Journal of Commerce,* May 11, 15, 1849; Henry W. Bellows, *A Sermon Occasioned by the Late Riot in New York* (New York, 1849), 13–15.

50. Wilentz, *Chants Democratic,* 264–66 (on 1834 Bowery Theater inci-

dent); Bellows, *A Sermon Occasioned by the Late Riot,* 14; Allan Nevins, ed., *Diary of Philip Hone* (New York, 1936), 876; on working-class culture and the Bowery of the 1840s, see Wilentz, *Chants Democratic,* 326–57.

51. Henry W. Bellows, *The Christian Merchant* (New York, 1848), 7–8; Nevins, ed., *Diary of Philip Hone,* 727. For other instances of New York's Whig (and Republican) elite criticizing Democratic filibustering, see John Jay, *The Rise and Fall of the Pro-Slavery Democracy, and the Rise and Duties of the Republican Party* (New York, 1861), 40; *Strong Diary,* I:227–31, 250; on Whig attitudes toward expansionism, see especially Daniel Walker Howe, *The Political Culture of the American Whigs* (Chicago, 1979), 93–95, and Thomas R. Hietala, *Manifest Design: Anxious Aggrandizement in Late Jacksonian America* (Ithaca, 1985), 186–87.

52. *Strong Diary,* III:509; and see esp., Edwin L. Godkin, *Reflections and Comments* (New York, 1895), 53, for the patrician critique of mass markets and mass parties.

53. Frederick Law Olmsted to Charles Loring Brace, South Side, July 4, 1848, in Charles C. McLaughlin, ed., *The Papers of Frederick Law Olmsted* (Baltimore, 1977), I:320.

54. F. L. Olmsted to C. L. Brace, Cumberland River, Dec. 1, 1853, in McLaughlin, ed., *Olmsted Papers,* II:232–36.

55. *Ibid.;* Charles Loring Brace, *Address upon the Industrial School Movement Delivered at a Union Meeting of the Ladies of the Industrial Schools at University Chapel* (New York, 1857); Matthew Arnold, *Culture and Anarchy: An Essay in Political and Social Criticism* (Indianapolis, 1971), 79.

56. On merchants' post-Astor Place interest in Central Park, see Ian R. Stewart, "Politics and the Park," *NYHSQ* 61 (July/Oct. 1977), 124–55; on Peter Cooper's career and the merchant support for Cooper Union, see Edward C. Mack, *Peter Cooper: Citizen of New York* (New York, 1949), 243–73, and Olmsted to Brace, Dec. 1, 1853, in McLaughlin, ed., *Olmsted Papers,* II:232–36; Henry W. Bellows, *The Relation of Public Amusements to Public Morality, Especially of the Theatre to the Highest Interests of Humanity* (New York, 1857), 7, 12, 17, 37.

57. Bellows, *Relation of Public Amusements,* 17.

58. Henry W. Bellows, "Cities and Parks: With Special Reference to the New York Central Park," *Atlantic Monthly* 7 (April 1861), 419, 21.

59. On Olmsted's use of "religion" as the cultural opposite of Southern "materialism," see F. L. Olmsted to C. L. Brace, Dec. 1, 1853, in McLaughlin, ed., *Olmsted Papers,* II:232–36.

60. Charles Loring Brace, *The Dangerous Classes of New York, and Twenty Years' Work among Them* (New York, 1872), 446; Bellows, "Cities and Parks," 429 (on Central Park as a "mission").

61. *Strong Diary,* II:391. On the Wall Street religious revival of 1858, see Timothy L. Smith, *Revivalism and Social Reform in Mid-Nineteenth Century America* (New York, 1957), 63–79, and Perry Miller, *The Life of the Mind in America from the Revolution to the Civil War* (New York, 1965), 88–95.

62. On Dodge and Phelps's evangelical activity, see Richard Lowitt, *A Merchant Prince of the Nineteenth Century,* 191–203; *Strong Diary,* II:218.

63. Quoted in Henderson, "Young America and the Astor Place Riot," 100.

64. *Strong Diary,* III:124–26; Henry W. Bellows, *The State and the Nation—Sacred to Christian Citizens* (New York, 1861), 7; H. W. Bellows to his sister, Dec. 12, 1860, quoted in Fredrickson, *The Inner Civil War,* 54; John Jay to Salmon Chase, New York, April 4, 1861, in American Historical Association, *Annual Report for the Year 1902* (Wash., D.C., 1903) ("Diary and Correspondence of Salmon P. Chase"), 493–94; Henry W. Bellows, *Unconditional Loyalty* (New York, 1863), 4; on the new preoccupation with the "doctrine of loyalty" in early 1863, see Fredrickson, *The Inner Civil War,* 130–50.

65. *Strong Diary,* III:343, II:140, III:343; Union League Club, *Banquet Given by Members of the Union League Club of 1863 and 1864* (New York, 1886), 24–25.

66. F. L. Olmsted to C. L. Brace, Dec. 1, 1853, in McLaughlin, ed., *Olmsted Papers,* II:236; *Strong Diary,* III:325; F. L. Olmsted to Edwin L. Godkin, St. Louis, April 4, 1863, Edwin L. Godkin Papers, HU.

67. F. L. Olmsted to O. W. Gibbs, N.Y., Nov. 5, 1862, in Safe File Box Mss., Union League Club, N.Y.; August Belmont to S. J. Tilden, N.Y., Jan. 27, 1863, Samuel J. Tilden Papers, NYPL.

68. *Strong Diary,* III:333, 452.

69. *Ibid.,* 301, 319.

70. For the complete Union League Club membership in 1863, see below, Appendix H. Eight (or possibly nine) men who had attended the Belmont farewell dinner in 1853 became members of the Union League Club in 1863 (John L. Aspinwall, Henry A. Coit, Charles A. Heckscher, Sheppard Knapp, Alfred Pell, George S. Robbins, Paul Spofford, Moses Taylor, and possibly O. De Forrest Grant). Note that leading Democrats like Moses Taylor and John A. Dix were members of the Club. By January, 1864, the Club's membership numbered over five hundred. Union League Club of New York, *Report of Executive Committee, Constitution, By-Laws, and Roll of Members. January, 1864* (New York, 1864), 41–47. The sheer size of the Club alone suggests that it had become the majority voice in the merchant community by the mid-sixties. On the Club and the loyalty of "the intelligent, cultivated, gentlemanly caste," see *Strong Diary,* III:321–22.

71. Frederick Law Olmsted, "The Economical, Moral and Political Relations of Slavery" (Letter Number 48), in the *Times,* Feb. 13, 1854; on Northerners who shared Olmsted's perception of the free black community, see Fredrickson, *The Inner Civil War,* 151–65; on the campaign to raise the black Twentieth Regiment in New York City, merchants' relief efforts for black riot victims, and Sturges's speech on the worthiness of the black poor, see Chapter 2, above.

72. Fredrickson, *The Inner Civil War,* 165.

73. Bellows, *Unconditional Loyalty,* 4.

CHAPTER 5. INDUSTRIALISTS

1. Frances W. Gregory and Irene D. Neu, "The American Industrial Elite in the 1870's," in William Miller, ed., *Men in Business, Essays in the History of Entrepreneurship* (Cambridge, 1952); material on Allaire, Roach, Robert Hoe and Richard M. Hoe, Delamater and Allen drawn from *Dictionary of American*

Biography; Norman Brouwer, "New York Shipbuilding in the Industrial Age," *Seaport* 21 (Spring 1988), 36 (Delamater and the *Monitor*); Charles H. Davis, *Organization and Operation of an Industrial Establishment* (New York, 1900), 1 (Stephenson); U.S. Comm. of the Sen. upon Relations between Labor and Capital, *Report and Testimony,* II:1104 (Britton). The pattern of mobility among New York City industrialists coincides with Herbert G. Gutman's findings for mid-century Paterson, New Jersey. See "The Reality of the Rags-to-Riches 'Myth': The Case of the Paterson, New Jersey, Locomotive, Iron, and Machinery Manufacturers, 1830–1880," in Herbert G. Gutman, *Work, Culture, and Society in Industrializing America: Essays in American Working-Class and Social History* (New York, 1977), 211–33.

2. Montgomery, *Beyond Equality,* 12–13; Thomas H. O'Connor, *Lords of the Loom: The Cotton Whigs and the Coming of the Civil War* (New York, 1968); Stephenson's and Brewster's Carriage Factories were both partnerships; Hogg & Delamater was conducted by a single proprietor, Cornelius Delamater, who took over the Phenix Foundry, owned by the engineer James Cunningham. See J. L. Bishop, *A History of American Manufactures* (Phila., 1868), III:126–30; the Morgan Iron Works were owned and managed by the sole proprietor George W. Quintard. See Bishop, *History,* III:131; Roach's Etna Works seems also to have been run as an individual proprietorship until his move to Chester, Pennsylvania, in the early seventies, when he set up a joint stock concern owned entirely by himself, his sons, and two "partners," one of whom was William Rowland, a master joiner "well known in shipbuilding circles for his consummate taste and ingenuity in the combination of hard woods." See Bishop, *History,* III: 134, and *Coal and Iron Record* II (Dec. 18, 1872), 194–95; James Allaire ran his works himself until 1842, when he "associated others with him, who formed a joint-stock company . . . [and] elected him their first President." See Bishop, *History,* III:123–25; for Stephen D. Tucker and the Hoe Works, see Stephen D. Tucker, *History of R. Hoe and Company New York,* R. Hoe and Company Collection, CU. It is important to note that heavy industry located in Newark and other cities on the metropolitan periphery *outside* Manhattan was often owned by New York City merchants. On this point, see Bishop, *History,* III:218. For John Roach's strenuous advocacy of protective tariffs, see *Times,* Jan. 11, 1887.

3. For a thorough discussion of the exodus of heavy industry from lower Manhattan between 1845 and 1870 and the new ascendancy of labor-intensive enterprise in the downtown manufacturing center, see Richard B. Stott, "The Worker in the Metropolis: New York City, 1820–1860" (Ph.D. diss., Cornell Univ., 1983), 18–26; on the uptown migration of different factories, see Tucker, *History of R. Hoe and Company* (Hoe); Stott, "Worker in the Metropolis," 25 (Delamater); William R. Bagnall, *Sketches of Manufacturing Establishments in New York City, and of Textile Establishments in the United States* (Wash., D.C., 1908), 37, 51 (Singer); *Times,* Jan. 11, 1887 (Roach).

4. Forney, *Memoir of Horatio Allen,* 45; U.S. Sen. Comm., *Report and Testimony,* II:1113–19.

5. *Times,* Oct. 23, 1869.

6. Albion, *The Rise of New York Port,* 149; James S. Brown, *Allaire's Lost Empire: A Story of the Forges and Furnaces of the Manasquan* (Freehold, N.J.,

1958), 61; also see Halsted H. Wainwright, *The Howell Iron Works and the Romance of Allaire* (Freehold, N.J., 1925).

7. Haswell, *Reminiscences,* 332 (Allaire's tenement); also, see "Allaire, James Peter," *Dictionary of American Biography,* I:182; *Herald,* April 9, 1853.

8. See Edgar M. Hoover, *The Location of Economic Activity* (New York, 1948), 72, on the increasing intensity of "land use" (in this case, production), as land cost or "rent" increases.

9. Horace Greeley and Henry J. Raymond, *Association Discussed, Or the Socialism of the Tribune Examined* (New York, 1847); Horace Greeley, *Industrial Association: An Address to the People of the United States* (Boston, 1850). On François-Marie-Charles Fourier (1772–1837), see G. D. H. Cole, *A History of Socialist Thought* (London, 1967), vol. I, ch. 6; James Joll, *The Anarchists* (Cambridge, 1980), 34–37; and Charles Gide, ed., *Selections from the Works of Fourier* (London, 1901).

10. Greeley and Raymond, *Association Discussed,* 27; Greeley, *Industrial Association,* 9, 14–15, 18.

11. Horace Greeley, *Hints toward Reforms* (New York, 1857), 38–39; Greeley and Raymond, *Association Discussed,* 16, 19; see Howe, *The Political Culture of the American Whigs,* 184–97, for an interpretation of Greeley similar to mine.

12. Greeley, *Hints toward Reforms,* 189–90; *Tribune,* Aug. 5, 1845 (debate with *Express*), Feb. 22, 1844 (on founding of AICP).

13. For Greeley's distinction between "attractive industry" and "isolated labor," see Greeley, *Industrial Association,* 14–15; *Tribune,* April 13, 1853.

14. *Herald,* April 7, 9, 11, 16, 1853; also, see David Montgomery, *The Fall of the House of Labor: The Workplace, the State, and American Labor Activism, 1865–1925* (Cambridge, 1987), 180–93, on assumptions about workplace practices shared by journeymen machinists and industrial employers during the nineteenth century.

15. *Times,* June 1, 1872 ("Printing Press Makers").

16. Bagnall, *Sketches of Manufacturing Establishments in New York City,* 654–55.

17. *Ibid.,* 655–60, and esp. 657.

18. *Constitution of the Brewster & Co. Industrial Association* (New York, 1869), 3.

19. *Ibid.,* 4–6.

20. *Ibid.,* 4–8.

21. *Times,* June 24, 1881; U.S. Sen. Comm., *Report and Testimony* (Wash., D.C., 1885), I:1010–12.

22. "Delaware River Iron and Shipbuilding Works," *Coal and Iron Record* II (Dec. 18, 1872), 194; American Iron and Steel Association, *Bulletin* VII (1873), 426 (interview with Roach); *Herald,* June 6, 1872; U.S. Senate, *Report and Testimony,* I:1008–9.

23. American Iron and Steel Association, *Bulletin* VII (1873), 425–26. On emerging middle-class notions of "sincerity" and "hypocrisy" in antebellum America, see Karen Halttunen, *Confidence Men and Painted Women: A Study of Middle-Class Culture in America, 1830–1870* (New Haven, 1982); on the same

issues in the English context, Walter E. Houghton, *The Victorian Frame of Mind, 1830–1870* (New Haven, 1957).

24. See Appendix I, "Social Composition of the Association for Improving the Condition of the Poor, 1863"; George Opdyke to Commissioners of the Metropolitan Police, July 16, 17, 18, 21, 22, 1863, Opdyke Letterbook, Mayors' Papers, MARC. Horatio Allen to Maj. Gen. John E. Wool, Novelty Iron Works, July 16, 1863, Letters Received, Dept. of the East, Entry 1403, R.G. 393, NA.

25. R. H. Tawney, *Religion and the Rise of Capitalism* (New York, 1926), 210–26; on the history of attitudes toward poverty in New York City, see Raymond Mohl, *Poverty in New York, 1783–1825* (New York, 1971).

26. Charles E. Rosenberg and Carroll Smith Rosenberg, "Pietism and the Origins of the American Public Health Movement: A Note on John H. Griscom and Robert M. Hartley," *Journal of the History of Medicine and Applied Sciences* 23 (1968), 16–35; Carroll Smith Rosenberg, *Religion and the Rise of the American City: The New York City Mission Movement, 1812–1870* (Ithaca, 1971); on the pietism of Phelps and Dodge, see Lowitt, *A Merchant Prince of the Nineteenth Century.*

27. See, for example, the report of J. B. Horton, Missionary of the 7th Ward, New York, Aug. 23, 1844, quoted in John H. Griscom, *The Sanitary Condition of the Laboring Population of New York. With Suggestions for Its Improvement* (New York, 1845), 27–28.

28. John H. Griscom, *Annual Report for 1842,* 173 n. 16, as quoted in Rosenberg and Smith Rosenberg, "Pietism," 24; Griscom, *The Sanitary Condition of the Laboring Population of New York,* 22–23.

29. See Smith Rosenberg, *Religion and the Rise of the American City,* 163–85, on reformers' changing perceptions of the poor and the city after 1837, and 154–59, on the failure of the Protestant City Mission Society. On the early years of the AICP, see Roy Lubove, "The New York Association for Improving the Condition of the Poor: The Formative Years," *NYHSQ* 43 (July 1959), 307–27.

30. Rosenberg and Smith Rosenberg, "Pietism," 28–35.

31. AICP, *Seventh Annual Report* (New York, 1850), 29; on the system of visiting, see AICP, "Constitution and By-laws," in *First Annual Report* (New York, 1845), 9, 11; AICP, *Seventh Annual Report,* 32; AICP, *Thirteenth Annual Report* (New York, 1856), 25.

32. AICP, *Seventh Annual Report,* 25.

33. Gareth Stedman Jones, *Outcast London: A Study in the Relationship between the Classes in Victorian Society* (Oxford, 1971), 251–52.

34. AICP, *Twelfth Annual Report* (New York, 1855), 38–41 (abstracts of visitors' returns), and 37 (visitors' "unquestioned access").

35. AICP, *Seventh Annual Report,* 18.

36. Robert M. Hartley, *Removal of Almshouse from Bellevue, Clippings from the Journal of Commerce* (New York, 1845) (NYPL Collection).

37. AICP, *Twelfth Annual Report,* 23–24; AICP, *Fifteenth Annual Report* (New York, 1858), 30–32 (Wood), 22 (the operation of AICP "machinery" during the 1857–58 depression), 24 (percentages of applicants aided during the depression).

38. AICP, *Thirteenth Annual Report* (New York, 1856), 45–51 (Working Men's Home Association); Lawrence Veiller, "Tenement Housing Reform in

New York City, 1834–1900," in Robert W. DeForest and Lawrence Veiller, eds., *The Tenement House Problem* (New York, 1903), 86–87; James Ford, *Slums and Housing with Special Reference to New York City* (Cambridge, 1936), 137 (degeneration of conditions in model tenements); Veiller, "Tenement Housing Reform," 87 (1856 State Committee). For the arguments of the 1856 Committee justifying proposed "violations" (the term is mine) of laissez-faire, see Select Committee Appointed to Examine into the Condition of Tenant Houses in New-York and Brooklyn, *Report* (Assembly Doc. No. 205, Mar. 9, 1857), 5, 53. On the passage of a metropolitan housing law after the Civil War, see below.

39. AICP, *Seventh Annual Report,* 19 (breaking up families of the unworthy poor), *Thirteenth Annual Report,* 41 (legislation against truants), *Fifteenth Annual Report,* 42 (Count Rumford), *Seventeenth Annual Report* (New York, 1860), 71 (Superintendent Kennedy's arrests of vagrant poor).

40. On the New York City metalworking industry as "open shop" in the 1860s, see Frank Roney, *Frank Roney: Irish Rebel and California Labor Leader* (Ira Cross, ed.) (Berkeley, 1931), 179–81; also, above, Chapter 3.

41. *Times,* Jan. 11, 1887.

42. On the brief service of the merchant Peletiah Perit as a Commissioner of Police during the "emergency" of summer 1857, see *Times,* March 10, 1864; Governor John A. King, who appointed the largely Republican Metropolitan Police Commission, was brother of James Gore King, the New York City merchant-banker. See Spann, *The New Metropolis,* 388–89.

43. James C. Mohr, *The Radical Republicans and Reform in New York during Reconstruction* (Ithaca, 1973), 21–114; James Richardson, *The New York Police: Colonial Times to 1901* (New York, 1970), 91–108. For an excellent discussion of the Civil War-era political struggle over commission rule in another Eastern city, see Douglas V. Shaw, "The Making of an Immigrant City: Ethnic and Cultural Conflict in Jersey City, 1850–1877" (Ph.D. diss., Univ. of Rochester, 1972); on the Tenement House Law of 1867, see Veiller, "Tenement House Reform," 94–97; Emmons Clark, "Sanitary Improvement in New York during the Last Quarter of a Century," *Popular Science Monthly* 39 (1891), 324–26.

44. *Times,* Oct. 30, 1857 ("German Republicans in Mass Meeting"). In the fall 1863 city mayoral race, Gunther's plurality over Union (Republican) Party candidate Orison Blunt in the four *Kleindeutschland* wards (10, 11, 13, 17) was, respectively, 624, 1858, 936, and 1784. Gunther's plurality over Tammany Democrat candidate Francis I. A. Boole in the four German wards was 830, 716, 865, and 1660. *The Tribune Almanac and Political Register for 1864* (New York, 1864), 57.

45. AICP, *Nineteenth Annual Report* (New York, 1862), 14; Forney, *Memoir of Horatio Allen,* 41, 44–45; also, see the discussion of industrial workers' views of their employers in Chapter 3.

46. The Republican Party controlled City Hall from December 1861 through November 1863.

On Peter Cooper and the Citizens' Association, see Mack, *Peter Cooper,* 344–56; on the Association and opposition to Tammany, see Citizens' Association, *Work Is King! A Word with Working Men in Regard to Their Interest in Good City Government. Read, Think, and Act* (New York, n.d.); George F.

Noyes, *Argument in Favor of the Metropolitan Board of Public Works, before Senate and Assembly Committees* (Albany, 1867) (fraud in Tammany contract system); on the Association, the draft riots, and sanitary reform, see Citizens' Association, *Report of the Council of Hygiene and Public Health,* xiv–xv; Mohr, *Radical Republicans and Reform in New York,* 67–69.

47. New Yorker *Arbeiter-Zeitung,* Sept. 3 (painters and German cigar makers), Sept. 10 (Citizens' Association workers' meeting), Oct. 1 (Painters), 1864; *Times,* Sept. 2, 1864 (woodworking trades, Printers and Citizens' Ass'n); also see Leonard Newman, "Opposition to Lincoln in the Elections of 1864," *Science and Society* 8 (Fall 1944), 319.

What kinds of employers attended Citizens' Association rallies? The evidence is scant. Two of the 49 "Workers in Wood" at a September 1, 1864, meeting in support of the Citizens' Association were traceable in the 1860 U.S. Federal Manuscript Census Industrial Schedules. Frederick Germann, a chair manufacturer, employed 25 male workers, with $6000 invested in the business. Francis Hayek, a cabinet manufacturer, employed 30 male workers, with $10,000 invested in the business. Both of these shops were moderate-sized to large by New York City woodworking shop standards. U.S. Census Office, *Census of the United States, Industrial Schedule, New York County, 1860,* 13th and 14th Wards, State Archives, NYSL.

Though working-class interest in the Citizens' Association ebbed after fall 1864, it should be noted that some metal-trades workers did ask Peter Cooper and the Citizens' Association to mediate the eight-hour strikes of June 1872. See below, Chapter 6.

48. Mohr, *Radical Republicans and Reform in New York,* 146–47. I am especially indebted to Elizabeth Blackmar for her insights into the history of nineteenth-century housing and housing reform.

49. John William Bennett, "Iron Workers in Woods Run and Johnstown: The Union Era, 1865–1895" (Unpubl. Ph.D. diss., Univ. of Pittsburgh, 1977), 181.

50. On cultural conflict between Republican Protestant reformers and immigrant workers in the 1850s, see especially Weinbaum, "Temperance, Politics and the New York City Riots of 1857"; Rorabaugh, "Rising Democratic Spirits"; and M. J. Heale, "Harbingers of Progressivism: Responses to the Urban Crisis in New York, c. 1845–1860," *Journal of American Studies* 10 (April 1976), 17–36.

51. AICP, *Twentieth Annual Report,* 33; Eric Foner, *Free Soil, Free Labor, Free Men,* 11–39, and esp. 22.

CHAPTER 6. THE RISE AND DECLINE OF TWEED'S TAMMANY HALL

1. On Tammany's resolution of the political crisis during and immediately following the draft riots and its successful campaign against the Peace Democracy, see above, Chapters 1 and 2, and the discussion below, esp. note 38.

2. For the Tweed Ring as evidence of "the perversion and corruption of democratic government in great American cities," see James Bryce, *The American Commonwealth* (New York, 1913), II:397–405; on the extent of Tweed

Ring fraud, see Alexander B. Callow, Jr., *The Tweed Ring* (New York, 1965); on Tweed's "program of official social insurance" and public aid, see John W. Pratt, "Boss Tweed's Public Welfare Program," *NYHSQ* 45 (Oct. 1961), 396–411; for an examination of the Tweed Ring as an experiment in "administering a complex environment," see Seymour J. Mandelbaum, *Boss Tweed's New York* (New York, 1965); on Tweed's New York as a "heralded failure" in municipal administration in contrast with the more general "unheralded triumph" of Gilded Age city government, see Jon C. Teaford, *The Unheralded Triumph: City Government in America, 1870–1900* (Baltimore, 1984), esp. 4; for an attempt to debunk the evil "myth" of Boss Tweed and portray him instead as a "pioneer spokesman for an emerging New York," see Leo Hershkowitz, *Tweed's New York: Another Look* (Garden City, 1977), esp. 348–49.

3. Morton Keller, *Affairs of State: Public Life in Late Nineteenth Century America* (Cambridge, 1977), 238–88.

4. New York State Constitutional Convention, *Proceedings and Debates of the Constitutional Convention . . . Held in 1867 and 1868* (Albany, 1868), IV:2954–55 (speech of M. I. Townsend).

5. New York *Leader,* July 9, 1864; William L. Riordon, *Plunkitt of Tammany Hall* (New York, 1963), 69.

6. Man, "The Church and the New York Draft Riots of 1863," 33–50; David J. Alvarez, "The Papacy in the Diplomacy of the American Civil War," *Catholic Historical Review* 69 (April 1983), 227–48.

7. *Strong Diary,* IV:171. The Dead Rabbits were an Irish gang involved in the 1857 riots against the Republican-sponsored Metropolitan Police. Wood, a favorite of the Irish working class, was waging his own campaign against the Metropolitans; hence, Wood as "King of the Dead Rabbits."

8. Croswell Bowen, *The Elegant Oakey* (New York, 1956), 5, 70–71, 76–77, 83; Matthew P. Breen, *Thirty Years of New York Politics* (New York, 1899), 272–73.

9. Chandler, *Visible Hand,* 92–94.

10. Edward D. Durand, *The Finances of New York City* (New York, 1898), 107–9, on 1852 law instituting assessment bonds; see Jeremiah O'Donovan, *A Brief Account of the Author's Interview with His Countrymen, and of the Parts of the Emerald Isle Whence They Emigrated, Together with a Direct Reference to Their Present Location in the Land of Their Adoption, during His Travels Through Various States of the Union in 1854 and 1855* (Pittsburgh, 1864), 93–94, 161–68, on emerging Irish middle class in New York City during the 1850s, 87–88, on Irish contractors in other American cities; Dennis Clark, *The Irish in Philadelphia: Ten Generations of Urban Experience* (Phila., 1973), 57–60, on Philadelphia's emerging Irish middle class in the 1850s.

11. On Belmont and the Rothchilds' involvement in American railroad development, see Chandler, *Visible Hand,* 146, 165, 170, 183; on Belmont, European investment bankers, and the Tweed regime, see Seymour J. Mandelbaum, *Boss Tweed's New York* (New York, 1965), 78, and below.

12. Board of Supervisors of the County of New York, *Proceedings* (1863), II:213; see Durand, *Finances of New York City,* 102–7, for an excellent discussion of the size and significance of New York City's war debt and the role of the draft riots in legitimating the large expenditure on bounties and exemption

fees; on Tweed's Exemption Committee, see Hershkowitz, *Tweed's New York,* 93, and above, Chapter 2.

13. On Gustavus A. Conover and the People's Union movement, see *Times,* Nov. 16, 1857; "Miscellaneous Contractors," in "Finance Committee, Democratic Republican General Committee, 1863," Misc. Manuscripts, NYPL.

See Map 5, "Names and Geographic Distribution of Tammany Contractors, 1863." The Tammany contractors were an uptown group. Note that most lived and worked above Fourteenth Street, and many above Thirtieth Street.

14. *Daily News,* July 23, 1863, on Peter Masterson's opposition to the midweek uprising, and Chapter 2, above; Case of Richard Lynch, Indictments, Aug. 1863; Case of John Halligan, Martin Hart, and Adam Schlosshauer, Indictments, Aug. 1863, MARC, on Jacob Long; William C. Gover, *The Tammany Hall Democracy of the City of New York and the General Committee for 1875* (New York, 1875), 97, 114, on John Masterson and Jacob Long as Tammany leaders of the 1870s.

15. Henry James, *Autobiography* (Fred. W. Dupee, ed.) (New York, 1956), 38, 39–42, as quoted in Jean-Christophe Agnew, "The Consuming Vision of Henry James," in Richard Wightman Fox and T. J. Jackson Lears, eds., *The Culture of Consumption: Critical Essays in American History, 1880–1980* (New York, 1983), 79.

16. Isaac G. Kendall(?), *The Growth of New York* (New York, 1865), 43, 28.

17. *Times,* Oct. 23, 1869 (Iron Works of New York); Bagnall, *Sketches of Manufacturing Establishments,* 37, 51 (Singer); *Times,* Jan. 11, 1887 (Roach); *Real Estate Record and Builders' Guide,* I (May 30, 1868), Forney, *Memoir of Horatio Allen,* 45 (Novelty Works); also, on decline of iron ship-building in the 1860s, see John H. Morrison, *History of New York Ship Yards* (New York, 1909), 153–65.

18. Blackmar, "Housing and Property Relations in New York City," 386–492.

19. *Real Estate Record and Builders' Guide,* I (May 9, 1868), on the history of the West Side Association, and I (June 27, 1868) on the incorporation and membership of the East Side Association. See Appendix J, "Names and Occupations of Members of Uptown Property Associations, 1866–71."

20. *Real Estate Record and Builders' Guide,* I (May 9, June 13, 1868), and Appendix J, "Names and Occupations of Members of Uptown Property Associations, 1866–71."

21. *Real Estate Record and Builders' Guide,* I (May 9, 1868), on "East and West Side Associations."

22. William Martin's brother Howard had assisted Olmsted as chief clerk of Central Park prior to the Civil War and on Olmsted's recommendation was appointed accountant to the United States Sanitary Commission. Charles E. Beveridge and David Schuyler, eds., *The Papers of Frederick Law Olmsted* (Baltimore, 1983), III:39.

23. West Side Association, *Proceedings of Six Public Meetings* (New York, 1871), Document 1, pp. 13–14, Document 5, p. 7; Callow, *The Tweed Ring,* 225–27; Citizens' Association of New York, *An Appeal by the Citizens' Association . . . Against the Abuses in the Local Government, to the Legislature of the State of New York, and to the Public* (New York, 1866); Union League

Club, *The Report of the Committee on Municipal Reform Especially in the City of New York* (New York, 1867), esp. 24–51.

24. East Side Association, *To the Friends of Rapid City Transit* (New York, 1871), esp. 7–9 (West Side and East Side Association petitions); West Side Association, *Address of William R. Martin on Rapid Transit* (New York, 1871); West Side Association, *Proceedings of Six Public Meetings,* Document 5, pp. 18–19.

25. *Times,* April 29, 1869 (letter of A. Oakey Hall).

26. *Real Estate Record and Builders' Guide,* I (May 9, 1868), on East River Improvement Association's interest in Hoffman and Cooper; West Side Association, *Proceedings of Six Public Meetings,* Document 5, pp. 20–23, on James F. Ruggles; on endorsement of Tammany financial records by Moses Taylor, E. D. Brown, J. J. Astor, George K. Sistare, Edward Schell, and Marshall O. Roberts, see esp. *Times,* Nov. 7, 1870; John Foord, *The Life and Public Services of Andrew Haswell Green* (Garden City, 1913), 165; Hodas, *Business Career of Moses Taylor,* 195–97.

27. *Merchants' Magazine and Commercial Review* 58 (April 1868), 320–21, on spring 1868 fluctuations in the money market; *Sun,* June 22, 1868, *Real Estate Record and Builders' Guide,* I (July 25, 1868), and "The Eight Hour Strikes," in *Merchants' Magazine and Commercial Review* 59 (Aug. 1868), 94, on merchant capital, the building industry, and the bricklayers' strike.

The only extended treatment of the 1868 bricklayers' strike is Irwin Yellowitz, "Eight Hours and the Bricklayers' Strike of 1868 in New York City," in Irwin Yellowitz, ed., *Essays in the History of New York City: A Memorial to Sidney Pomerantz* (Port Washington, N.Y., 1978), 78–100.

28. *Times,* Sept. 2, 1864 (woodworkers and the Citizens' Association); Mohr, *Radical Republicans and Reform in New York during Reconstruction,* 120–39.

29. *Sun,* June 23, 1868; *Real Estate Record and Builders' Guide,* I (July 4, 1868) on "Labor."

30. *Sun,* June 23, 1868.

31. Yellowitz, "Eight Hours and the Bricklayers' Strike of 1868," 88, 91, 94; *Merchants' Magazine and Commercial Review* 59 (Aug. 1868), 92 (Canadian strikebreakers); *Real Estate Record and Builders' Guide,* I (Sept. 5, 1868), on "front-men," and I (Sept. 19, 1868) on German "Ten Hour Mason Society"; also see Harry C. Bates, *Bricklayers' Century of Craftsmanship* (Wash., D.C., 1955), 27–30.

32. United States Senate Committee, *Report . . . upon the Relations Between Labor and Capital* (Wash., D.C., 1885), I:817.

33. *Sun,* June 25, 1868; *Herald,* July 16, 1868.

34. *Herald,* July 16, 18, 21, 1868.

35. *Herald,* July 28, Aug. 15, 1868; see above, Chapter 5, for discussion of John Roach's taxonomy of good and bad workmen.

36. *Real Estate Record and Builders' Guide,* I (Sept. 19, 1868), on dismissal of Tweed Court House bricklayers; *Herald,* Nov. 26, 1869, for interview with Peter Barr Sweeny.

37. *Address of the Liquor Dealers and Brewers of the Metropolitan Police District to the People of the State of New York* (New York, 1868), 17; *Herald,* Nov. 26, 1869, Sweeny on the Excise Question.

38. The shift in the political loyalties of the downtown manufacturing district occurred in the following manner: In 1862, the Sixth and Fourteenth wards elected to Congress, respectively, the Peace Democrat brothers Benjamin and Fernando Wood. But by fall 1863, the Tammany nominee Francis I. A. Boole was able to carry these wards in the mayoral race, and Tammany men won the Sixth and the Fourteenth against Peace candidates in the 1864 Congressional election. By the mid-sixties, these neighborhoods were Tweed strongholds; Tweed himself took the two wards in his victorious run for the state senate in 1867. Fernando Wood ran (and probably had to run) for Congress in 1866 from a new district in Upper Manhattan and Harlem. *The Tribune Almanac and Political Register for 1863* (New York, 1863), 51; *Tribune Almanac . . . for 1864* (New York, 1864), 57; *Tribune Almanac . . . for 1865* (New York, 1865), 49; *Tribune Almanac . . . for 1867* (New York, 1867), 52; *Tribune Almanac . . . for 1868* (New York, 1868), 50.

Tammany led anti-riot brigades in the Sixth and Fourteenth wards in July 1863; also see Chapter 1 on the confirmation of the downtown manufacturing district, and particularly the Sixth and Fourteenth wards, as Tammany territory after 1863.

On the rise of luxury goods consumption during the Civil War, see esp. Basil Leo Lee, *Discontent in New York City, 1861–1865* (Wash., D.C., 1943).

39. *Times,* Oct. 14, 1865 (Occupations of New York City Inspectors of Registry and of Elections for November 1865, by Ward); see Callow, *The Tweed Ring,* 208–13, on inspectors and Tammany management of elections in the sixties; Timothy L. Smith, "New Approaches to the History of Immigration in Twentieth-Century America," *AHR* 71 (July 1966), 1270.

Of course, no political organization seeking to reach out to the Protestant middle and upper classes would care to advertise its connection to saloonkeepers. Nonetheless, the incidental nature of the *Times*'s remark regarding "liquor stores" gives it added credence as a hint of a changing social and political reality. Tammany's new interest in "respectability" doubtless reflected changes in its constituency.

40. *Herald,* Oct. 23, 1851; Lee, *Discontent in New York City,* 2–3; *Times,* Oct. 21, 1869 (Hall on Cuba); Hershkowitz, *Tweed's New York,* 142–43 (Garvin); *Herald,* Nov. 26, 1869 (Sweeny).

41. *Leader,* Oct. 3, 1863.

42. See William Bigler to Samuel J. Tilden, Feb. 3, 1868 (letter mislabeled by editor) and S. L. M. Barlow to S. J. Tilden, June 21, 1868, in John Bigelow, ed., *Letters and Literary Memorials of Samuel J. Tilden* (New York, 1908), I:216–17 and 231–32; Montgomery, *Beyond Equality,* 351–52, on New York State Democrats' posture toward Chase candidacy; Jerome Mushkat, *The Reconstruction of the New York Democracy, 1861–1874* (Rutherford, N.J., 1981), 133–36, on movement for Chase and Chase's position on the Fourteenth Amendment and black suffrage; *Herald,* Nov. 26, 1868, Sweeny interview. Mushkat cites a *Leader* masthead for Seymour as evidence of Tammany's rejection of the Chase candidacy. Tweed lieutenant Sweeny's recollection of Tammany's support for Chase as the means of effecting a political compromise seems to be the better evidence.

On Chase's Democratic and antislavery antecedents, see Foner, *Free Soil, Free Labor, Free Men,* 73–102.

43. For William E. Dodge, Jr.'s advocacy of martial law, see D. Stuart Dodge, ed., *Memorials of William E. Dodge* (New York, 1887), 90, and Chapter 2, above; Dodge, ed., *Memorials,* 130, 134, Dodge speech to Congress; "The Crisis of Reconstruction," in *The Merchants' Magazine and Commercial Review* 58 (Feb. 1868), 122–23; for a discussion of Northern business interests' position in the reconstruction debates, see Kenneth M. Stampp, *The Era of Reconstruction, 1865–1877* (New York, 1965), 206–8.

44. William S. McFeely, *Grant: A Biography* (New York, 1981), 338; *Herald,* Nov. 26, 1868.

45. Alexander Delmar, *The Great Paper Bubble; Or, the Coming Financial Explosion* (New York, 1864), 52–53, 88; also, Alexander Delmar, *Gold Money and Paper Money* (New York, 1863); also see Joseph Dorfman, *The Economic Mind in American Civilization, 1606–1865* (New York, 1946), II:975–76.

46. Simon Sterne, *Report to the Constitutional Convention of the State of New York on Personal Representation Proposed at the Request and Printed under the Auspices of the Personal Representation Society* (New York, 1867), 8–10, 23, 39–40. The Society officers were David Dudley Field, Pres., Francis George Shaw, V.P., Robert B. Minturn, Rec. Sec., Sidney Howard Gay, Corr. Sec., Edward Cooper, Treas.; the Executive Committee was composed of Field, Shaw, Minturn, Cooper, David G. Croly, J. Francis Fisher, Alfred Pell, Jr., Mahlon D. Sands, and Simon Sterne.

47. See "Del Mar, Alexander," and "Sterne, Simon," *Dictionary of American Biography,* III:225–26 and IX:592–93.

48. See Francis Lieber, *Notes on Fallacies of American Protectionists* (New York, 1870), for an 1869 list of members of the American Free Trade League. David Dudley Field was president of both the Personal Representation Society and the Free Trade League. Francis George Shaw was vice president of both groups. Alfred Pell, Jr., Mahlon Sands, and Simon Sterne were officers in both groups. Samuel J. Tilden was also a vice president of the Free Trade League.

49. See above, Chapter 4.

50. Horatio Seymour to Samuel J. Tilden, Albany, Jan. 12, 1863, and Horatio Seymour to S. J. Tilden, Utica, Sept. 5, 1871, in Samuel J. Tilden Papers, NYPL.

51. William C. Gover, *The Tammany Hall Democracy,* 49, on Gunther campaign of 1863 and the formation of the McKeon Democracy; *Weekly Day Book,* Jan. 2, 1864, on Berckmann; see Map 6, "Election Returns, New York City Mayoral Race, 1863, by Ward," on Gunther's areas of political strength.

52. New Yorker *Staats Zeitung,* Dec. 1, 1863 ("Die Wahl").

53. Montgomery, *Beyond Equality,* 110–11; *Times,* Oct. 13, 1869; *Staats Zeitung,* Oct. 15, 1869 ("Einige Worte uber den Plan einer 'Arbeiter-Partei' ").

54. *Times,* Oct. 13, 1869 (Matthews); *Times,* Oct. 30, 1869 (Ennis).

55. *Times,* Nov. 4, 1869 (election results); *Herald,* Nov. 26, 1869 (Sweeny on Young).

56. On Peter Cooper's relationship with the Tweed Ring in 1870–71, see Edward C. Mack, *Peter Cooper: Citizen of New York* (New York, 1949), 344–56.

57. Peter B. Sweeny, *On the "Ring Frauds" and Other Public Questions* (New York, 1894), 51–57 (Sweeny contended that if Ottendorfer had accepted the Tammany nomination, the Tweed administration would have survived); *Die*

Arbeiter Union, Oct. 21, 1869 (Arbeiter Clubber 17 Ward); *Times,* Sept. 1, 1869 (17th Ward German-American Club).

58. On Adolph Douai's involvement in the German free soil movement of the fifties, see Levine, " 'In the Spirit of 1848': German-Americans and the Fight over Slavery's Expansion," esp. 142–43.

59. *Herald,* Nov. 26, 1869 (Sweeny on "the Negro Question"); *Die Arbeiter Union,* April 1, 1870 ("Das 13. Amendment und unsere Zukunft"). The most compelling evidence for this shift in working-class racial opinion appears in the discussion of the anti-coolie movement and Orange riot, below.

60. *Sun,* July 1, 1870; *Sun,* May 20, 1871, on Young's arrangement with Tammany; *Irish American,* July 1, 1870; *Die Arbeiter Union,* July 1, 1870 ("Die Chinesen"); John Swinton, *The New Issue. The Chinese-American Question* (New York, 1870), 6, 14.

61. Siegfried Meyer to Karl Marx, Hoboken, N.J., Jan. 10, 1871, File D3426, IISH; Samuel Bernstein, *The First International in America* (New York, 1962), 31–34; on the IWA's American branches and platform, see also Hermann Schlüter, *Die Internationale in Amerika: Ein Beitrag zur Geschichte der Arbeiter-Bewegung in den Vereinigten Staaten* (Chicago, 1918); George McNeill, *The Labor Movement: The Problem of Today* (Boston, 1887), 141–42, and below, Chapter 8.

62. *Sun,* May 19, 20, 1871; *Herald,* Nov. 26, 1869.

63. Montgomery, *Beyond Equality,* 323–34.

64. Mandelbaum, *Boss Tweed's New York,* 77–80; *Times,* June 13, 20, 1871.

65. *Sun,* May 2, 3, 4, 6, 18, 1871; see Chapters 1 and 2, above, on laborers' activities during the draft riots.

66. Bowen, *Elegant Oakey,* 88–89; Bernard Cook, "A Report from Friedrich Sorge to the General Council of the I.W.A.: The New York Riot of 1871," *Labor History* 13 (1972), 415–18.

67. Bowen, *Elegant Oakey,* 87–95; *Times,* July 10, 11, 12, 1871; *Herald,* July 12, 1871 (interviews with street pavers, quarrymen, and longshoremen on the eve of the Orange riot).

68. Joel T. Headley, *The Great Riots of New York, 1712 to 1873* (New York, 1873), 299–304; *Sun,* July 13, 1871, on participation of quarrymen and pipe men in Orange riot; *Civil Rights. The Hibernian Riot and the "Insurrection of the Capitalists." A History of Important Events in New York, in the Midsummer of 1871* (New York, 1871), 25; Montgomery, *Beyond Equality,* 377–78.

69. *Civil Rights,* 3, 27, 46, 70–71; Peter B. Sweeny to John T. Hoffman, July 21, 1871, John Thompson Hoffman, Pers. Misc., NYPL.

70. Peter B. Sweeny to John T. Hoffman, July 21, 1871, Hoffman, Pers. Misc., NYPL; *Times,* July 4, 1924 (George Washington Plunkitt on 1871 Orange riot).

71. *Times,* July 4, 1924. In 1902, Hoe Press workers attacked a huge Jewish funeral parade for a chief rabbi. When Jews counterattacked the factory, mounted police assaulted them brutally. Also, on New York's Kosher Butchers' Riot of 1902—arguably an intrareligious conflict—see Gutman, *Work, Culture & Society in Industrializing America,* 61–63.

72. *Times,* Oct. 29, 1869, on "Labor Organizations among the Colored Men"; see Charles H. Wesley, *Negro Labor in the United States, 1850–1925: A Study*

in American Economic History (New York, 1927), 175, 199, on the black labor movement in New York City during the sixties and New York's black delegation to the National Labor Convention; Bernstein, *The First International,* 39, on Section 1 of the IWA and black workers. Section 1's cordial relations with the black community may account for the National Colored Labor Convention's December 1869 decision to send a delegate to the 1870 Paris Congress of the First International.

73. *Times,* Sept. 14, 1871; *Tribune,* Sept. 14, 1871; *Herald,* Sept. 14, 1871; Montgomery, *Beyond Equality,* 327; Bernstein, *The First International,* 66.

74. See above, Chapter 2, for discussion of the Union League Club's recruitment of black regiments in 1863–64.

75. *Times,* June 7, 1871.

76. *Ibid.*

CHAPTER 7. 1872

1. *Herald,* April 6, 1866 (Greeley and the 1866 shipyard strike).

2. C. L. Vallandigham to Horace Greeley, Jan. 10, 1863; James W. White to Horace Greeley, Feb. 9, 1863; Benjamin Wood to Horace Greeley, April 22, 1863; Horace Greeley to Lemuel Smith, Aug. 16, 1864, Horace Greeley Papers, NYPL; Peter Cooper to J. A. Dix, N.Y., Sept. 4, 1872, John A. Dix Papers, CU.

3. *Times,* June 22, 1872. New York City "Independents" may also have rejected Greeley because he was protectionist and they were tariff reformers.

4. Siegfried Meyer to Karl Marx, Joliet, Ill., April 14, 1872, D 3431, IISH.

5. On the influence of the IWA in New York City, see above, Chapter 7.

6. Quoted in Montgomery, *Beyond Equality,* 420–21; see Lewis Masquerier, *Sociology: Or, the Reconstruction of Society, Government, and Property, upon the Principles of Equality, the Perpetuity, and the Individuality of the Private Ownership of Life, Person, Government, Homestead and the Whole Product of Labor, by Organizing All Nations into Townships of Self-Governed Homestead Democracies—Self-Employed in Farming and Mechanism, Giving All the Liberty and Happiness to Be Found on Earth* (New York, 1877).

7. Siegfried Meyer to Karl Marx, Hollidaysburg, Pa., Aug. 30, 1871, D 3428, IISH.

8. *Ibid.*

9. Mohr, *Radical Republicans and Reform in New York during Reconstruction,* 135–36; Montgomery, *Beyond Equality,* 323–25; *Herald,* June 21, 1872.

10. *Daily News,* Oct. 8, 1864; *Sun,* Oct. 8, 1864, June 5, 1865; Lawrence Costello, "The New York City Labor Movement, 1861–1873" (Unpubl. Ph.D. diss., Columbia Univ., 1967), 343–46.

11. *Times,* Oct. 13, 1869; *Herald,* Nov. 26, 1869; Montgomery, *Beyond Equality,* 323–25.

12. Costello, "The New York City Labor Movement," 253–56; on the structure and activities of the United Labor Party, see David Scobey, "Boycotting the Politics Factory: Labor Radicalism and the New York City Mayoral Election of 1886," *Radical History Review* 28–30 (1984), 303–14.

13. *Ibid.,* 220–21; *Tribune,* Dec. 11, 1869 (People's Cooperative Association); *Herald,* June 21, 1872 (Blissert on "republic of labor.")

14. [Executive Committee of Citizens and Taxpayers], *Appeal to the People of the State of New York* (New York, 1871); New York *Herald,* June 21, 1872.

15. Siegfried Meyer to Karl Marx, Mill City, Colo., July 26, 1870, D 3424, IISH; see Scobey, "Boycotting the Politics Factory," 280–325, on the significance of the Central Labor Union of the 1880s.

16. *Times,* May 21, 1872.

17. Montgomery, *Beyond Equality,* 234.

18. *Herald,* May 21, 1872.

19. *Ibid.*

20. *Ibid.; Tribune,* June 8, 1872.

21. Montgomery, *Beyond Equality,* 238–40.

22. On the distinction between "rough" and "respectable" as a Victorian construct which did not describe the complicated realities of Victorian workers' lives, nor their penchant for role-playing in the anonymous urban setting, see Peter Bailey, " 'Will the Real Bill Banks Please Stand Up?' Towards a Role Analysis of Mid-Victorian Working-Class Respectability," *Journal of Social History* 12 (Spring 1979), 336–53; see Brian Harrison, *Drink and the Victorians: The Temperance Question in England, 1815–1872* (London, 1971), on non-deferential working-class interpretations of temperance within the pan-class British temperance movement.

23. *Tribune,* June 3, 6, 8, 1872; Montgomery, *Beyond Equality,* 329.

24. *Times,* June 6, 1872.

25. *Times,* June 9, 1872; *Tribune,* June 11, 1872; Montgomery, *Beyond Equality,* 328.

26. *Times,* June 6, 8, 16, 27, 1872; *Herald,* June 8, 1872.

27. Montgomery, *Beyond Equality,* 330.

28. *Herald,* June 18, 1872; *Times,* June 18, 1872; *Tribune,* June 18, 1872.

29. *Times,* June 19, 1872; *Tribune,* June 19, 1872; A. S. Cameron, *The Eight Hour Question* (n.p., n.d.); see J. W. Orr Engraving Co. to McCormick Bros. and Co., June 19, 1872, in William T. Hutchinson, *Cyrus Hall McCormick,* II: 485–86, n. 117, for another employer's comments on the paucity of "good men" among the city's skilled workmen during the third week of June.

30. *Times,* June 25, 1872.

31. AICP, *Twenty-ninth Annual Report* (New York, 1872), 60–65.

32. E. L. Godkin, "Rich Men in City Politics," *Nation* 13 (Nov. 16, 1871), 316.

33. Henry W. Bellows, "The Battle of Civilization," in *The Chicago Pulpit* (Chicago, 1873), III:207–14; Bellows, *Historical Sketch of the Union League Club of New York,* 142–43 (falling away of political reform activities), and 162 (Bellows wrote the *Historical Sketch* to remind the Club membership that the "Union League Club should be an animated protest against [an] anti-national, anti-American, listlessness, and waste of political power, and failure in duty"); Union League Club of New York, *Report of Executive Committee* (New York, 1872), 9–10, Club Library, Union League Club, New York.

EPILOGUE: THE DRAFT RIOTS' LOST SIGNIFICANCE

1. Edward E. Pratt, *Industrial Causes of Congestion of Population in New York City* (New York, 1911), 114–15; David C. Hammack, *Power and Society: Greater New York at the Turn of the Century* (New York, 1982), 36; Christopher Gray, "The Lost Skyscrapers," *Times,* May 15, 1988 (the first Manhattan skyscraper, the Equitable Building at 120 Broadway, was completed in 1870); Moses Rischin, *The Promised City: New York's Jews, 1870–1914* (New York, 1970), 64–65; for twentieth-century efforts to resolve the problem of a divided metropolitan elite, see Robert A. Caro, *The Power Broker: Robert Moses and the Fall of New York* (New York, 1974); Sam Roberts, "Who Runs New York Now?," *Times,* April 28, 1985.

2. Bender, *New York Intellect,* 174; Stephen Skowronek, *Building a New American State: The Expansion of National Administrative Capacities, 1877–1920* (Cambridge, Eng., 1982), for a discussion of Roosevelt's and Root's efforts at state-building in the Progressive era; on the far-reaching national influence of Tilden and his elitist "educational politics," see Michael E. McGerr, *The Decline of Popular Politics: The American North, 1865–1928* (New York, 1986), 42–106; on the similarly extensive influence of Godkin's *Nation,* see Bender, *New York Intellect,* 181–91.

Bibliographical Essay

It would be difficult or impossible to attempt to review the extensive range of sources which appear in the annotation for this study or which were used in its preparation. What follows is a brief survey and evaluation of some of the most important primary sources, and of secondary works I have found helpful in nineteenth-century social, cultural, and political history.

DRAFT RIOTS

The Civil War context enables the historian to develop a rather fine-grained sociological portrait of the draft riots. The political climate of July 1863 was so charged with considerations of national loyalty and the stakes of involvement in the treasonous violence were so high that even the most minute circumstances of participation and non-participation were noticed. The newspapers—especially the *Herald, Irish American, Daily News, Evening Express, Daily Tribune,* and *Times*—were extremely thorough in their reportage of rioters' identities, words, and activities. No less attentive were the scores of eyewitnesses who published their impressions in pamphlets and recorded them in private correspondence, diaries, and memoirs. David Barnes, *The Draft Riots in New York, July 1863* (New York, 1863), a precinct-by-precinct account of police activities, William O. Stoddard, *The Volcano under the City. By a Volunteer Special* (New York, 1887), Ellen Leonard, *Three Days' Reign of Terror, or the July Riots in 1863, in New York* (New York, 1867), Lucy Gibbons Morse, "Personal Recollections of the Draft Riot of 1863," 2, Knapp/Powell, Trunk 1, New-York Historical Society, and the descriptions of attacks on the black community in Committee of Merchants for the Relief of Colored People, Suffering from the Late Riots in the City of New York, *Report,* in James M. McPherson, ed., *Anti-Negro Riots in the North, 1863* (New York, 1969), are a few of the more useful among the scores of first-hand accounts. The extraordinary self-consciousness of New Yorkers during the bloody week—their sense that every skirmish affected the future of the city and the nation—allows us to reconstruct the draft riots in very intimate detail.

No study of the riots could be written without consulting the records of the National Archives. To reconstruct the response of government and military officials to the violence, I found particularly useful R.G. 94 (Letters Received by the Secretary of War), R.G. 110 (Records of the Provost Marshal General's Bureau) (especially Entry 1360, Letters Sent to A.A.P.M.G. Robert Nugent), R.G. 107 (Registers of Letters Received by the Secretary of War), R.G. 108 (Headquarters of the Army, Letters Sent), and R.G. 393 (Department of the East, Letters Received and Letters Sent). The Papers of Abraham Lincoln, Manuscripts and Archives Division, Library of Congress, are an important source, as are the Samuel L. M. Barlow Papers at the Huntington Library, which provide fascinating commentary on both the view from the White House and the perspective of New York's wealthy Democrats. A crucial source for the history of state and federal conscription during the Civil War and the riot-week activities of government and military officials is U.S. War Dept., *The War of the Rebellion: A Compilation of the Official Records of the Union and Confederate Armies,* Series III, Volume III and Series I, Volume 27, Part II (Washington, D.C., 1899). For events pertaining to the enforcement of New York State conscription after the draft riots, I found the John Adams Dix Papers, Manuscripts Division, Butler Library, Columbia University, to be especially important. Civil War soldiers' diaries are still a largely ignored means of access to the perceptions of soldiers "occupying" New York City during and after the insurrection.

Fruitful material was found in what is probably any New York City social historian's greatest resource, the Municipal Archives and Records Center. Nineteenth Century Indictments, Proceedings of the Court of General Sessions (with Police Court Proceedings frequently attached), Draft Riot Claims on the Comptroller's Office, Coroners' Reports, Mayors' Papers, all part of the Municipal Archives collection, are essential to any study of the draft riots. The Indictments from mid-1863 to early 1864 are rich in eyewitness and participant discussion of riot-week events. Adrian Cook's *The Armies of the Streets: The New York City Draft Riots of 1863* (Lexington, Ky., 1974), was extremely helpful in directing me to the most useful sources at the Municipal Archives. Cook's is an exhaustive piece of research; many of the conclusions I draw regarding the social constituencies of the riots are derived from his thorough lists of those killed, arrested, wounded, and suspected of participation in the riots (pp. 213–68).

Court records can tell us much about the mentality and identity of the rioters, but the evidence they offer must be handled with care and corroborated by other sources wherever possible. There are two issues to bear in mind. First, how truthful were defendants and witnesses in their narratives of riot-week events? Here the historian has to evaluate testimony affidavit by affidavit and become a "finder of fact" in the manner of a jury. One discovers that the recounting of episodes in the draft-riot affidavits was surprisingly elaborate and revealing. Even where details about individual participation may have been disguised or omitted, the "framework" of the stories told—assumptions about social relations and practices and information about the social makeup of the neighborhood—usually remained consistent from witness to witness. Though the Irish poor were skilled in deceiving authority from long practice on both sides of the Atlantic, Democratic control of the city judiciary may have encouraged riot participants and witnesses to speak with greater freedom and less fear of punishment. There is some evidence that witnesses who considered testifying were discouraged by the intimidation of neighbors (see Cook, *Armies of the Streets,* 177), but it must have become clear before too long that, as Cook observes, "the judicial authorities proved somewhat less than dedicated"—after the first "couple of dozen" rioters were prosecuted, the district attorney "lost all interest in bringing the rest to justice" (pp. 177, 183). Of the 443 people arrested as suspected rioters, 221 were released without charges being brought against them; 10 were discharged by the judge at a preliminary hearing because of insufficient evidence; 74 were indicted but never brought to trial; of the 67 who were ultimately convicted, many were allowed to plead to a lesser charge and few were sentenced to long terms (Cook, *Armies of the Streets,* 177–87).

Second, the policeman's grab did not always distinguish the rioter from the bystander. The solution, I believe, is not altogether to ignore indictments and court proceedings as evidence of the social identity of the rioters. One can place more interpretative weight upon cases where a guilty plea was entered or a conviction obtained, and this I have tried to do, especially in instances where the surviving court papers provided no supporting affidavits from which to evaluate the episode. Most important, one can supplement and substantiate court records with the many other available sources. Crucial here is the information about social and industrial geography derived from the U.S. Census Office, *U.S. Eighth Federal Census* (1860), Industrial Manuscript Schedules, New York County, New York State Library, Albany, New York, which tells us where in the city we are (and are not) likely to find workers of certain occupational groups. But ultimately it is the diversity and abundance of other materials—the newspaper accounts, pamphlets, correspondence, diaries, and memoirs discussed above—that allow the historian to evaluate arrest records and court testimony against the background of a dense fabric of evidence.

NEW YORK CITY

Any historian of New York City or of the nineteenth-century American labor movement must refer first to Sean Wilentz's magisterial *Chants Democratic: New York City & the Rise of the American Working Class, 1788–1850* (New York, 1984). Wilentz's conclusions about artisan culture and consciousness and the uneven quality of metropolitan industrialization are a *terra firma* from which all other historians of nineteenth-century New York can now proceed. Also indispensable are Elizabeth S. Blackmar, "Housing and Property Relations in New York City, 1785–1850" (Ph.D. diss., Harvard Univ., 1980), John Jentz, "Artisans, Evangelicals, and the City: A Social History of Abolition and Labor Reform in Jacksonian New York" (Unpubl. Ph.D. diss., City Univ. of New York, 1977), Brian J. Danforth, "The Influence of Socioeconomic Factors upon Political Behavior: A Quantitative Look at New York City Merchants, 1828–1844" (Unpubl. Ph.D. diss., New York Univ., 1974), Lawrence Costello, "The New York City Labor Movement, 1861–1873" (Unpubl. Ph.D. diss., Columbia Univ., 1967), and Carl N. Degler, "Labor in the Economy and Politics of New York City, 1850–1860" (Unpubl. Ph.D. diss., Columbia Univ., 1952). Degler's splendid discussion of labor politics in the 1850s has been surprisingly neglected by subsequent labor historians. On New York City society and politics during the Civil War, see Basil Leo Lee, *Discontent in New York City, 1861–1865* (Wash.,

D.C., 1943). Paul Migliore, "The Business of Union: The New York Business Community and the Civil War" (Unpubl. Ph.D. diss., Columbia Univ., 1975), introduces some of the lines of inquiry regarding New York merchants' attitudes toward the war and reconstruction that are developed here. The starting point for any study of the nineteenth-century New York City German labor movement is Hermann Schlüter, *Die Anfänge der deutschen Arbeiterbewegung in Amerika* (Stuttgart, 1907); also very valuable are Wilhelm Weitling, *The Utopian Communist: A Biography of Wilhelm Weitling, Nineteenth Century Reformer* (Baton Rouge, 1950), Stanley Nadel, "*Kleindeutschland:* New York City's Germans, 1845–1880" (Ph.D. diss., Columbia Univ., 1981), and Bruce Carlan Levine, "'In the Spirit of 1848': German-Americans and the Fight over Slavery's Expansion" (Ph.D. diss., Univ. of Rochester, 1980). On New York City's immigrant experience in general, see Robert Ernst's thorough *Immigrant Life in New York City, 1825–1863* (New York, 1949); Carol Groneman [Pernicone], "'The Bloody Ould Sixth': A Social Analysis of a New York City Working-Class Community in the Mid-Nineteenth Century" (Unpubl. Ph.D. diss., Univ. of Rochester, 1973). No one can begin to understand the nineteenth-century urban immigrant experience, and the central place that women occupied within it, without consulting Christine Stansell, *City of Women: Sex and Class in New York, 1789–1860* (New York, 1986). Discussion of Irish-American attitudes can be found in Carl Wittke, *The Irish in America* (New York, 1956), and Florence E. Gibson, *The Attitudes of the New York Irish toward State and National Affairs, 1848–1892* (New York, 1951). For the transatlantic social and ideological context of Irish immigration to the United States in the mid-nineteenth century, see Kerby A. Miller, *Emigrants and Exiles: Ireland and the Irish Exodus to North America* (New York, 1985). On Catholic institutions and social thought in the nineteenth century, I found John Talbot Smith, *The Catholic Church in New York: A History of the New York Diocese from Its Establishment in 1808 to the Present Time* (New York, 1905), 2 vols., to be especially helpful. A general history of New York City's mid-nineteenth-century Irish community still remains to be written.

On the economic and industrial history of New York City in this period, I have profited from Edward D. Durand, *The Finances of New York City* (New York, 1898); Allan Pred, "Manufacturing in the Mercantile City, 1800–1840," *Annals of the Society of American Geographers,* 56 (1966), 307–25; the essays in David T. Gilchrist and W. David Lewis, eds., *Economic Change in the Civil War Era* (Greenville, Del., 1965), and Alfred D. Chandler, Jr., *The Visible Hand: The Managerial Revolution in American Business* (Cambridge, 1977), esp. 89–95. Robert G. Albion's classic *The Rise of New York Port, 1815–1860* (New York, 1939) is still without equal as an overview of the economic emergence of the commercial seaport.

The discussion of "the emergence and expansion of warehouse districts" in David Ward, *Cities and Immigrants: A Geography of Change in Nineteenth-Century America* (New York, 1971), 89–103, much influenced my understanding of the growth and boundaries of New York City's downtown manufacturing district during the middle decades of the century. David Harvey's provocative essay, "Paris, 1850–1870," in *Consciousness and the Urban Experience: Studies in the History and Theory of Capitalist Urbanization* (Baltimore, 1985), raises important questions about the relation between urban space and class relations in the Parisian context, which have great relevance to an inquiry into the ordering of social and political life in mid-nineteenth-century New York City.

LABOR AND CLASS RELATIONS

John R. Commons et al., *History of Labour in the United States* (New York, 1916), Vol. I, and Norman Ware, *The Industrial Worker, 1840–1860* (Boston, 1924) were both basic resources for this study, though the analysis of 1850s labor politics offered here revises the interpretations of Commons and Ware.

David Montgomery's *Beyond Equality: Labor and the Radical Republicans, 1862–1872* (New York, 1967) was in many ways the theoretical starting point and *vade mecum* for this work. Questions of social "taxonomy" were treated by the nineteenth-century metal-trades employer Denis Poulot in *Le sublime: Ou le travailleur comme il est en 1870, et ce qu'il peut être* (Cottereau, intro.) (Paris, 1980). David Montgomery, "Workers' Control of Machine Production in the Nineteenth Century," in *Workers' Control in America: Studies in the History of Work, Technology, and Labor Struggles* (Cambridge, 1979), and Benson Soffer, "A Theory of Trade Union Development: The Role of the 'Autonomous' Workman," *Labor History* 1 (Spring 1960), 141–63, have also had a strong influence on my thinking about the Civil War-era New York City trades. The essays in Theodore Hershberg, ed., *Philadelphia: Work, Space, Family and Group Experience in the 19th Century* (New York, 1981), raise many germinative questions about working-class experience in mid-century New York City.

Reading Leon Fink, *Workingmen's Democracy: The Knights of Labor and American Politics* (Urbana, 1983), made me think about the ways in which nineteenth-century workers' notions of the future polity might have shaped not merely their alternative political choices (the Knights of Labor, for instance), but also their terms of involvement with the major political parties. Gareth Stedman Jones, *Outcast London: A Study in the Relationship between Classes in Victorian Society* (Oxford, 1971), explores in the London context many of the issues I have examined in Civil War New York. I have also learned from Trygve Tholfsen, *Working Class Radicalism in Mid-Victorian England* (New York, 1977), Dorothy Thompson, *The Chartists: Popular Politics in the Industrial Revolution* (New York, 1984), Peter Bailey, " 'Will the Real Bill Banks Please Stand Up?' Towards A Role Analysis of Mid-Victorian Working-Class Respectability," *Journal of Social History* 12 (Spring 1979), 336–53, and I. J. Prothero, "London Chartism and the Trades," *Economic History Review* 24 (May 1971), 202–19. My discussion of working-class "association" is influenced by P. H. Noyes, *Organization and Revolution: Working-Class Associations in the German Revolutions of 1848–1849* (Princeton, 1966). On the subjects of the consciousness of emerging and declining social groups, and the relationship between organization and consciousness, I have found instructive Eric Hobsbawm, "Notes on Class Consciousness," in I. Meszaros, ed., *History and Class Consciousness* (London, 1971). On the problem of the relative possibilities of artisan and industrial worker radicalism in the mid-nineteenth century, see G. M. Stekloff, *History of the First International* (London, 1928).

SECTIONAL CRISIS, CIVIL WAR, AND RECONSTRUCTION

The point of departure for my thinking about the Civil War-era has been C. Vann Woodward's classic, *Reunion and Reaction: The Compromise of 1877 and the End of Reconstruction* (Boston, 1951). David Potter's *The Impending Crisis, 1848–1861* (New York, 1976) has taught me much about the complicated relationships between nationalism, the economics and politics of territorial expansion, and the sectional conflict. See Eric Foner's definitive study, *Free Soil, Free Labor, Free Men: The Ideology of the Republican Party before the Civil War* (New York, 1969), on the social bases and complex political inheritance of the antebellum Republican Party. Many of Foner's conclusions about the outlook of the national Republican Party in the fifties have amplified my sense of what made New York City Republicans distinctive in the pre-war and war decades. Roy F. Nichols, *The Disruption of American Democracy* (New York, 1948), Michael F. Holt, *Forging a Majority: The Formation of the Republican Party in Pittsburgh, 1848–1860* (New Haven, 1969), and William E. Gienapp, *The Origins of the Republican Party, 1852–1856* (New York, 1987) are essential reading on the fast-shifting party coalitions of the fifties.

Leonard L. Richards's discussion of elite conflicts over abolitionism in the 1830s and the "atypical" nature of New York City anti-abolitionist rioting in *"Gentlemen of Property and Standing": Anti-Abolition Mobs in Jacksonian America* (New York, 1970) stimulated many of the ideas about middle- and upper-class conflicts over anti-slavery and social change in this work. A model study of state social and political conflict in the middle decades of the century is Barbara J. Fields, *Slavery and Freedom on the Middle Ground: Maryland during the Nineteenth Century* (New Haven, 1985). On racism and the social conflicts of the Civil War era, see Fields's provocative essay "Ideology and Race in American History" in J. Morgan Kousser and James M. McPherson, eds., *Region, Race, and Reconstruction* (New York, 1982), 143–78. Also suggestive on issues of lower-class racism and the Democratic Party is Alexander Saxton, "Blackface Minstrelsy and Jacksonian Ideology," *American Quarterly* 27 (1975), 3–28. George M. Fredrickson, *The Inner Civil War: Northern Intellectuals and the Crisis of Union* (New York, 1968), explains much about New York City cultural leaders' views on the war and social change; in many ways his book was the starting point for my own analysis of merchants' perspectives on society and polity in the period 1846–65. Philip Foner, *Business & Slavery: The New York Merchants and the Irrepressible Conflict* (Chapel Hill, 1941), has worn surprisingly well with time, though must be supplemented with the observations I offer in Chapter 4, above. Finally, I have found John Higham, *From Boundlessness to Consolidation: The Transformation of American Culture, 1848–1860* (Ann Arbor, 1969), to provide useful terminology to help conceptualize the process of social change in New York City between 1850 and 1872, though I have interpreted Higham's essay in my own way, discussed above in Chapter 3.

POLITICS

My understanding of New York City politics in the 1840s and 1850s has been much shaped by a number of older essays and works: Merle Curti, " 'Young America,' " *AHR* 32 (Oct. 1926), 34–55; Herbert D. A. Donovan, *The Barnburners: A Study of the Internal Movements in the Political History of New York State and of the Resulting Changes in Political Affiliation, 1830–1852* (New York, 1925); and Louis Dow Scisco, *Political Nativism in New York State* (New York, 1901). Marvin Meyers, *The Jacksonian Persuasion* (New York, 1960), and Daniel Walker Howe, *The Political Culture of the American Whigs* (Chicago, 1979), are also essential resources for any discussion of antebellum New York City politics. On New York City and State politicians, see Alexander C. Flick, *Samuel J. Tilden: A Study in Political Sagacity* (New York, 1939), Edward C. Mack, *Peter Cooper: Citizen of New York* (New York, 1949), and Stewart Mitchell, *Horatio Seymour of New York* (Cambridge, 1938). We still lack biographies that fully place Horace Greeley and James Gordon Bennett, two of the city's most influential figures, in their Civil War-era social and political context.

Amy Bridges, *A City in the Republic: Antebellum New York and the Origins of Machine Politics* (Cambridge, Eng., 1984), explains much about the social and ideological origins of political nativism and Democratic Party allegiance, especially in the period before 1850; Jerome Mushkat, *Tammany: The Evolution of a Political Machine, 1789–1865* (Syracuse, 1971), maneuvers through the complexities of the local Democratic Party's relationship to the state and the nation in the 1850s. I rely on Sidney David Brummer, *Political History of New York State during the Period of the Civil War* (New York, 1911), for a discussion of wartime political factions in the state; Seymour J. Mandelbaum, *Boss Tweed's New York* (New York, 1965), and James C. Mohr, *The Radical Republicans and Reform in New York during Reconstruction* (Ithaca, 1975), were invaluable resources in my analysis of the post-war political settlement in New York City and State. John G. Sproat, *The Best Men: Liberal Reformers in the Gilded Age* (New York, 1968), remains an excellent overview of the elitist reform of the seventies and eighties, though it now must be read alongside Michael E. McGerr, *The Decline of Popular Politics: The American North, 1865–1928* (New York, 1986).

Index